THE LIBRARY OF
ISAAC NEWTON

THE LIBRARY OF
ISAAC NEWTON

―――

JOHN HARRISON

Senior Under-Librarian
University Library, Cambridge

CAMBRIDGE UNIVERSITY PRESS

Cambridge
London   New York   Melbourne

CAMBRIDGE UNIVERSITY PRESS
Cambridge, New York, Melbourne, Madrid, Cape Town, Singapore, São Paulo, Delhi

Cambridge University Press
The Edinburgh Building, Cambridge CB2 8RU, UK

Published in the United States of America by Cambridge University Press, New York

www.cambridge.org
Information on this title: www.cambridge.org/9780521218689

First published 1978
This digitally printed version 2008

A catalogue record for this publication is available from the British Library

Library of Congress Cataloguing in Publication data
Harrison, John R
The library of Isaac Newton.
Includes index.
1. Newton, Sir Isaac, 1642–1727 – Library. I. Title.
Z997.N56H36   018'.2   77–83994

ISBN 978-0-521-21868-9 hardback
ISBN 978-0-521-10145-5 paperback

CONTENTS

CONTENTS

ILLUSTRATIONS

FOREWORD

—————

'Some books', wrote Francis Bacon, 'are to be tasted, others to be swallowed, and some few to be chewed and digested.' And yet others, one might add, suffer the fate of never being put to the mouth at all. But for whatever complexity of reasons and circumstances one forms a collection of books of one's own, the quality and range of content in a man's personal library must broadly reflect the depth and extent and variety of his intellectual interests and pleasures, and individual works within it may by their rarity, annotation, or known context cast their unique light upon facets of his thought and character. That has long been acknowledged as a truism in the areas of literary and general historical scholarship, where the value of approaching a person's mind and indeed imagination through the books which he or she has read, or had opportunity to read by possessing them, is well appreciated. In the specialist field of historical bibliography there has come to be a whole new scholarly art and science whereby one seeks to 'feel', through the touchstone of his books, the intellectual attitudes and predilections of their owner, extrapolating the care and forethought with which he has chosen and handled them and stored them away on his shelves into wider comment upon the habits and idiosyncrasies of his mental activity and awareness. In the relatively youthful discipline of scientific history the lesson has taken longer to learn. Even so, the gathering flood of publication of the book holdings of many scientific worthies of the past is a mark of the increasing interest there is in such once ignored documentary records. And indeed, even at the lowest level, how revealing to compare the libraries of a Robert Hooke and a John Flamsteed, seeing how many works they owned in common in such unexpected subjects as pure mathematics and the Latin classics, as well as noting their differences! A whole generation of students of the multifold genius of that greatest of all English mathematical scientists, Isaac Newton, has hitherto been served by the publication in 1931 of Colonel de Villamil's transcriptions of the list of books in his library at his death which was made for their purchaser John Huggins, and of the more complete catalogue of these compiled some thirty years afterwards by James Musgrave. Particularly since the generous gift by the Pilgrim Trust to Trinity College, Cambridge, in 1943 of the remnant (nearly half) of Newton's books which had not then been dispersed by piecemeal sale, 'de Villamil' has been a valued if far from wholly trustworthy tool of research; no serious Newton scholar would be without his personal copy, suitably corrected and marked up with present locations and call-numbers of those items which interest him. But de Villamil, for all his unflagging enthusiasm and the guiding mentor's hand of the learned Henry Zeitlinger of Sotheran's, was neither trained bibliographer nor experienced historian of Newton's period. The time is long overdue for his pioneering efforts to be given firmer foundation, the deficiencies of his lists repaired, their inadequate titles elaborated, present locations of books (as these are known) cited, the library as a unit set in the broad context of Newton's intellectual experience, and its general character outlined. In his present work Mr John Harrison does this with efficiency and despatch, and much, much more.

To the many who have had contact with him either personally or professionally Mr Harrison needs no introduction. He has spent his life working in one of the world's great institutional collections of books, the Library of Newton's own University of Cambridge, where he is now Senior Under-Librarian in charge of cataloguing. Beyond such confines most will know him for his careful, infinitely painstaking catalogue of *The Library of John Locke* which he first brought out a dozen years ago in collaboration with Mr Peter Laslett – being responsible, in the latter's generous words, for 'all the really difficult and exacting things'. Here now he brings to fruition his ambition to be a parallel authority upon the greatest of Locke's acquaintances: one who was (if I may betray my own partisanship) yet more catholic in his intellectual interests. During the half-dozen years Mr Harrison has laboured to produce this new catalogue it has been my onlooker's pleasure and privilege to watch from close by as he pursued his intricate detective work in identifying and collating individual titles in the Huggins and Musgrave lists, tracking copy after copy of the books themselves down to their present locations in public and private ownership, and gleaning all he can of their condition, pedigree, annotation, and such singularities as the 'dog-earing' of page corners which is so often found in books which Newton read. His hunt and his other corroborative foot-sloggings have, entirely at his own expense (how refreshing to be reminded that scholarship does not of necessity demand either sponsorship or subvention!), taken him all over this country and to the United States, and are still not ended – cannot be ended till he has satisfied himself of the fate of every last one of the books which Newton once possessed. The fruits of that investigation, insofar as they have accrued to date, are set out by him in the detailed catalogue which follows. Its enduring usefulness as a tool of bibliographical reference and research I need not emphasize. The professional scholar and librarian has in it an authoritative, precisely

documented record of Newton's known holdings of books, listing and identifying their titles, and in the case of that majority of these whose originals
have been traced adding a wealth of ancillary information in concisely coded
form. To a yet wider circle of casual consultants and booklovers, from anti-
quarian booksellers to private owners and prospective purchasers of rare
books, it will serve as an eye-opening bible, allowing them for the first time
to distinguish the works with authentic Newtonian association from the
chaff of the many spurious 'books from Newton's library' with which the
market is so often filled. As one who has himself a few times in his life had the
ioy of finding an unidentified volume with the Huggins and Musgrave bookplates within, let me hope with Mr Harrison that his catalogue will flush out
into public light some good many more of the thousand or so of Newton's
books which at present still lurk in the limbo whither they were dispersed after
the 1920 Thame Park sale.

There will of course be very few who will read Mr Harrison's catalogue
through, item by item, with clerkish delight in its precision. He himself
would, I know, prefer me to stress the broader themes which he treats in his
splendid introductory essay, placing the books in their contemporary context, sketching in lightly but expertly the background of Newton's contacts
with his contemporaries and of the practices of the late-seventeenth-century
printing and publishing world, and also tracing the separate history of
what happened to Newton's collection of books in the two and a half centuries
after he died. His discussions there of Newton's habits and preferences in
reading and buying books, his seemingly carefree lack of system in shelving
them (or even putting marks of his ownership within), and his usual way of
annotating their content on separate sheets of paper bring vividly to life the
user rather than the lover of books in themselves which was Newton. One
can learn a great deal merely from looking at Mr Harrison's tabular analyses
of the works in Newton's library by topic, provenance, and year of publication; and also rightly be put on one's guard against any simplistic supposition that any such unweighted book numbers are an exact index of
relative significance. The subtleties of the novel sidelight which he throws in
so many unexpected ways on Newton's developing mind in reaction to
reading the printed works available to him are well worth studying in themselves, in divorce from his itemization of their titles. The cliché of being
a mine of information here regains its freshness with so rich an exploitation
of Newtonian lode.

To say more would be to steal Mr Harrison's thunder. Let him forthwith,
with my commendation and my blessing, speak for himself.

D. T. WHITESIDE

ACKNOWLEDGEMENTS

This book has been more than five years in the making, and were I to delay its publication yet longer it is possible that some additional details about Newton's books would be forthcoming. But the author has eventually to decide when the moment has arrived to number the catalogue entries, to work out and analyse totals and percentages, and to present his findings. The material which I have been able to discover and the steady flow of information which I have been fortunate enough to receive, both here in Cambridge and from many parts of the world, have now virtually come to an end. Yet I am, of course, aware of the possibility that someone, somewhere, will unearth a cache of Newton's books. Should this happen I hope that this volume will have had some part in tracking down and authenticating such items.

Among the unexpected pleasures encountered in compiling this catalogue have been the kindness and the help which I have received from many people, most of them previously unknown to me. This list is a long one and includes librarians, booksellers, several distinguished Newton scholars, and others. If I name only some of them I trust that the rest will accept this warm expression of gratitude.

I owe a great deal to my friend Peter Laslett, since it was due to his prompting that I turned (though not for ever) from Locke to Newton. He continued to urge and encourage me to work systematically on the latter's library, even after I knew quite well that I had become a Newton addict!

From the world of librarianship I wish to acknowledge the assistance I have had from colleagues in the University Library, Cambridge: A. J. C. Bainton, Peter Fox, Peter Gautrey, Wilfrid Lockwood, John Oates, and George Rawlings (who took the photographs). Elsewhere in Cambridge, Dr Philip Gaskell, Trevor Kaye, and other members of the staff of Trinity College Library have all most generously smoothed my path, while the Master and Fellows of the College permitted me to reproduce the items shown in the plates of this book. Peter Croft, Librarian of King's College, has not only provided ready access to the Keynes Collection there but has also given me the benefit of his knowledge of the London second-hand book market. From the United States I wish to record the valued assistance I have had from Miss Dorsey Fiske, Mrs Virginia Harrison, formerly Curator of the Newton Collection at Babson College, Massachusetts, Dr Kenneth A. Lohf of Columbia University, and Dr John Neu of the University of Wisconsin – Madison. In addition to these, several other librarians of American institutions have replied promptly and patiently to my enquiries.

I have benefited much from the vast experience of the book trade of Dr H. A. Feisenberger of Sotheby's, David Low of Emmington, Chinnor, and Jacob Zeitlin of Los Angeles. Sir Geoffrey Keynes kindly gave me permission to read and to quote from his brother's unpublished papers, as well as showing me the two Newton items he owns. The Rector of Chinnor, the Rev. R. W. Horner, readily made available the Registers of his Parish for my examination.

From the world of Newton scholarship, Professor Bernard Cohen, Professor B. J. T. Dobbs of Northwestern University, Dr Karin Figala of the Technische Universität, Munich, Dr John Rogers of Keele University, and Dr Peter Spargo of the University of Cape Town have all given me freely of their knowledge and advice on particular aspects of Newtonian study.

But my greatest indebtedness by far is to Dr D. T. Whiteside. Although we were on little more than nodding terms when I first began to gather material for this book, no one could have been kinder or more generous with his time. He made available to me his unparalleled knowledge of matters Newtonian to be drawn upon, widely and regularly. And I did. He read my introductory essay in draft, corrected errors of fact, made suggestions for improving its content and style, pointed out further areas for investigation. In addition (and equally important to me), Tom Whiteside's invaluable help and guidance firmly restrained my early tendency to jump to speculative conclusions or to offer facile hypotheses without supporting evidence. For any blemishes that remain the responsibility rests squarely with me.

J.H.

University Library, Cambridge
23 November 1977

CHAPTER I

ISAAC NEWTON: USER OF BOOKS

If the level of a man's reputation and importance in the scientific and intellectual thought of his time were reflected in the size of his library, then the total of 2100 volumes which Isaac Newton owned when he died is a lower total than might have been anticipated. But Newton acquired his books, insofar as they were not gifts and presentations, in order to read them, not just to own them. The delights of mere collecting did not appeal to him. His library was a set of working books most of which he came to know well and in some cases use extensively: the well-thumbed books are evidence of this. For Newton everything had its place in a regulated state of apparent disorder and so presumably he saw no advantage in making extensive book-lists or catalogues – nor in asking anyone else to compile them for him. Among the vast amount of manuscript material left behind by Newton only two short book-lists of any significance in his hand have come to light.[1]

John Dee, Samuel Pepys, John Evelyn, John Locke, or other 'professional' book-collectors may have regarded their personal libraries partially as show-pieces: Newton did not. Misleading references have been made elsewhere by booksellers and others to Newton's shelf-marks. None of the books from his library which I have examined so far carries a shelf-mark in Newton's writing, nor one written in the book by anyone else during Newton's life-time.[2] If he wished to consult a book he knew just where to lay his hands on it. Problems of shelf-space or book-storage leave no mark in his papers. If he had any difficulty in finding room for his books, either in Trinity College, Cambridge or later in London he apparently did not consider that it called for any written comment. In his correspondence Newton wrote about individual authors and their works, sometimes generous in his appreciation, and sometimes expressing his gratitude for presentation copies. In a long letter to Henry Oldenburg dated 23 June 1673 Newton described his copy of Huygens's *Horologium oscillatorium*, 1673 (no. 820 in the catalogue below)

---

[1] Bodleian Library, Oxford, MS New College 361 (Ekins Papers), Vol. II, fol. 47 (see p. 9 below), and Babson MS 418 (see pp. 41–2 below).

[2] For the shelf-marks in Newton's books see pp. 38–41 below.

as 'M. Hugens kind present...full of very subtile and useful speculations very worthy of ye Author',[1] and on 17 September of the same year he wrote to John Collins, thanking him 'for ye little but ingenious tract of P. Pardies',[2] showing his judgement of *La statique*...1673 (no. 1245). Such expressions of opinion, however, refer to individual authors or books, and not to their relationship with other works in his library.

Most young people are anxious to demonstrate their ownership of a book by writing their names inside the front cover at the earliest opportunity, and many of these inscriptions tend to be fuller and sometimes more pompous than those used later in life. So it was with Newton. His 1653 Greek Testament (no. 199), now in Trinity College, Cambridge, bears the inscription of the serious-minded eighteen-year-old 'Isaac Newton hujus libri verus est possessor. Pretium – $\overset{£}{0}$–$\overset{s}{3}$–$\overset{d}{0}$. Aprilis 3 die An° Dni 1661'. There are two candidates for the claim to bear the earliest known Newton signature on a printed book. His copy of Sabinus's *P. Ovidii Metamorphosis*...1593 (no. 1224), now in the Newton Collection at Babson College, Massachusetts, carries the legend 'Isaci Newtoni liber Octobris 15 1659. prætium –0–1–6', while his 1560 Pindar (no. 1317), now in King's College, Cambridge, has on the verso of the title-page 'Isaacus Newton hunc librum possidet. Pret. 8^d. 1659'. Whether the Pindar inscription preceded the Ovid one in time can only be conjecture, but both certainly came from the pen of a sixteen-year-old proud of the ownership of his books.

When Newton wished to indicate his ownership of a book he would usually, though not invariably, write his name in the book, sometimes adding the price he paid (or a donor's name), occasionally also the date of acquisition – early habits of which he soon tired, as the summary below shows. The Newton books in Trinity (862 volumes) and elsewhere, together with others not so far located but described in detail by Heinrich Zeitlinger in his Sotheran catalogues,[3] altogether 953 of the overall total of 2100, show the following:

	Total
Newton's signature with price and date	8

(Dates: 1659, 15 October 1659, 29 March 1661, 3 April 1661, 1680, 1682 (three separate works). See respectively nos. 1317, 1224, 1264, 199, 552, 1102, 1515, 1579.)

[1] Isaac Newton, *Correspondence*, ed. H. W. Turnbull, J. F. Scott, A. R. Hall and L. Tilling (vols. 1–7, Cambridge, 1959–77), I, 290. (Later references to this edition read: Newton, *Correspondence*, followed by the editor(s) of the volume cited.)

[2] *Ibid.* p. 307. See also Isaac Newton, *Mathematical papers*, ed. D. T. Whiteside (vols. 1– , Cambridge, 1967–), III, 391. (Later references to this work read: Newton, *Math. papers*.)

[3] Henry Sotheran & Co., *Bibliotheca chemico-mathematica*, ed. H. Zeitlinger and H. C. Sotheran (2 vols. and Suppl. 1–2, London, 1921–37). (Later references to this work read: Sotheran, *Bibl.*)

Newton's signature and date 8
(All eight books have the same inscription: 'Isaac Newton.
Trin: Coll: Cant: 1661'. See nos. 181, 335, 609, 629, 793,
1442, 1559, 1640.)

Newton's signature with price 15
(See nos. 55, 76, 337, 347, 377, 557, 589, 770, 839, 1489, 1561,
1562, 1687, 1714, 1763. Details of the prices are given at these
entries.)

Newton's signature showing the book to have been a gift 16
(The catalogue nos. and donors are 1688 (Étienne Baluze), 122
(Isaac Barrow), 261, 266, 271, 273 (all from Robert Boyle),
315 (Thomas Burnet), 678 (John Collins), 605 (Oliver
Doyley), 822 (Christiaan Huygens), 314 (Richard Mead), 49,
485, 1115, 1116 (all from Henry More), 1705 (William Walker).
The inscriptions are reproduced in full at the catalogue
entries.)

Newton's signature only[1] 10
(See nos. 75, 112, 322, 584, 869, 1006, 1106, 1156, 1362, 1574.)

Price-note and date in Newton's hand but without his signature 2
(See nos. 764, 1401. The prices and dates are 14s. 6d., 1680 and
6s. 6d., 1682 respectively.)

Price-note only 8
(See nos. 230, 476, 546, 704, 1076, 1206, 1209, 1210. The prices
paid range from 7s. for no. 1210 to £2. 15s. 0d. for the
7-volume set of no. 546.)

A survey of the style Newton adopted for his signature on these books may
help to clear up some of the confusion now spread by a possible early-nine-
teenth-century forger.[2] At least four institutional libraries have books which
Newton might well have owned – they were published in 1642, 1668, 1678,
and 1686 respectively – and they all carry an inscription in the same hand
reading 'I. Newton. A.M.' In addition to the fact that the handwriting does
not resemble Newton's and that none of the titles appears in the book-list

[1] One further work (no. 869) was described in the catalogue as being 'With the Autograph
of Isaac Newton' in the Thame Park sale, 13–15 January 1920 – when possibly as many as
1000 books from Newton's library were sold. The present whereabouts of this book is not
known, nor is it traceable in the Sotheran catalogues, and it is not therefore included in
the above count. For the Thame Park sale see pp. 48–50 below.
[2] The mathematician and historian of science Augustus De Morgan, writing in 1852–3,
reported that 'books are about the world with Newton's signature, known to have been
in them before the time at which forgeries commenced'; see his posthumously published
Newton: his friend and his niece (London, 1885; facs. repr. 1968), p. 153.

and catalogue discussed below, the mere presence of 'A.M.' after his name virtually rules out their being Newton's copies. None of his autograph inscriptions on the books which I have examined show any degree after his name, or an 'R.S.S.' [Regiæ Societatis Socius]. Newton's signature has often, I may add, been confused with that of his distant kinsman Sir John Newton who, when Newton submitted a pedigree to the College of Arms in 1705, made an affidavit in its support.[1] I know of three books[2] bearing John's signature which have been attributed with some confidence to Isaac. Comparison of photocopies of these with that in an autograph letter of Sir John Newton written to Isaac Newton on 13 April 1714[3] clearly demonstrates their common authorship. The use of a personal bookplate had not become a widespread practice among private book-collectors of Newton's time. Pepys had one designed and printed for insertion in his books, but Locke did not. It will not be surprising that Newton, plain man that he was, did not concern himself with such fripperies. In the fifty-seven signatures written by Newton in his books as listed above, 'Isaac' is used twenty-six times, 'Isaacus' twice, 'Isaaco' (as part of an inscription indicating a presentation) once, 'Isaci [Newtoni liber']' once, 'Is.' twenty-five times, 'Is°' (part of a note on a presentation) once, and 'I' once. The surname 'Newton' is only Latinized once, in the case shown immediately above.

There are few regular purchasers of second-hand books who do not derive much satisfaction from the thought that occasionally they pick up a bargain. Newton was no exception. While most of the books referred to above as listing the prices he paid have that price recorded by him without further comment, in the case of two works Newton was sufficiently pleased with the purchase to add a comment upon it. In his copy of Richer, *Historia Conciliorum generalium*...1680 (no. 1401) Newton wrote '1682. pret. 6^s 6^d, valet 10^s', while in Elmacinus, *Historia Saracenica*...1625 (no. 552) he added 'E libris Is. Newton 1680, Pret 9^s, valet 25^s' – an even greater bargain. On the title-page, this latter work has scored through 'Trin: Coll. Cant. A° Dñi 1668. Ex dono Mgri Thomæ Gale huius Collegij Socij'. John Laughton, Librarian of Trinity 1679–83, and subsequently University Librarian, was willing to sell this particular book to a Fellow of the College because the library had

---

[1] C. W. Foster, 'Sir Isaac Newton's family', *Reports and papers of the Architectural Societies of the County of Lincoln, County of York* [etc.], xxxix (1928), 10, 62. In an earlier letter of April 1707 to Sir John, Isaac styled himself 'Your affectionate Kinsman and most humble Servant' (Newton, *Correspondence*, ed. J. F. Scott, iv (1967), 488–9). For further information on the Newton family see Newton, *Correspondence*, ed. A. R. Hall and L. Tilling, vii (1977), 485–8: Appendix ii, 'Newton's genealogy'.

[2] I gratefully acknowledge the kindness and courtesy of the librarians at the Babson College, Columbia University, and Pierpont Morgan Libraries in providing me with Xerox copies of the title-pages which carry these signatures.

[3] Cambridge University Library, MS.Add.3968(41), fol. 113/14, printed in Newton, *Correspondence*, ed. A. R. Hall and L. Tilling, vi (1976), 100.

received another copy in 1679.[1] Newton took a close interest in the library at Trinity. In 1675, 1679, and 1681 he presented it with *Sancti Irenæi Adversus Valentini...libri V...Cum scholiis & annotationibus J. Billii* [etc.], 1675, Huet's *Demonstratio Evangelica*, 1679, and Grew's *Musæum Regalis Societatis*, 1681, as well as a copy of his *Principia* in 1687.[2] Furthermore, in 1676 he subscribed £40 towards the new College Library and loaned a further £100 four years later. It is not surprising then, even if these were all the gifts Newton made to it, that the Librarian was prepared to sell the Elmacinus at a favourably cheap price.

In his schooldays Newton had books available for him to read. There was the library of his late stepfather Barnabas Smith with '2 or 300 books in it, chiefly of divinity and old editions of the fathers...These books Sir Isaac gave to his relation Dr. Newton of Grantham, who gave some of them to me [William Stukeley], when I went to live there.'[3] Such other boyhood books as Newton had he also gave away in later life. While he was a pupil at the King's School at Grantham, Newton lodged with an apothecary named Clark and here he had the opportunity of reading the books that were in the house as well as pursuing his interests in making mechanical models.[4] In addition he might also, while at school, have climbed the narrow stairs up to the chained library at St Wulfram's Church in the cramped little room over its south porch to escape the rush and noise of the country town around and to read in comparative quiet. This, Grantham's original public library (established in 1598 and one of the first in England), was accessible to all burgesses and residents of the town. Since the church is only a hundred yards or so from the grammar school, Newton would have passed it every time he went to school and home again to his lodgings next to the George Inn. Of the library's early holdings there still exist some three hundred volumes (almost all, predictably, on theological subjects).[5]

The size of Newton's own library while he was at Trinity College from 1661 to 1696 is not known, and although only a rough guess can be made

[1] The Trinity Catalogue of 1667 (Add.MS.a.101) does not contain a copy of no. 552. The copy which Newton acquired is in the shelf-list of the 1670–90 Catalogue (Add.MS.a.104) at B.α.17: the '17' is deleted and the entry annotated 'Vid. Dup:'. The second copy, still in Trinity Library (W.17.45²), was bequeathed to the College by James Duport, who died in 1679.

[2] Trinity College Add.MS.a.106, fol. 13v. See also J. Edleston (ed.), *Correspondence of Sir Isaac Newton and Professor Cotes*...(London, 1850), pp. xxvi–xxix *passim*.

[3] W. Stukeley, *Memoirs of Sir Isaac Newton's life*... ed. A. Hastings White (London, 1936), p. 16.

[4] For an account of Newton's opportunities for reading as a schoolboy and of his early years in Trinity see D. T. Whiteside, 'Isaac Newton: birth of a mathematician', *Notes and Records of the Royal Society of London*, XIX (1964), 53–62, esp. 54–5.

[5] Of the volumes still in the library, 85 are yet secured with their original chains, and 153 others bear traces of having once been so fastened, but the rest have been rebound over recent years, so losing all signs of their earlier chaining. See A. A. Markham, *The story of Grantham and its church*, 18th ed. (Gloucester, 1973), pp. 19–20.

there is good reason to think that a very modestly sized personal library might well have satisfied his requirements. He had immediate access to Trinity Library, which over the period of Newton's residence there comprised 3000–4000 volumes.[1] Evidence that Newton used the books in his College Library is to be seen in his Common Place Book now in the Keynes Collection in King's College, Cambridge (K.MS.2). Folio 1 of the manuscript consists mainly of a list of historical authors and their works, and at the side of the entry for five of these writers Newton has added 'Trin. Coll.' and by another, 'Tr. C.' In the case of two of these works Trinity shelf-marks are also given, and these marks are still present in the books. Another entry reads 'Trin. Coll. Bib. S.[ub] P.[relo]', showing that the item was not yet catalogued and so had no press-mark at the time. The book in question is known to have been bequeathed to Trinity by James Duport, Regius Professor of Greek at Cambridge 1639–54, and later Dean of Peterborough, who died in July 1679, and Newton may have consulted the book before it was catalogued.

Though we do not know exactly where Newton lived before he moved into his rooms north of the Great Gate of Trinity in 1678, it is unlikely that his residence was ever more than a few hundred yards from the Public Library of the University of Cambridge. In the second half of the seventeenth century it had between 10,000 and 12,000 volumes, all of which Newton had the right to consult. In addition, there were other Cambridge colleges whose libraries contained valuable collections. Together these covered virtually the whole range of what scholarly material was then in print.

Even more immediate and important to Newton was his free access to Isaac Barrow's considerable private library. On 27 September 1670, Newton wrote to John Collins, 'I have hitherto deferred writing to you, waiting for Dr Barrows returne from London that I might consult his Library about what you propounded.'[2] Though Barrow and Newton may not have been the close personal friends which tradition has it, they were certainly professional colleagues with a common interest in mathematics and other branches of science, and in theology. The ready use Newton was evidently allowed to make of Barrow's library made it the less necessary for him to build up a library of his own until after the latter's death in May 1677. The contents of Barrow's library are set out in the Bodleian MS.Rawl.D.878, fols. 39–59: 'A Catalogue of the bookes of Dr Isaac Barrow sent to S. S.[3] by Mr Isaac Newton, Fellow of Trin: Coll: Cambs. July 14. 1677. Obiit Dr Barrow, Maii 4. 1677'.[4] This lists 992 separate titles (1099 volumes), at least 151 of which

---

[1] P. Gaskell and R. Robson, *The library of Trinity College, Cambridge: a short history* (Cambridge, 1971).

[2] Newton, *Correspondence*, ed. H. W. Turnbull, I (1959), 16.

[3] 'S. S.' may be one or other of the London Stationers Samuell Sprint or Samuel Smith; see Newton, *Math. papers*, III (1969), xiv n. 12.

[4] The catalogue is written in four different hands, none of which is Newton's. Only two of the compilers give place and date of publication of the works.

are mathematical and 24 on astronomy, but otherwise including numerous classical, theological, philosophical, and scientific works, and it provides a clear guide to many of the books which were accessible to Newton until the middle 1670s. Where, therefore, the absence from Newton's library of works by a particular author or on a certain subject has hitherto aroused comment or speculation, the reason for the gap may well be that when, up till 1677, Newton had needed a book Barrow's copy was available for him to consult.

As an undergraduate Newton was by no means a struggling, impoverished scholar. A notebook of his, now in the Fitzwilliam Museum, Cambridge, which contains his accounts for the period May 1665 to April 1669, has an entry dated 12 February 1667/8: 'Received of my Mother. 30.0.0.', and another later, undated: 'Received from my Mother. 11.0.0.' An earlier entry dated 22 April 1667 reading 'Received 10.0.0.' may indicate money from the same source. Life for Newton was not wholly devoid of pleasures: thus the Notebook records, also on 22 April 1667: 'At yᵉ Taverne severall other times &c. 1.0.0.', and later, probably also in 1667: 'At yᵉ Taverne twice. 0.3.6.', and again: 'Lost at cards at twice [?] 0.15.0.' Newton had, then, enough money at his disposal even in his twenties to be able to lend not inconsiderable amounts to his more needy young contemporaries.[1]

A pocket-book now in Trinity College, Cambridge,[2] which is inscribed on the fly-leaf 'Isaac Newton, Martij, 1659', contains the record of his accounts in his first years at Trinity, from 1661 on. This lists a few minor extravagances, though no drinking or gambling, and it also shows that he lent amounts ranging from a shilling to a pound to fifteen different people. There was money enough available, also, to buy books. In the Trinity Notebook Newton records that he bought 'Sleidan's 4 Monarchies' for a shilling (though only a later, 1686, edition (no. 1521) was in his final library) and 'Schrevelius his lexicon' (no. 1472) for five shillings and fourpence. The Fitzwilliam Notebook has an entry on 22 April 1667 recording the purchase of 'Ye Hystory of yᵉ Royall Soc.' (by T. Sprat, no. 1549) for which Newton paid seven shillings, followed by 'Gunters book & sector &c' (no. 728), which cost him five shillings from 'Dr [Francis] Fox'. In February 1667/8 Newton recorded 'Bacons Miscelanys' bought for one shilling and sixpence, but this book too was not listed in his library. Other purchases known to have been made by Newton at this period were recorded in an 'accompt' of his expenses in 1664 which seemingly no longer exists.[3] When Newton became a Fellow of Trinity

---

[1] The Fitzwilliam Notebook shows that over the period 1667 to 1669 Newton lent sums varying from a single shilling to one pound six shillings to six named individuals.

[2] Trinity College MS.R.4.48ᶜ.

[3] In Cambridge University Library MS.Add.4000, fol. 14v Newton wrote on 4 July 1699 that 'By consulting an accompt of my expenses at Cambridge in the years 1663 & 1664 I find that in yᵉ year 1664 a little before Christmas I being then senior Sophister [i.e. a third-year undergraduate], I bought Schooten's Miscellanies (no. 1471) & Cartes's Geometry' (no. 506 or 507).

College in 1667 he had a deal of income of his own to add to that received through his mother.

One of Newton's favourite and active areas of study from his early years at Trinity and continuing throughout his residence there was the field of chemistry and alchemy. Newton himself provides direct evidence of this interest in the late 1660s. The Fitzwilliam Notebook records the purchase by him of chemical materials and equipment in April 1669, together with Lazarus Zetzner's *Theatrum chemicum* (no. 1608) for one pound eight shillings. Newton's copy of this very important six-volume work (Strasburg, 1659–61) was sold for fifty pounds by Sotheran of London in 1926, when it was described by Zeitlinger as being copiously annotated by Newton.[1] Unfortunately for Newton scholars, and in particular those researching in his alchemical interests and ideas, the present whereabouts of this set is not known. The list of Barrow's library shows that he had collected very few books on this subject; hence Newton would either have had to look elsewhere in Cambridge, or buy such books himself. The indications are strong that he followed the latter course. This would certainly account for the unusually high number of chemical and alchemical books in Newton's final library – 169 or 9.5 per cent of a total of 1752 known titles on all subjects. Of these titles 99 are in that section of Newton's library now housed in the Wren Library at Trinity College. Some of them carry the names and annotations of their earlier owners, and their general appearance, together with the early date of publication of many of them, suggests that Newton acquired them second-hand. Several of his alchemical books in Trinity, and elsewhere, have notes by Newton in them, and the presence of such notes is indicated in the individual catalogue entries below.

Among the mass of Newton manuscript material still surviving there is very little directly indicative of his book-buying activities.[2] Reference has already been made to comments on individual books and authors in Newton's correspondence and in the Trinity and Fitzwilliam account books. The Stanford University Newton manuscript, headed 'De scriptoribus chemicis',[3] consists of five closely written pages of authors and titles of alchemical works, sometimes with details of the contents of the volumes. The majority of these works were published in the late sixteenth or early seventeenth century, and only a few of them were apparently in Newton's library. This manuscript

---

[1] See Sotheran, *Bibl.* (1937), Suppl. 2, I, 757 (Catalogue 800 (1926) 12098); see also *Library of Sir Isaac Newton: presentation by the Pilgrim Trust to Trinity College Cambridge 30 October 1943. With Appendix: Newton's library and its discovery*, by H. Zeitlinger (Cambridge, 1944), pp. 16, 17, 19 *passim*. (Later references to the Appendix by Zeitlinger read: Zeitlinger, *Newton's library*.)

[2] Unlike his friend John Locke, who carefully saved many of the lists of his purchases at book auctions together with his bookseller's bills, which can be examined in the Bodleian Library MS Locke.b.2.

[3] Stanford University Library, MS, Container 2, Folder 13.

appears to be either a collection of notes for an alchemical bibliography or a detailed reading-list for Newton's own use at some time after 1692 – the latest publication date of the works included. On the other hand, the Babson MS 418, Newton's 'Lib. Chem' list, which is discussed further on pages 41–2 below, is a document showing books which Newton owned in late 1696 or in 1697. But neither the Stanford nor the Babson manuscript throws any direct light on Newton's actual book-buying activities.

There exists one document, however, bound up with a wide variety of miscellaneous material, mainly in Newton's hand, in Bodleian Library MS New College 361 (Ekins Papers), vol. II, fol. 78, headed 'Books for Mr Newton'. We may readily suspect that this list was kept by Newton not so much for the details of what he paid for the sixteen works tabled there (bought at a cost of altogether three pounds and seventeen shillings) as because he wished to preserve the notes on Trojan history which he subsequently wrote on the back of the bill. There is no clue to the bookseller's identity, no date on the bill, no endorsement on it by Newton, and no indication as to when it was paid. Booksellers' accounts then, as now, carried tantalizingly short and sketchy entries, but it is possible to identify all sixteen books listed there by title with certainty, and all but two were present in Newton's final library.[1] Fourteen works were in French, two in Latin; twelve of them were on alchemical subjects. Their publication dates range from 1626 to 1701, and as the bill is headed 'Books for Mr Newton' (it would presumably have been 'Sir Isaac' after 1705), its probable date is 1702 to mid-1705. Twelve of these books are in the Trinity collection today.

The same Bodleian manuscript contains on its folio 47 a book-list written on both sides of the leaf in Newton's hand. It is difficult to know how significant this is, and there is no heading or other clue to the particular purpose for which it was compiled. The list sets out thirty-four separate works, comprising thirty-seven volumes, and gives for each item author, title, and place and date of publication. Their titles are individually grouped by size: 'Folio' (twelve works), 'Quarto' (ten), and 'Octavo' (twelve), with the earliest date of publication, 1477, correctly given. The works concern mainly classical literature, history, and mythology. A dash is present at the left of each of twenty-six of the entries, but this does not necessarily mean that Newton acquired the books so distinguished, for only seventeen of the total of thirty-four are traceable in Newton's library, and three of these have no such line alongside. This book-list can be dated only as after 1697 – the latest date of publication recorded.

---

[1] The catalogue numbers, with prices paid for each book are: 237 (5s. 0d.), 511 (1s. 6d.), 531 (2s. 6d.), 539 (5s. 0d.), 540 (3s. 6d.), 619 (3s. 0d.), 635 (1s. 6d.), 1242 (£1. 5s. 0d.), 1263 (2s. 6d.), 1316 (3s. 0d.), 1372 (5s. 0d.), 1397 (4s. 0d.), 1607 (2s. 6d.), 1675 (3s. 6d.). The two works missing from the final library were: *La physique des anciens.* [By D.R.] 12°, Paris, 1701 (3s. 6d.) and *Traité de perspective.* [By B. Lamy.] 8°, Paris, 1701 (6s. 0d.).

Newton's scientific eminence and his reputation and influence particularly in his own country were recognized, among other things, by his having a number of authors dedicate their books to him.[1] There were many more who for varying motives considered him a worthy recipient of their books. How frequently such presentations were made we can but guess. A few such are documented in his correspondence, though even here in some cases it is not wholly clear whether the book mentioned was a gift or had merely been bought for Newton. In some instances Newton himself specified in the book so donated that it was a gift; in others the donor himself, usually the author, entered an inscription therein which makes this clear. But who can tell whether Newton bothered to keep all the books that were given to him? He may have glanced at the donation, perhaps read a bit of it, and scoffed at it before quickly discarding it. I have examined just over nine hundred volumes, 44 per cent of the total number of volumes in the catalogue below: twenty-nine of these are clearly marked as gifts, and my firm impression is that the number of such gifts is a deal lower than might reasonably be expected.[2] On the other hand, since, as outlined above, Newton hardly ever bothered in his later years to indicate his ownership of a book purchased by him by entering his signature in it, it could well be that he then took correspondingly less trouble to record the fact when a book was given to him. Some of the books in Trinity, particularly a few in an especially handsome binding, have all the appearance of being presentation copies; many of the early-eighteenth-century mathematical and scientific tracts also now there doubtless came likewise to Newton directly as gifts from their authors.

Fifteen inscriptions of the sixteen so far seen in which he added that the book so endorsed was a gift were made before Newton left Cambridge. The lone exception is a presentation copy of Thomas Burnet's *De statu mortuorum*... 1720 (no. 314) which the donor, Dr Richard Mead, physician and Fellow of the Royal Society, had had privately printed in a limited edition. Newton endorsed his beautifully bound copy with the legend 'Is. Newton Ex dono D^{ris} Ri. Meade'. On the fly-leaf of his copy of Isaac Barrow's *Lectiones XVIII*...1669(–1670) (no. 122) Newton's note reveals that his Trinity colleague gave him the book on 7 July 1670.[3] Barrow formally presented Newton with none of his other publications. Newton had, indeed, an 'author's copy' of Barrow's 1675 epitome of the works of Archimedes and Apollonius (no. 76), but its inscription shows that Barrow had given it to Edmund

---

[1] For example John Freind (no. 637), Willem Jacob 's Gravesande (no. 689), Colin Maclaurin (no. 1012), Abraham de Moivre (no. 1094), William Whiston (no. 1726), John Woodward (no. 1751).

[2] John Locke, far more concerned than Newton with registration and arrangement of his books, also appears to have had a surprisingly small number of books given to him. J. Harrison and P. Laslett, *The library of John Locke* (Oxford, 1965; 2nd ed. 1971).

[3] Presentation inscriptions, etc. are given in full at the entry for the individual works in the catalogue below.

Matthews, Fellow of Sidney Sussex College, from whom Newton subsequently bought it (his own note in the book tells us) for three shillings and sixpence.[1] It is not known how Newton acquired the full 1653 Greek Bible
(no. 196) which has 'Isaac Barrow' – in Barrow's hand – on the fly-leaf, as
well as many notes made by its original owner, but one may hazard a guess
that it came to Newton after Barrow's death as part of the 'fee' for going
through the library and despatching it for sale (see p. 6 above).[2] Newton's
frequent correspondent John Collins gave him a copy of Abraham de Graaf's
De beginselen van de algebra of stelkonst...1672 (no. 678). If Newton's choice
of the words he wrote in a book can be taken as a true sign of his regard for
the donor, then 'Ex dono Viri amicissimi', describing Collins, would mean
that Newton placed a high value on his friendship, even if his frequent refusal
to answer Collins's letters for months on end may appear to belie this.

Robert Boyle and Isaac Newton shared a number of common interests,
notably in physics, chemistry, and alchemy. They each made a deep study of
these subjects, they occasionally met in London, at the Royal Society and
privately, and they corresponded in February 1679 and August 1682.[3] The
esteem in which Newton held Boyle and the importance he attached to
Boyle's publications are highlighted by the large number of noticeably well-
used copies of the latter's works in his library, by far the largest number of
any single author there represented. John Locke, also a close associate of
Boyle, owned thirty-four of Boyle's works, which was the highest number of
any individual writer in his library.[4] These totals are not surprising when
account is taken of the closeness of the interests of the three men, and more
especially of Boyle's output – he published thirty-nine separate works over the
years 1655 to 1691. Of the twenty-four Boyle books which Newton possessed,
the present locations of sixteen are known. They show clear signs of having
been minutely studied by Newton – some of them have notes by him in their
margins.[5] Of these at least six came to Newton as gifts: in four of them (nos.
273, 271, 261, 266), published in 1673, 1675, 1680, and 1683/4 respectively
Newton has written 'Isaac Newton Donum Authoris' (or something very
similar); while two others (nos. 259 and 263) of 1673 and 1675 carry the
inscription 'For Mr Isaac Newton from the Authour', in the hand of Henry
Oldenburg. (Oldenburg had, at Boyle's behest, sent the former of these two

[1] The Newton books show that he also acquired nos. 1189 and 1357 either directly from
Matthews or from a sale of his books after the latter's death in 1692.

[2] Newton also 'acquired' his copy of Cosin's *Historia transubstantiationis papalis*...1675 (no.
444) from Barrow's library. The book has notes by Barrow, and the title was included in
the list made after his death, though there can be no proof that this was necessarily the
listed copy.

[3] Newton, *Correspondence*, ed. H. W. Turnbull, II (1960), 288–95 (28 Feb. 1678/9, Newton to
Boyle), 379 (19 Aug. 1682, Boyle to Newton).

[4] Harrison and Laslett, pp. 23–4, 91–3.

[5] Details of present locations and indications of the presence of notes by Newton are given at
the entries for these works in the catalogue below.

to Newton on 14 September 1673, observing in his accompanying letter that 'he [Boyle] desired me to present [it] to you in his name, with his very affectionat service, and assurance of ye esteem he hath of your vertue and knowledge'.[1] Oldenburg added that he was also including two other copies of the book in the same parcel, one for Dr Barrow and one for Dr Henry More.) Robert Boyle seems to have been generous with presentation copies of his works, and it is likely that Newton received more books as gifts from him than the six already referred to.

Christiaan Huygens presented Newton with at least two of his works. In June 1673, using Oldenburg as an intermediary, the eminent Dutchman sent Newton (among others) a copy of his work on dynamical theory, the *Horologium oscillatorium* (no. 820), which has on the lower right-hand corner of the title-page, in Huygens's writing, 'Pour Monsieur New', the rest of the inscription having subsequently been clipped away by a bookbinder.[2] Newton owned two copies of Huygens's *Traité de la lumière*, 1690 (nos. 822 – a special large-paper copy – and 823), both now in Trinity, and on the fly-leaf of the former Newton has written 'Is. Newton Donum Nobilissimi Authoris'. Both copies show signs of having been used by the owner.

Newton registered the gift of William Walker's *Modest plea for infants baptism...* 1677 (no. 1705) with the inscription 'Is. Newton. Ex dono R^{ndi} Authoris'. Walker knew Newton and became headmaster of the grammar school at Grantham some while after Newton left. The influence of his friend Dr Henry More on Newton during his earlier years at Cambridge has often been discussed by Newton's biographers and others.[3] Newton wrote in his copy of More's *Tetractys anti-astrologica...* 1681 (no. 1116) 'Donum Reverendissimi Auctoris'; his *Answer to Several remarks upon Dr Henry More...* 1684 (no. 49) bears a similar inscription. Newton's copy of the works of Pope Damasus I (no. 485), published in 1672 and now in Stockholm, has 'Is. Newton Donum R^{ndi} amici D. Moor S.T.D.', and here Newton was almost certainly yet again referring to Henry More.

In 1687 the French historian and man of letters Étienne Baluze edited a work of Saint Vincent de Lérins (no. 1688) which was published at Cambridge, where presumably he met Newton. In any event Newton indicated his regard for the writer of its foreword by inscribing 'Is. Newton. Donum amicissimi Authoris Præfationis' in his copy of the book.

Thomas Burnet, a Fellow of Christ's College (Henry More's college) from 1657 to 78 and Senior Proctor of the University when Newton took his M.A. degree in 1668, held a vigorous correspondence with Newton on cosmogony over the period December 1680 to January 1680/81. In volume I of his

---

[1] Newton, *Correspondence*, ed. H. W. Turnbull, I (1959), 305.

[2] Newton's appreciation of the book and its author is given on p. 2 above.

[3] See, particularly, Marjorie H. Nicolson's edition of the *Conway letters: the correspondence of Anne, Viscountess Conway, Henry More, and their friends, 1642–84* (London, 1930).

copy of Burnet's *Telluris theoria sacra*...2 vols., 1681–9 (no. 315) Newton wrote 'Isaac Newton Ex dono Authoris', thought it is uncertain whether this was done in 1681 when the first volume appeared or in 1689 when the work was completed or even later. In his copy of *The fame and confession of the Fraternity of the R:C:*...1652 (no. 605), a book which has considerable evidence of use by him,[1] Newton inserted 'Is. Newton. Donum M^{ri} Doyley', but without adding a date; this seems likely to have been the gift of Dr Oliver Doyley, Fellow of King's 1637–93, an acquaintance of Henry More.[2]

In surveying the books presented to Newton which carry inscriptions made by the authors themselves (or on their behalf) we might well begin with more fulsome ones. The Swiss physician and naturalist (and member of the Royal Society) Johann Jacob Scheuchzer had presented Newton with a copy of his *Piscium querelae et vindiciae*, 1708 (no. 1467) inscribed 'Illustr. D. Newtono Societ. Regiæ Præsid.', but to cap this in 1723 Scheuchzer's son Johann Caspar sent Newton a copy of his father's two-volume work on Alpine natural history (no. 1466) with the inscription 'Perillustri Viro Isaaco Newtono Equiti aurato & Societatis Regiæ Præsidi Itinera hæcce Alpina nomine Authoris Ea quâ par est, observantiâ humillimè offert Joh. Casparus Scheuchzerus Fil. Londini Calendis Julij MDCCXXIII'. This somewhat over-fulsome dedication may, I suspect, have had something to do with the fact that the plates opposite pages 1, 65, and 147 of the first volume of this lavishly illustrated book were produced 'Sumptibus D. Isaaci Newton Equitis Aurati'. Edward Bernard, Professor of Astronomy at Oxford, linguist, and theologian, was a close runner-up to the younger Scheuchzer in volubility, inserting 'Isaaco Newtono Mathematico præstantissimo & Philosophiæ instauratori perfelici E. Bernardus ex merito' on the end-paper of his gift copy of his *De mensuris et ponderibus*...1688 (no. 166). Nor was Robert Greene far behind. The copy of his presentation of his *Principles of natural philosophy*... 1712 (no. 699) Greene endorsed 'To the Honoured & Much Esteemed Sir Isaac Newton These Principles of Natural Philosophy are with the Greatest Deference & Submission Presented'. In 1723 Pieter van Musschenbroek sent Newton a copy of his *Oratio de certa methodo philosophiæ experimentalis*...(no. 1134) on whose title-page he put 'Viro Amplissimo, Nobilissimo, Geometrarum Principi, Solidæ Philosophiæ Instauratori Isaco Newtono mittit Auctor'.

Continental authors are well represented among those who sent Newton gifts of their works. The French mathematician Pierre Rémond de Monmort presented Newton with his *Essay d'analyse sur les jeux de hazard*, 2e éd....1713

---

[1] I. Macphail, *Alchemy and the occult: a catalogue of books and manuscripts from the Collection of Paul and Mary Mellon given to Yale University Library* (2 vols., New Haven, 1968), II, no. 102.

[2] See Nicolson (ed.), *Conway letters*, p. 270 n. 6 and esp. pp. 322–3, where in a letter to Lady Conway dated 13 Oct. 1670 Henry More writes that '[Francis Mercury] Van Helmont... dined with me...with Mr Doyly and Mrs Foxcrofts son [Ezekiel]'.

(no. 573). Rémond de Monmort, who here conventionally styles himself Newton's 'tres humble serviteur', had already sent him a copy of the 1708 edition of this work (no. 572).[1] The yet more celebrated mathematician Johann I Bernoulli sent his *Essay d'une nouvelle théorie de la manœuvre des vaisseaux* ...1714 (no. 173), inscribing it 'Illustr. Newtono Mathematico D.D. Auctor'. Bernard Le Bovier de Fontenelle, the French 'philosophe' and 'Secrétaire perpétuel' of the Paris Académie wasted few words when he gave Newton a copy of his two-volume *Histoire du renouvellement de l'Académie Royale des Sciences*...1717 (no. 624). A curt 'Pour Monsieur Newton' sufficed.[2] Frederick Slare, physician and F.R.S., was even briefer: on the fly-leaf of his *Experiments*...*upon Oriental and other Bezoar-stones*...1715 (no. 1520) he merely penned 'Ab Authore'.

If the facts presented therein or the opinions expressed provoked Newton to any personal reaction he rarely carried this on to the printed page of the book he was reading. Approval or disagreement on his part is rarely signified and he was not usually given to underlining or sidelining the text. In the margins of pages 34, 37, and 38 of Sir Norton Knatchbull's *Annotations upon some difficult texts*...*of the New Testament*, 1693 (no. 889) the word 'Error', in Newton's hand, is present. But such strictures are exceptional. His copy of the two-part edition of Descartes's *Geometria*...1659–61 (no. 507), in Trinity College, but not stemming from the Barnsley Park sale, bears Newton's marginal annotations of 'Error' on pages 11 (twice), 12 (twice), 24, 25, 84, and 97, 'non probo' on pages 7 and 17, 'Non Geom[etricum est]' on pages 38, 67, and 85, and 'Imperf.' on page 65.[3] These notes, however, are not an emotional reaction or snap judgement on Newton's part, but rather a series

---

[1] On 16 Feb. 1708/9. See Newton, *Correspondence*, ed. J. F. Scott, IV (1967), 533. Newton's acknowledgement to Rémond de Monmort for the gift of the 2nd edition is interesting and enlightening. The extant draft (Cambridge University Library [MS.Add. 3968(41), fol. 99v [Nov. ?] 1714) of his letter reads:
'Sr I received your Book of the second Edition & return you my hearty thanks for the favour of such a present. Its now about eighteen years [Easter 1696] since I left off the study of Mathematicks & the disuse of thinking upon these things makes it difficult to me to take them into consideration: but however I have run my eyes over your perform^{ce} & cannot but applaud it much. I wish heartily that France may always flourish with men who improve these sciences.'
(I am grateful to Dr D. T. Whiteside for drawing my attention to this draft.)

[2] Books are known to have been sent as gifts to Newton by Jacques Cassini in 1722 (no. 1252), Felix Antonio de Alvarado, ?1718 (no. 35), Guido Grandi in 1704 (nos. 684 and 685), and Caspar Neumann, ?1712 (no. 1148) (Newton, *Correspondence*, ed. A. R. Hall and L. Tilling, VII (1977), 209, 377–8, 434–5, 481 respectively). All these items are now in Trinity, but carry no presentation inscriptions.

[3] David Brewster in his *Memoirs of the life, writings, and discoveries of Sir Isaac Newton* (2 vols., Edinburgh, 1855), I, claims (p. 22 n. 1) to have seen 'Newton's copy of Descartes' Geometry...among the family papers. It is marked in many places with his own hand, *Error, Error, non est Geom.*' Brewster here repeats word for word a memorandum by John Conduitt, who reports the information from Robert Smith (then the book's owner). See also Newton, *Math. papers*, I (1967), 17; VII (1976), 194 n. 46.

of catchwords for a more developed criticism on intrinsic mathematical grounds which he elaborated in a contemporary piece on 'Errores Cartesij Geometriæ'.[1] It was unusual too for Newton to betray the sort of testy impatience to be seen in a note added at the end of the dedication of *The works and life...of Charles, late Earl of Halifax...*1715 (no. 733). The writer of this dedication, possibly William Pittis, there admits that 'he is sensible he has been guilty of many omissions, thro' want of intelligence, from persons who might have oblig'd him with proper informations; particularly, in leaving out the mention of his Lordship's care for inspiring future ages... by causing medals to be struck on the most memorable achievements of the two last reigns'. Newton, who maintained a lasting contact (begun in Cambridge and continued in London) with Charles Montague, Earl of Halifax, could not, knowing Halifax and being Master of the Mint when the medals were produced,[2] allow this innuendo to go unchallenged, and in consequence appended the observation 'Sr Rich. Steel had opportunities of informing himself that my Ld Halifax had no hand in causing those Medalls to be struck.'[3]

Compared with the considerable amount of crucially important autograph manuscript material which still survives, the number of the (not always significant) notes which Newton made in his books is small. The survey which follows is, I repeat, based on my examination of nine hundred of Newton's library books, which may not be entirely typical of the whole. By 'note' I mean any manuscript addition to the volumes, varying from a single word or number to a whole page or more of carefully written material, but books with signatures, price-records and donor-inscriptions only are not included. The number of volumes I have found to have such notes by Newton is eighty-four. The Trinity collection has fifty-two of these, twenty-four other volumes are held elsewhere, and there are yet eight more, whose present location is unknown, which are reliably documented to have notes by Newton.[4] The subjects of these eighty-four volumes are chemistry and alchemy (thirty-two volumes), mathematics (twenty-four), physics (eight), astronomy (six), theology (seven), together with bibliography, chronology, classical literature, English literature, Judaism, medals, and Persian history (one each). The dates of the notes range from some made in late 1664

---

[1] Cambridge University Library MS.Add.3961(4), fols. 23r–24r. See Newton, *Math. papers*, IV (1971), 336–45.

[2] See Sir John Craig, *Newton at the Mint* (Cambridge, 1946), pp. 51–5.

[3] Newton here is taken to imply that the author of the *Life* of Halifax (attributed to William Pittis in a manuscript note in the Cambridge University Library copy of the work) was in a position to consult Sir Richard Steele on this point. Though Halifax had been a friend and patron of Steele, who had dedicated *Tatler* IV (1711) and *Spectator* II (1712) to him, the part that Steele had, direct or indirect, in the preparation of the book is not known.

[4] Five of these (no. 1608) are described in detail in the Sotheran catalogues (see p. 16), and three are shown in the Huggins List (see p. 33) as with notes by Newton.

or early 1665 in Schooten's *Exercitationum mathematicarum libri V*...1657 (no. 1471) to a correction made by Newton in the first English edition of his own *Universal arithmetick*...(no. 1155), published in 1720. The fact that twenty of the volumes with notes by Newton were published after 1700 suggests that he persisted in the practice – though never extensively – for most of his working life.[1]

Newton's annotations in his books can be considered to fall into four main categories, though with a degree of overlap between the groupings: (1) books containing notes and comments of a general nature, some of which do not even relate directly to the book itself, (2) books where the notes are mainly of a bibliographical nature, (3) books with manuscript corrections, alterations, and errata-list insertions, and finally (4) those copies of books written by Newton himself, or ones to which he contributed, which contain emendations or additions (some on interleaved sheets), many of which were incorporated in later editions of the works. Detailed descriptions and analyses of the contents and subtleties of these notes, and an evaluation of their potential value to scholars, are beyond the scope of the present work. Many of the more important and substantial annotations have already been discussed in print, and the intention here is to give some examples which will illustrate their variety and range. In the catalogue below the presence of Newton's notes in particular volumes of his personal library is indicated in the individual entries for the works, where reference is also given to pertinent published material; a list of their catalogue numbers is given in Appendix C.

General notes and observations were usually made by Newton in the margins or in any other vacant space on the printed page (see plates 1 and 2). Occasionally they were written on fly-leaves or on end-papers, or on loose inserted slips of paper left in the book as in Schooten's *Exercitationum mathematicarum libri V*...1657 (no. 1471) where the notes are now inserted between pages 178 and 179 and 360 and 361, or Wing's *Harmonicon cœleste*...1651 (no. 1744).[2] Illustrative sketches and diagrams by Newton are present in his copies of Ridley's *Short treatise of magneticall bodies*...1613 (no. 1403) and Oughtred's *Trigonometria*...1657 (no. 1220). When Messrs Sotheran offered for sale in 1926 Newton's copy of Zetzner's *Theatrum chemicum*... 1659–61 (no. 1608), for £50, their catalogue[3] gave a list of the authors annotated, with textual corrections and references, all in Newton's hand, in five of the six volumes. Of the books in his library containing notes of a general nature added by Newton the Zetzner set was probably the most comprehensively annotated. *Secrets reveal'd*...*by*...*Eyrœneus Philaletha Cos-*

---

[1] For a bibliographical note on the mathematical works annotated by Newton see his *Math. papers*, I (1967), 19–24.

[2] Now in Columbia University Library. See Zeitlinger's description in Sotheran, *Bibl.*, Suppl. 2, I (1937), 299, no. 3976.

[3] *Ibid.* p. 757, no. 12098.

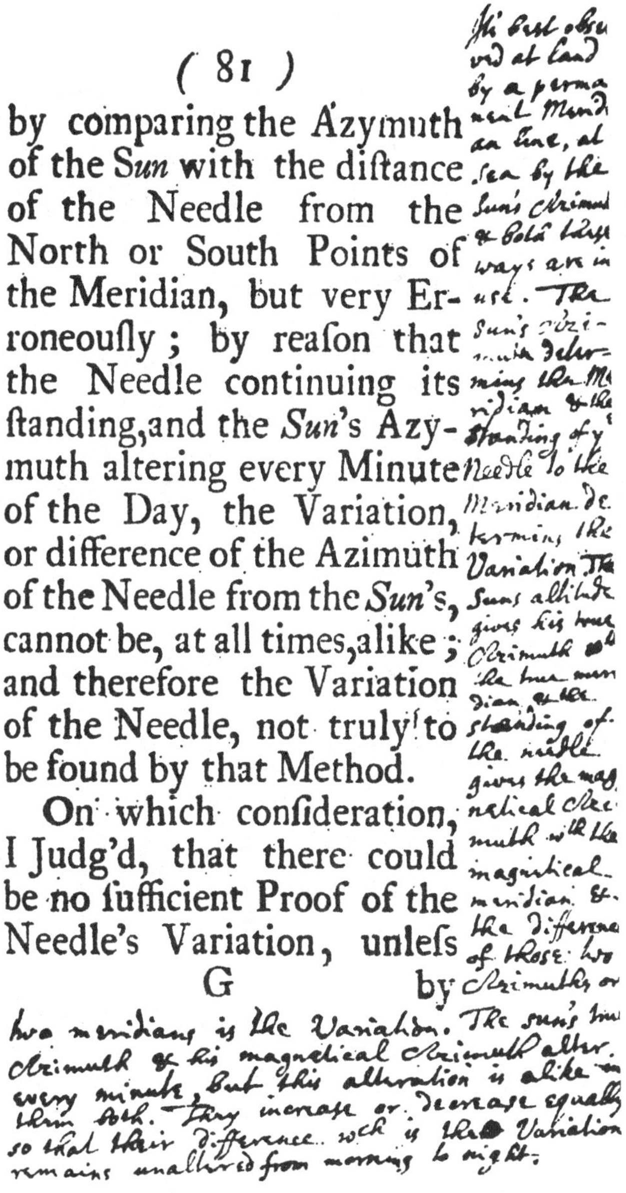

(81)

by comparing the Azymuth of the Sun with the diſtance of the Needle from the North or South Points of the Meridian, but very Erroneouſly; by reaſon that the Needle continuing its ſtanding, and the Sun's Azymuth altering every Minute of the Day, the Variation, or difference of the Azimuth of the Needle from the Sun's, cannot be, at all times, alike; and therefore the Variation of the Needle, not truly to be found by that Method.

On which conſideration, I Judg'd, that there could be no ſufficient Proof of the Needle's Variation, unleſs

G by

His best observ'd at land by a permanent Meridian line, at sea by the Sun's azimuth & Pole. Both ways are in use. The Sun's azimuth determines the Meridian & the standing of ye Needle to the Meridian determines the Variation. The Suns altitude gives his true azimuth wth the true meridian & the standing of the needle gives the magnetical azimuth with the magnetical meridian & the difference of those two azimuths or two meridians is the Variation. The sun's true azimuth & his magnetical azimuth alter every minute, but this alteration is alike in them both. They increase or decrease equally so that their difference wch is the Variation remains unaltered from morning to night.

Plate 1 Newton's notes in Howard, *Copernicans of all sorts, convicted*. . .1705 (no. 810).

ad vesperam 19, cùm viderentur propinquæ limbo, dies sunt 6 ½, & adhuc diffabant tantùm à limbo; ut facile effet collesu, eas non exituras isto die. Nubes & nox deinde obstabant observantibus, at die 20 orto Sole, cùm nondum completi dies septem, jam disparverant. Apparens velecitas prope centrum talis erat, ut si manifest eadem, maculæ quatriduo propemodum perventuræ fuissent ad limbum disci, sed fecundum hanc hypothesin apparens velocitas imminui debuit, prout maculæ à centro discessissent longius, id quod re ipsâ ita evenit. Sed & figura Coronæ illius nebulosæ, quæ maculas ambierat, & prope centrum oblonga fuerat instar auris humanæ, cùm ad limbum pervenerint erat; contrahebatur, ut quæ illic fuerat maxima, hic videretur minima, quo motus circa axem Solis circularis similiter adstruitur.

Tertiò, *De Saturno & ejus satellitibus.* Anno 1610. Galilæus primùm deprehendit Saturnum tricorporeum, cujus nativam faciem postmodum Hugenius Anno 1655 accuratiùs determinavit, & Systema Saturni in lucem edidit, quale hìc verbis describere conabimur. Adhibito igitur globo vulgari, elevato polum ejus boreum ad ipsum usque Zenith, tum globus ipse refert annulum Saturni, dummodo ligneus, planus & latus, refert annulum Saturni, dummodo spatium inter globum & Horizontem ligneum concipias multò amplius, & ipsum marginem extendas eousque, dum ejus diameter ad diametrum globi eam circiter rationem obtineat, quæ est 9 ad 4, & excessum semidiametri marginis supra semidiametrum globi bipartiaris, ita ut semis unus cedat latitudini annuli, & alter interstitio, quod est inter annulum & corpus Saturni. Tum in plano Horizontis, vel annuli, undique continuato concipias circulum quendam satis longo intervallo ipsum annulum ambientem, in quo feratur Comes Saturni ab Hugenio deprehensus primùm anno 1655; periodum suam absolvens diebus proxime 16. Ita concipias jam Systema Saturni. Deinde finge tibi globulum quendam minutum, cum annulo, & interstitio ejus à corpore Saturni

materialem cogita esse sphæram, in qua movetur Saturnus; cujus centrum occupet Sol, & non longè ab hoc divagatur tellus. Cæterùm adjunges principium Arietis cardini occiduo, & initium Libræ ortivo; ita ut Cancri initium sit sub Meridiano. Tum globulum Saturni, cum suo annulo & orbita comitis, suspende filo ipsi Zenith alligato, ita ut annulus Saturni, & orbita comitis, dum globulus secundum eclipticæ ductum circumferetur, maneant semper Horizonti parallela. Videbis jam secundum hanc hypothesin explicari posse varia ista phænomena, quæ circa phases Saturni, & motum ejus Comitis observata fuere. Nam circa initium Arietis & Libræ, cùm annulus Saturni, & orbita Comitis in idem cum Horizonte, vel Æquatore, planum incidunt, spectantur & annulus & orbita satellitis, quasi in lineam rectam porrecta, quare tum annulus ipse (utpote tenuis, vel saltim extremâ orâ lucem exiguam aut nullam refundens, ut sunt corpora quædam repercutiendo lumini minùs apra) visum effugiens, solum corpus Saturni relinquit rotunditate suâ conspicuum, ita tamen, ut medium nigricante quadam lineâ distinctum appareat, quod ab ipso, de quo loquimur, annulo procedere facilè credas. Provecto deinde paulatim Saturno versus initium Cancri adnascantur brachia vel anfæ, exiguæ primùm, mox grandiores, donec circa 20 ½ grad. Geminorum latissimus appareat annulus figurâ ellipticâ, cujus longior diameter æquatori parallela est, ut semper, minima vero ad eundem perpendicularis. Illud verò ita contingere necesse est, quia circulus Horizontis plano parallelus, quò magis supra Horizontem elevatur, eo majorem offendit latitudinem, adeò, ut si ad ipsum usque Zenith attolleretur, perfectum omnino circulum visuri essemus, qui in Horizontis plano positus non nisi rectam lineam mentiebatur, Sed & orbita Comitis, quò magis ab Æquatoris (vel Horizontis) plano attollitur, eo latiorem ellipsin porrigit, unde fit, ut Comes tum in utraque cum Saturno conjunctione à nobis conspiciatur, qui prope Æquatorem non nisi rectam lineam per corpus Saturni, vel prope, excensam describere

Plate 2 One of Newton's notes in Mercator, *Institutionum astronomicarum libri II*...1676 (no. 1072).

mopolita [i.e. George Starkey]...1669 (no. 1478), now at the University of Wisconsin at Madison, is heavily annotated and corrected by Newton. Also in the field of alchemy, Geber's *Chimia*...1668 (no. 657) has two closely written pages of notes in Newton's hand on the fly-leaves, while *The fame and confession of the Fraternity of R:C:*...1652 (no. 605) has a long note by Newton on 'R.C. the founder of the supposed Rosy Crucian society'.[1] On the rear end-papers of Wing's *Astronomia Britannica*...1669 (no. 1743) Newton made extensive notes in about 1670 explaining, among other things, 'the disturbance of the moon's orbit from its theoretical elliptical shape through the action of the solar vortex (which "compresses" the terrestrial one bearing the moon by about $\frac{1}{43}$ of its width)'.[2] Another astronomical book, Bayer's *Uranometria*...1655 (no. 142), now at Babson College, has the names of the constellations supplied in Newton's hand on every map. Among the mathematical works annotated, the 1655 first edition of Barrow's Latin epitome of Euclid's *Elements* (no. 581) has copious notes by Newton; there are annotations by him also in his copy of William Jones's edition of his minor mathematical works, *Analysis per quantitatum series*...1711 (no. 1153). Other mathematical books with Newton's alterations and amendments will be referred to below.

The most heavily annotated theological volume by far which I have encountered among Newton's books is his English Bible (no. 188) printed in London by Henry Hills and John Field in 1660, bound together with a 1639 Book of Common Prayer (no. 240) and Sternhold and Hopkins's Metrical version of the Psalms, 1661 (no. 1560). Bearing clear signs of frequent use, it contains very many annotations and Biblical references written by Newton in the margins of the text, especially at the Books of Daniel and Revelation, together with a six-line note on the end-paper. In addition to providing material for the study of Newton's theological thoughts and inclinations, the book also offers a special personal and human – not to say sentimental – sidelight on Newton the man. A note attached to the volume (presented to Trinity in July 1878) states that 'it was given by Sir Isaac Newton in his last illness to the woman who nursed him'. I am prepared to believe that this is substantially, even if not literally, true.[3]

---

[1] Macphail, *Alchemy and the occult*, II, 349, has a facsimile of this note.

[2] D. T. Whiteside, 'Newton's early thoughts on planetary motion: a fresh look', *British Journal for the History of Science*, II (1964–5), 119 n. 11.

[3] A letter written by John Cox the donor, dated 3 Aug. 1878 and now pasted in the volume (Trinity College, Adv.d.1.10), refers to an accompanying statement (also attached to the book) written by his uncle Joseph Cox of Hurstbourne Priors setting out what the latter knew of the Bible's provenance. Joseph Cox's undated note reads:
   'The Bible formerly belonging to Sir Isaac Newton was in the possession of my father-in-law Mr Edward Golding of Hurstbourne Priors up to the time of his death. His statement respecting it was that it was given by Sir Isaac Newton in his last illness to the woman who nursed him – it was given either by her or some member of her family to

2-2

The category of notes here treated as bibliographical consists of material relating to the author or the subject of the book. These notes are usually listed on the fly-leaves, and most of the works containing this kind of note are on alchemical subjects. They show Newton to be well versed in the literature of alchemy. In his copy of Fabre's *Opera*, 2 vols., 1652 (no. 598) Newton carefully copied a long list of other works by Fabre, adding at the end 'Desumpta sunt hæc ex Libris ipsis impressis in Bibliotheca Bodleiana Oxonij', that is from Hyde's *Catalogus impressorum librorum Bibliothecæ Bodlejanæ*...2 vols. in 1, 1674, which Newton also had in his library (no. 824). The fly-leaf of *Philosophiæ chymicæ IV. vetustissima scripta*...1605 (no. 1301) carries a long manuscript addition where Newton gives 'Authores a Seniore citati' – a list of eighteen authors with page numbers given for each (twenty alone for 'Hermes') – as well as providing many short references in the margins of the pages of the book to other alchemists. His copy of Mariotte's *Œuvres*...2 vols. in 1, 1717 (no. 1031) shows that in that year, or possibly later, Newton was still concerned to make this particular kind of annotation. In this book, there, at the bottom of page **2, he added '†first printed in 1681' as a footnote to a citation of '*Le traité des couleurs*...de Paris de 1686' in lines 11 and 12 of the printed text (where a corresponding '†' is inserted after '*couleurs*'). The preface to *Tractatus aliquot chemici singulares*...1647 (no. 1623) is signed 'L.C.': here Newton has added 'id est Lud. Combachius. Vide Bibl. Chem. p. 64' in a reference to Borel's *Bibliotheca chimica*...1654 (no. 246), which he elsewhere quoted as an authority for similar notes made in other of his alchemical books. *Le Triomphe hermetique*...1689 (no. 1642) ends at page 153 with the author's name printed as the anagram 'DIVES SICUT ARDENS, S✱✱✱'. Newton offers the solution immediately below: 'S.E. Sanctus Didierus' (Limojon de Saint-Didier). Similarly we are able to see from his own annotations (plate 3) how Newton deciphered the code which Michael Mayer (or Maier) employed at the bottom of page 160 of his *Themis aurea; hoc est, De legibus Fraternitatis R.C. tractatus*...1618 (no. 1049). Here Newton returns the coded lines to Latin and shows us the key he used to do so.

The third main grouping of Newton's notes, consisting of manuscript corrections and additions, occurs for the most part in his mathematical books.

Mr Golding's mother and at her death came to him and was in his possession more than sixty years. He died two years ago and then it came to me.'
I have no doubt about the authenticity of the Newton notes and it is highly significant that Hurstbourne Priors is but three miles from Hurstbourne Park, the seat of the Portsmouth family. Even if Newton on his death-bed did not himself actually hand over the Bible to the nursing woman, then it would seem quite likely that John Conduitt, who looked after Newton's affairs after his death, may have offered her the book as a memento of devoted service. Furthermore, if the nurse was in service to the Conduitts, this could account for the book's presence at Hurstbourne Park (or Priors) when their grandson became the second Earl of Portsmouth. (See also p. 266 below.)

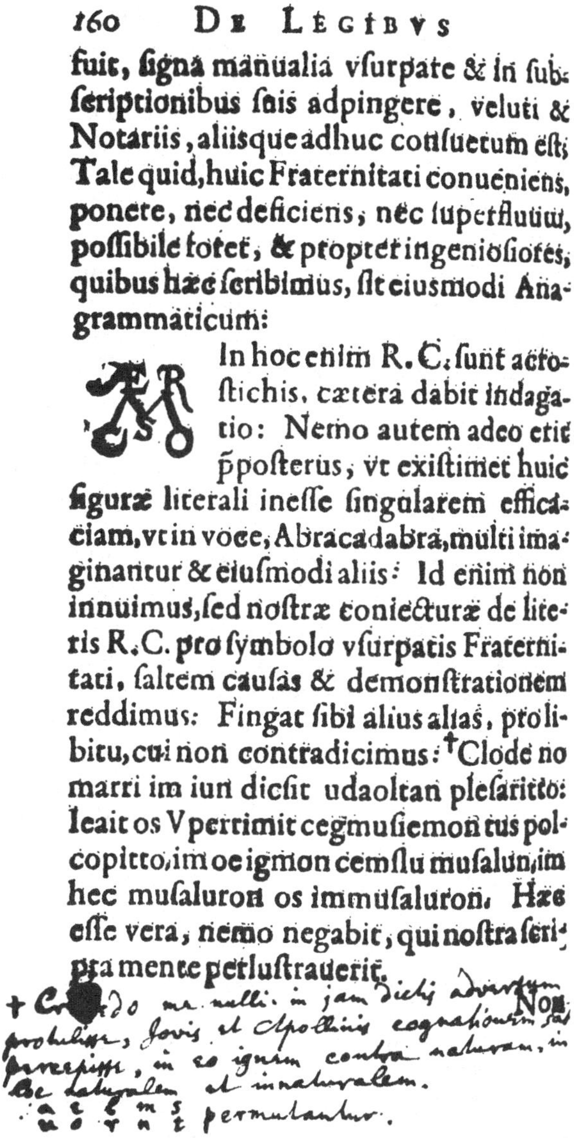

160   D ɪ L ᴇ ɢ ɪ ʙ ᴠ s

fuit, figna manualia vſurpate & in fub-
ſcriptionibus ſuis adpingere, veluti &
Notariis, aliisque adhuc cotſuetum eſt;
Tale quid, huic Fraternitati conueniens,
ponere, nec deficiens, nec ſuperfluum,
poſſibile fotet, & propter ingenioſiores,
quibus hæc ſcribimus, ſit eiusmodi Ana-
grammaticum:

In hoc enim R. C. ſunt acro-
ſtichis, cætera dabit indaga-
tio: Nemo autem adeo erit
ꝓpoſterus, vt exiſtimet huic
figuræ literali ineſſe ſingularem effica-
ciam, vt in voce, Abracadabra, multi ima-
ginantur & eiuſmodi aliis: Id enim non
innuimus, ſed noſtræ coniecturæ de lite-
ris R. C. pro ſymbolo vſurpatis Fraternitati, ſaltem cauſas & demonſtrationem
reddimus: Fingat ſibi alius alias, pro li-
bitu, cui non contradicimus: ⁺Clode no
marri im iun dicſit udaoltan pleſaritto:
leait os Vperrimit cegmuſiemon tus pol-
copitto, im oe igmon cemſlu muſalun, im
hec muſaluron os immuſaluron. Hæc
eſſe vera, nemo negabit, qui noſtra ſeri-
ꝓta mente petluſtrauerit.

[handwritten annotation:]
+ Credo me nulli... in jam dictis aduerſum
probibere, Jovis et Apollinis cognationem ſat
perceptiſſe, in eo ignem contra naturam, in
ec naturalem et innaturalem.
permutantur.

Plate 3   Newton's annotation in Mayer, *Themis aurea; hoc est, De legibus Fraternitatis R.C. tractatus*...1618 (no. 1049).

21

The notes vary from the alteration of a single number to substantial and significant additions. On page 14 of his copy of *Norwood's epitomie*...1645 (no. 1189) the misprinted number '46' has been corrected to '45' in a Newtonian hand, and the second volume of Wallis's *Opera mathematica*, 1693 (no. 1710) also has minute corrections by Newton on pages 392–3; while his copy of Raphson's English version of his own *Universal arithmetick*...1720 (no. 1155) is corrected by him in the margin of page 4. Similar annotations are present in the imperfect copy of Rudd's English version of Euclid's *Elements* ...1651 (no. 584) on page 22, and in *Dary's miscellanies*...1669 (no. 488) on pages 4, 6, 19, and 36. While Newton was reading his Lucretius, *De rerum natura libri VI*...1686 (no. 990) he seems to have decided to number the lines of text. He started by listing every tenth one in the margin (10, 20, 30 etc.); then the intervals became larger, widening to be every fifty lines (1000, 1050) on page 17, at which point he either saw no good reason to continue or ran out of patience. Robert Boyle in his *General history of the air*...1692 (no. 264) asserts on page 100, line 6, that the weight of a cubic inch of water is $254\frac{7}{16}$ grains, and in the margin opposite Newton has added 'Dr Wyberd found it $253\frac{1}{8}$ᵍʳ· Put it 254ᵍʳ· & a sphære of water of an inch in diameter will weigh 134ᵍʳ· [133 in fact]'.

The last category of books with Newton's notes, those of his own works containing his additions and revisions for inclusion in later editions, has an importance which has been recognized and discussed in a number of works published in recent years. Cambridge University Library has an interleaved copy of the first (1687) edition of Newton's *Principia* (no. 1167) with the author's annotations both on the printed pages and on interleaves. Unfortunately this volume was seriously damaged and stained by fire and water after Newton's death (though the volume is now restored), with the result that some of the notes were made difficult to read. A copy of the same edition in Trinity College (no. 1168) also contains many notes by Newton.[1] The same two libraries also each have an annotated copy of the second edition, 1713, of the work. In the interleaved volume in the University Library (no. 1169) there are notes by Newton in the printed text, on the interleaves, and on a number of additional inserted paper slips. Similar emendations by Newton occur throughout the Trinity copy of this edition.[2] Two bound sets of the first (1704) English edition of Newton's *Opticks*, one in Cambridge University Library (no. 1158) and the other in Trinity College (no. 1157) show the author at work, both as proof-corrector and as reviser of his book for the second English edition of 1717 (nos. 1159 & 1160). The former

---

[1] See I. B. Cohen, *Introduction to Newton's 'Principia'* (Cambridge, 1971), pp. 25–6 and plates 1–3, 13.

[2] The autograph annotations in both sets of interleaved and annotated copies are now reproduced in Isaac Newton, *Philosophiae naturalis principia mathematica, 3rd ed., 1726*..., ed. A. Koyré and I. B. Cohen, 2 vols. (Cambridge, 1972).

consists of the printer's proofs and contains the complete text of Books 1 (pp. [1] 1–144) and 2 (pp. [2] 1–137), but lacks the title-page, the plates, and the appended Latin treatises ('Enumeratio linearum tertii ordinis' and 'Tractatus de quadratura curvarum' which take up the remaining pages [2] 138–211 in the complete edition). The Trinity volume has the title-page, the same English text (except for pages [1] 1–17, which are missing), and the plates, but it too lacks the Latin section. A comparison of the many corrections and insertions made by Newton in the wide margins of each of the two volumes shows that they are identical (and very clearly written) for Book 1; Newton took particular care to add to each long additional passages in the ample margins of pages 73 and 80. In Book 2, however, many of the lengthy inserts which he made in the printer's proofs, particularly on pages 70, 117, 133–5, and 137, were not copied into the Trinity volume, though the straight-forward corrections and short additions are present in both volumes. In addition to its copious holograph annotations the University Library book carries the printer's markings in the text and margins showing that the text of the quarto 1704 edition was used as copy for the slightly smaller octavo format of the revised second English edition of 1717. These markings indicate exactly where the first signature of each gathering of the new edition was begun: thus, in the text of page 14 (B3v), line 2, the sign '[' was placed before the word 'Paper' and 'C.I.ᵐᵃ. Fol: 17' correspondingly written in the margin, and so on throughout. Examination of the second edition shows that these markings are indeed accurately keyed to the printed text and that all the annotations made by the author in the University Library volume were included. Newton also made many similar notes and loose inserts in his personal copy of the first edition of Samuel Clarke's Latin translation, *Optice*...1706 (no. 1162). In his copy of his *Arithmetica universalis*...1707 (no. 1154), now privately owned, Newton corrected several minor misprints, inserted more appropriate running heads, made a number of transpositions in its text, and marked a great many passages for deletion. Almost none of these latter changes were effected in the revised 1722 edition.[1]

Joseph Raphson's *Historia fluxionum*...1715 (no. 1370) reasserted Newton's sole priority in discovering the method of fluxions against the counter-claims made for the differential calculus of Leibniz. Newton was perhaps aware of the proposed contents of this book before its appearance; he certainly wrote and published (anonymously) in 1717 an appendix which set out correspondence during 1715–16 between himself, Conti, and Leibniz vindicating his claim to priority. Newton's copy of this Latin version of Raphson's work carries an inserted printed 'Præmonitio lectori' which makes it clear that the author had died before the work was published. It is a first issue, lacking Newton's 1717 appendix, but has manuscript Latin notes by him, some of which were introduced into the English version *The history of fluxions*...

---

[1] Newton, *Math. papers*, v (1972), 14.

1715 [–1717]. The title-page of this work has it as being 'By (the late) Mr. Joseph Raphson'; Newton's copy (no. 1371) of this edition contains the 1717 appendix at pages 97–123, and in it he has made a number of still unpublished corrections and alterations. (The appendix was simultaneously added by Newton to later issues of both the Latin and English versions.)[1] When Pierre Des Maizeaux was preparing his edition of *Recueil de diverses pièces*... 2 vols., 1720 (no. 1379), which contains correspondence by Newton, Leibniz, Samuel Clarke, and others, he sent Newton a set of preliminary proof-sheets of pages 1–88 (signatures A–E4) of the opening section of volume 2, entitled 'Lettres de M. Leibniz et de M. le Chevalier Newton, sur l'invention des fluxions & du calcul différentiel'.[2] These proofs (no. 1380) contain on their pages 71–88 (in the final published version pages 75–93) 'Remarques de M. le Chevalier Newton sur la lettre de M. Leibniz' on the invention of fluxions. Newton's 'Observations', which had been published in their original English on pages 111–17 of his 1717 Appendix to Raphson, were put into French by Des Maizeaux, and had already been profusely corrected by the translator when they reached Newton in June 1720. Though Newton added two small inserts to the sheets – on pages 16 and 17 (these are not in the heavily annotated part) – they were not included in the published edition.

When scrutinizing the pages of the books from Newton's library for any trace of a manuscript note there is the ever-present risk of wishfully attributing to him an annotation which is in the handwriting of someone else. It is not surprising that such misbegotten identifications have often occurred, for though eighty-four volumes of the nine hundred here examined contain (in my estimation) notes by Newton himself, there are another ninety-seven volumes in the collection with notes in alien hands. There can, however, be no absolute guarantee that in my present examination of his books I have not overlooked some obscure note made either by Newton or by another's pen. Most of the non-Newtonian manuscript additions seem already to have been present when Newton acquired the books second-hand. The remainder, entered after Newton's death I would presume, consist mainly in spelling out the full names and titles for contemporary historical, political, or literary figures who are indicated in the printed texts (in the early-eighteenth-century fashion) only by initial letters accompanied by dashes or asterisks. (The handwriting here bears a close resemblance to that of the Rev. Charles Huggins, Rector of Chinnor, Oxford, who acquired Newton's library in July 1727.) Of the ninety-seven volumes, twenty-eight are on theology and religious history, fourteen on the classics, eleven are mathematical, ten alchemical, eight historical, and the rest cover a variety of other subjects. A Greek Old Testament of 1653 (no. 196) and Cosin's *Historia transubstantiationis papalis*...1675 (no. 444) both contain notes by their pre-

[1] *Ibid.* III (1969), 11 n. 29.
[2] See Newton, *Correspondence*, ed. A. R. Hall and L. Tilling, VI (1976), 454–63.

vious owner Isaac Barrow, and Halley's *Catalogus stellarum Australium*...1679 (no. 735) has a series of numbers written by the author in the margins and in the text, evidently referring to some earlier catalogue of stars.

It is rare to find lists of page numbers written by Newton for reference purposes on the (front or rear) end-papers of his books. Though Wing's *Astronomia Britannica*...1669 (no. 1743) has a page-list of topics made by Newton on the first fly-leaf of the book (as well as a number of notes and longer related comments by him on its rear sheets) it was not his normal practice to do this. He more often preferred to use his own characteristic method of marking passages that had some special significance or interest for him by turning back (up or down) the nearer corner of the page. This was no casual act: such 'dog-earing' was executed with precision. The upper (or lower) corner was turned down (or up) so that its tip should pinpoint exactly a previously ordained part of the printed text – a sentence, phrase, or even a single word. If required, both corners of the same page were used. Ricard's *Traité général du commerce*...1700 (no. 1399) has the corner of page 187 turned down (see plate 4) and that of page 204 turned up with their tips resting on references to the French 'Loüis d'or', while in volume 1 of Desaguliers's *La science des nombres*...2 vols., 1701 (no. 503) the lower corner of page 252 is turned up to point directly to another comment on this same coin (plate 5). In the section headed 'Traité des couleurs' in volume 1 of Mariotte's *Œuvres*...2 vols. in 1, 1717 (no. 1031), the lower corner of page 226 has been turned back with its tip resting at lines 20–21 in the centre of the page where the text reads 'Le savant Mr. *Newton* a fait une hypothèse nouvelle & fort surprenante pour expliquer tous ces effetz'.

Some idea of the frequency of Newton's use of this marker may be gained from the fact that, of the 862 volumes in Trinity College, 122 still have some of their pages turned back while a further 152 exhibit clear signs of having been subjected to the same treatment in the past (though now restored to their pristine state). Of this latter category of anterior 'dog-earing' Newton's copy of the 1706 Latin version of his *Opticks* (no. 1162) provides an interesting example which has a significance previously unrecognized by those who have handled the volume. Of those pages once turned back, the tip of page 346, when its lower corner is reversed along the track of the still-visible fold, points to the sixth line from the bottom – alongside the word '*Sensorio*' in the phrase 'Voluntate sua corpora omnia in infinito suo *Sensorio* movere, adeoq; cunctas Mundi universi partes ad arbitrium suum fingere & refingere.' A recent study of the controversy which arose between Leibniz and Clarke over the precise meaning of the word 'Sensorium' as used by Newton in his *Opticks* calls attention to the two states of the preceding page 315 of the 1706 *Optice*,[1] in the first of which the text read 'Annon Spatium Universum,

---

[1] A. Koyré and I. B. Cohen, 'The case of the missing *tanquam*: Leibniz, Newton & Clarke', *Isis*, LII (1961), 555–66, esp. 563.

Sensorium est Entis Incorporei, Viventis, & Intelligentis' and was then in the second state altered to be 'Annon Sensorium Animalium, est locus cui Substantia sentiens adest'. Since the later citation of page 346 was there left unchanged and survives in all subsequent editions, it might well be assumed that Newton overlooked the phrase's existence on this later page, but the 'dog-earing' proves that he did not do so, either in 1706 or at some later date. Some tidy-minded librarian or bookseller might reasonably be presumed to have straightened out in this way the bulk of the page corners earlier turned back were it not that several individual volumes, in addition to having some page corners still bent back, also have others which were certainly once similarly 'dog-eared' and later returned to their original position. I conclude therefore that Newton came back to these pages, did with them whatever he had in mind to do, and then, having finished his business, tidied them up. 'Dog-earing' was no uniquely youthful or middle-aged characteristic of Newton: he seems to have used the device all his life. Gretton's *A vindication of the doctrines of the Church of England*. . . (no. 715), published as late as 1725, has three pages still turned back in typical style. There is considerable variation in the degree of 'dog-earing' in different volumes. Some of the books examined have only a single page turned down, but Newton's copy of Vossius, *De theologia Gentili*. . .1641 (no. 1697) has 112 of its 732 pages turned back, and Bochart, *Geographia sacra*. . .1681 (no. 231) has 58 pages similarly treated. Furthermore some pages in the two volumes have both their top and bottom corners folded to indicate separate passages in the printed text; the Vossius on seven occasions, the Bochart five. Eight other works have between eleven and twenty pages with this distinctive marker. Details of surviving 'dog-earing' and of still existing evidence of such anterior page-marking are given in the catalogue entries below. The 'dog-eared' works are mainly theological, geographical, historical, and classical, though there are quite a few in other fields. Boyle's *Sceptical chymist*. . .1680 (no. 270), for example, has nineteen passages thus precisely marked, and James Gregory's *Optica promota*. . .1663 (no. 713) has eight page corners turned, while Wallis's *Opera mathematica*. . .vol. 2, 1693 (no. 1710) has but one. But the nature and content of certain classes of books, especially the scientific and mathematical ones, did not well suit the purposes for which Newton used his page-marking method, and the internal evidence that these books were heavily used and frequently consulted is provided rather by the heavy wear and tear on their pages.

The references so pin-pointed by folding the pages themselves are often to place-names, to obscure proper names occurring in ancient history and theology, to uncommon words and definitions usually in Latin and Greek and to occurrences of Newton's name in a printed text. This is the case in his copies of Mariotte's and Wallis's works cited above as well as in Hermann's *Phoronomia*. . .1716 (no. 756), where page 394 is still turned down marking a reference to Proposition xxxvii, Book ii of the *Principia*, and in Maclaurin's *Demonstration des loix du choc des corps.* . .1724 (no. 1010), where the bottom corner of page 21 remains pointing up to 'Le celebre M. Newton'. In the 'Éloge de Monsieur Leibnitz' published in volume 2 of Fontenelle's *Histoire du renouvellement de l'Académie Royale des Sciences.* . .1717 (no. 624) the corner of page 43 shows a very distinct diagonal crease in the paper which when folded down along its track rests its tip unmistakably on line 7 where we read of 'l'admirable Livre de M. Neuton *Des Principes Mathematiques de la Philosophie naturelle*'.[1] I have come across further clear signs of pages having formerly been folded in his typical manner to mark mentions of Newton's name in his copies of Molyneux's *Dioptrica nova.* . .1692 (no. 1098) (the top corner of page 273), Rémond de Monmort's *Essay d'analyse sur les jeux de hazard*, 2e éd., 1713 (no. 573) (the bottom corner of page 396), and Des Maizeaux's *Recueil de diverses pièces.* . .vol. 2, 1720 (no. 1379), where the bottom corner of page 32 was once turned up to indicate a passage from a letter of Leibniz dated 18 April 1716 which reads, 'Etant à Londres, mais très-peu de jours, je fis connoissance avec M. *Collins*, qui me montra plusieurs Lettres de M. *Newton*, [the sentence continues on page 33] de M. *Gregory* & d'autres, qui rouloient principalement sur les *series*.'

Though booklovers may deplore this ugly habit – Heinrich Zeitlinger of Sotheran's termed it 'naughty' – it clearly reflects Newton's attitude that books are working tools to be used as convenient and to destruction. The possibility cannot be entirely ruled out that later users of his books may here have imitated him, but this is unlikely to have happened to any substantial, misleading extent, and there would seem to be much useful information to be gained by a detailed analysis of those parts of his books which Newton, unconsciously or deliberately, left marked for posterity in this way. Its potential importance to Newton scholars as an index to the direction of his mind as he read the books in his library has certainly not as yet been fully realized.

---

[1] In addition to thus marking the page Newton wrote to Fontenelle (?Autumn 1719): 'On this occasion I should give you due thanks for the honours which you have conferred upon me in the 'Éloge' of Mr Leibniz. . .' (Translation of the original Latin printed in Newton, *Correspondence*, ed. A. R. Hall and L. Tilling, vii (1977), 72.)

CHAPTER 2

DISPERSAL OF NEWTON'S LIBRARY
AFTER HIS DEATH

The two documents on which this present catalogue of Newton's library is based are the Huggins List, 1727, and the Musgrave Catalogue, *c.* 1766. They cannot be discussed adequately without some knowledge of the dispersal of Newton's estate following his death on 20 March 1726/7 and in particular of the immediate fate of his books.

Newton died intestate. This was not through ignorance, for he had been executor to his mother's will[1] and was also from time to time engaged in drafting various official documents for his mother's family, and was clearly familiar with all the legalities. As he lived to the age of eighty-four with (so it is reported) his faculties virtually unimpaired, it would seem that Newton deliberately refrained from making a will. Even had he done so, however, we might well doubt if he would have bothered to make any special provision for his books. In the event Newton's estate was divided equally among his eight (half-)nephews and (half-)nieces, though none of them was able to benefit from it until the complications ensuing from Newton's official position had been resolved. As Master of the Mint he was held personally responsible for all outstanding debts at the Mint, and immediately after his death Newton's assets were frozen by the Prerogative Court of Canterbury until a guarantor would stand bond for the total of £34,330 which he 'owed' the Crown. John Conduitt, who had married Newton's favourite niece, Catherine Barton, and was to succeed Newton as Master of the Mint, was the only relative wealthy enough to stand bond for this 'debt'. This he did in late May.[2] Before the bond was forthcoming, however, the Prerogative Court had already (on 18 April) appointed the seniors in each of the three groups of inheriting nephews and nieces – viz. Thomas Pilkington, Benjamin Smith, and Catherine Conduitt (for whom her husband John *de facto* acted) – as administrators of the estate, and the Court's Commission of Appraisement

---

[1] See C. W. Foster, 'Sir Isaac Newton's family', *Reports and papers of the Architectural Societies of the County of Lincoln, County of York* [etc.], xxxix (1928), 1–62 and particularly 50–53 where the probate copy of the will is reproduced in full.

[2] Conduitt's own account of the proceedings is contained in King's College, Keynes MS 127A. For further details see Newton, *Math. papers*, i (1967), xvii–xix.

of the same date required a detailed inventory to be made of Newton's personal effects, his house contents, his books, and his manuscripts. Accordingly 'A true and perfect inventory of all and singular the goods, chattels and credits of Sir Isaac Newton' was 'taken and appraised on the 21st, 22nd, 24th, 25th, 26th, 27th, days of April...1727' and the final declarations of the document were signed 'John Conduit, Catherine Conduit'.[1] This comprehensive record was written down on a series of vellum skins sewn together, making a document seventeen feet long and five inches wide. Included among the contents of Newton's house in St Martin's Street is 'Item 362 books in folio, 477 in Quarto, 1057 in Octavo, duodecimo and 24$^{mo.}$ together with above one hundred weight of pamphlets and Wast books valued at the sum of £270. 0. 0.' The printed books in Newton's library[2] soon after his death therefore were 1896 in total, together with an unspecified number of pamphlets.[3] An entry relating to Newton's manuscripts, loose papers, letters, and other material follows the item for the printed books.

By early May further steps had already been taken to dispose of Newton's library. A small vellum-bound manuscript notebook of nine leaves, now in the Huntington Library at San Marino, California, contains accounts of expenses incurred by Newton's administrators, including: 'May 6. – Paid to the booksellers for appraising the books etc., 21$^{l.}$ For taking a catalogue of the library, 4$^{l.}$ 4$^{s.}$...For 4 copies, 2$^{l.}$ 2$^{s.}$'. Other items relate mainly to Newton's funeral, and the notebook ends: 'These expenses were paid by Mrs. Conduitt and are signed, as examined, by Thomas Pilkington and Benjamin Smith...'[4] Pressure from the Prerogative Court, who appreciated both the need and the opportunity for a swift conversion of Newton's books into hard cash, most probably influenced the administrators in their decision to try to make a quick sale of the library as a whole, rather than have the books split up and traded off piecemeal to booksellers, or sold by auction. Before they benefited from his estate none of Newton's relatives themselves, apart from the Conduitts, appeared to be interested in acquiring the library, and the financial state of the Royal Society together with the lack of space in its

---

[1] The inventory, now in the Public Record Office, London (Inventory [Sir I. Newton] PROB.3/26/66), is transcribed and reproduced in full by R. de Villamil in his *Newton: the man* (London [1931]; repr. New York, 1972), pp. 50–61. It places a total valuation of £31,821. 16s. 10d. on the estate, that is, about £2500 short of Newton's legal 'debt' to the Crown.

[2] Though we assume here that all the books in the house really had belonged to Newton, it is possible that some of them may have been left behind by his niece Catherine Barton, who had lived with him there continuously for fifteen years before her marriage to Conduitt, and also at intervals afterwards. See pp. 68, 70.

[3] Thomas Maude in his *Wensley-Dale*...3rd ed. (London, 1780), p. 30, referring to Newton dying intestate, states that 'his library...consisted of 2000 volumes and 100 weight of pamphlets', but makes no mention of its subsequent history.

[4] Historical Manuscripts Commission, *Reports on the manuscripts of the late Reginald Rawdon Hastings, Esq....*, I (London, 1928), 415–16.

rooms at Crane Court did not encourage any move to be made towards buying the books of its distinguished (if not too well liked) President. Why Conduitt himself was not allowed to buy the books is not clear. In the event the purchaser of the books was as unexpected in his appearance on the scene as he was unworthy to possess them. Though his unsavoury reputation as a cruel and grasping Warden of the Fleet Prison hardly made John Huggins the most appropriate new owner of Newton's books, since he was, however, on the spot (living near by) and with ample ready money at his disposal, the administrators lost no time in speedily accepting the £300 he offered – £30 more, I may add, than the booksellers' valuation of the library's worth.[1] The motive which caused Huggins to snap up his chance of buying Newton's books was probably a determination to make the country gentleman he himself never could be out of his third son, Charles. The Huggins family, jointly with a Christopher Tilson, had already on 16 September 1723 bought the patronage of the living at Chinnor, near Oxford, for £2000,[2] and after the death of the incumbent John Pocock in February 1727/8 Charles Huggins became Rector there on 20 March 1727/8 (exactly a year after Newton's own death). The deal which led to the Newton library being housed in the Rectory at Chinnor was, however, clinched well before that date. Having at the beginning of the previous June shown his firm intention of buying the books and having agreed on a price, John Huggins had, by the end of July 1727, paid the £300 which concluded the sale.

The list now preserved in the British Library[3] may well be the original catalogue of Newton's library which was made at this time on behalf of the administrators (and whose costs were paid for by them on 6 May 1727) rather than one of the four copies then made from it. Since it records the books bought by Huggins and bears no title of its own it is usually referred to as

[1] Huggins was a moderately rich man. He had become Warden of the Fleet Prison in July 1713, having secured the appointment for his own life and that of his son (John junior) by (so it was reported) a gift of £5000 to Lord Clarendon, and his tenure of the offce became notorious for the cruelty shown to the inmates and for the corruption and extortion which then came to prevail. See J. Ashton, *The Fleet, its river, prison, and marriages* (London, 1888), pp. 265–9.

[2] See Bodleian Library, Wykeham–Musgrave Deposited Deeds, C.41, where both the indenture and the abstract of the sales show that the 'bargain' whereby Robert Gardiner and Catherine his wife sold the advowson of Chinnor for £2000 was made with John Huggins junior and Christopher Tilson. The abstract adds that 'The said John Huggins the Younger dyed intestate [13 Jan. 1736/7] and William Huggins Esq. claims title to the said premises as his next brother and Heir at Law.' In Feb. 1747/8 John Tilson (nephew of Christopher) sold his share of the advowson to William Huggins. (On both see de Villamil, p. 3, and *Victoria history of the counties of England: Oxfordshire*, VIII (London, 1964), 73; these, however, have the 'notorious' John Huggins as the part-purchaser of the living, whereas the documents indicate that it was his son John who actually carried through the transaction – though possibly Huggins senior put up the money for the sale – and who had the hard business head to obtain the good bargain which Huggins got.)

[3] Huggins List (British Library Reference Division, MS.Add.25, 424, bought from Mr W. J. Tait of Rugby on 21 October 1863).

the Huggins List.[1] Compiled within six weeks of Newton's death, it is the most reliable guide we have to the contents of his library, and every work shown there is included in the present catalogue. The document itself consists of twenty-one leaves. Of these, fol. 1 bears the word 'Catalogue' only; the listing itself starts at fol. 2 under the heading 'Folio', and the books are thereafter arranged in descending order of format. The entries, not in alphabetical order, give author and title (title only of course for anonymous works, periodicals, and the like) and the number of volumes (when more than one) together with place and date of publication. Its readily legible writing is in the same unknown hand throughout. As might be expected in a tabulation which shows every indication of having been completed in some haste, the titles are short and rarely run beyond a single line, but are very competently and accurately reproduced. The compiler occasionally even finds time to offer such comments on the condition of some of the books as 'In boards', 'In quires', 'damaged', 'spoild & damagd', 'Dirty & leaf wanting', 'Wants title' or (in Latin) 'Caret Titulo', 'Interleavd, imperfect', 'The last leaf written', 'Ch. Mag. a little staind' 'Large paper. Red Turkey [leather]', 'Gilt Back & Leaves'. There is only one significant error in citing author, namely on fol. 18 where 'I. Gaston' (instead of I. Gaston Pardies) is given as the writer of *La statique*...1673 (no. 1245); and no more than seven datings can be queried, either because, in one case, we can trace no edition of the work in question published in the year stated or because the remaining six books themselves survive (all as it happens in Trinity), allowing a direct check on the title-date to be made. It has been possible to identify all the titles listed with a high degree of certainty.

[1] The document ends:

'2 June 1727. Recd of Mr John Huggins in money five pounds five shillings wch with Three hundred pounds sterling more agreed to be paid by the sd Mr Huggins to the Administrators of Sr Isaac Newton Decd is the price agreed on for the Library of Books, Pamphletts &c. of the sd Sr Isaac Newton sett forth in this Catalogue. Save only and except that the Books [six in total which afterwards passed to the Conduitts] particularly mentioned in the Page to have Sr Isaac Newton's notes, are not be delivered to the sd Mr Huggins, but to be valued by two indifferent persons...and such valuation to be deducted out of the Three hundred pounds, wch is to be paid and the Books to be delivered on or before the first day of July next. I say recd as one of the Administrators ...By me Tho Pilkington. Witness. Robt Champion.

July ye 20th 1727. Recd the three Hundred Pounds above mentioned in full for the Books set forth in this Catalogue. [Signed] Tho Pilkington Catherine Conduitt. Witness J. Woodman.'

Except for the signatures of Thomas Pilkington and Robert Champion, and the receipt of 20 July, the above is in the hand of John Huggins senior. I am able to make this assertion by comparing the handwriting in this List with specimens of the writing in two letters in the British Library (MS.Add.36, 135, fol. 326, and 36, 137, fol. 204), penned in Newgate in 1729 and signed 'John Huggins' when he was imprisoned there awaiting trial for the murder of one of the inmates of the Fleet Prison (he was eventually acquitted). Even though his son John, therefore, had been responsible for the purchase of the living at Chinnor, it was Huggins himself who drew up the deed of transfer of the books.

The Huggins List has two main drawbacks, which will not be found surprising in a document hurriedly completed in response to the urgent dictate of the Prerogative Court. While without its demand no list of any sort (not to mention its four extra copies) would probably have been made, the Court basically required a sale catalogue, one taking note as pertinent of the physical condition of the books. The pressure to produce a list which was seemingly of little personal concern either to its compiler or to those who commissioned the work no doubt occasioned its failure to supply the titles of each separate item in volumes containing more than one work. Reference to these books themselves, which look still to be in the bindings in which they were sold in 1727, permits me to add these omitted titles to the catalogue. A much more serious ensuing defect, one detracting considerably from the real value of the document as a record, is the employment in it of blanket entries for certain groups. This is best illustrated by the following summary of its contents, set out in the sequence (ordered by format) in which the items appear:

Folio	204 separately listed titles comprising	314 volumes
	+'33 wast Folios' (untitled)	33 volumes
Quarto	245 separately listed titles comprising	361 volumes
	+'81 wast Quartos' (untitled)	81 volumes
Octavo	264 separately listed titles comprising	303 volumes
'Small 8° & 12°'	35 separately listed titles comprising	36 volumes
	+'3 Dozen of small [untitled] chymical books'	c. 36 volumes
'Livres François 8° & 12°'	82 separately listed titles comprising	263 volumes
	+'16 Dozen of Wast 8° & 12°' (untitled)	c. 192 volumes
	+'About a hundred & half of Wast Books & Pamphletts'	c. 150 volumes
'Sm. 8° & 12°'	133 separately listed titles comprising	159 volumes
'Books that has Notes of Sir Is. Newtons'	6 separately listed titles comprising	6 volumes
In all:	969 listed titles ⎫ comprising c. 492 untitled works ⎭	c. 1934 volumes

The 969 works whose titles are listed comprise 1442 volumes, and though such vague entries as '3 Dozen', '16 Dozen' and 'About a hundred & half' make it difficult to arrive at a precise count by volumes, the overall total of 1934 shown above does not differ seriously from the inventory count of 1896. The Huggins List shows that Newton possessed three copies of three particular works, namely his own *Analysis per quantitatum series*...1711 (no. 1153) together with his *Traité d'optique... Traduit par M. Coste...2ᵉ éd. françoise...* 1722 (no. 1166), and Varignon's *Nouvelle mécanique...* 2 vols., 1725 (no. 1670). He also had duplicate copies of twenty others, including the first three English editions of his *Opticks*, 1704 (nos. 1156–7), 1717 (nos. 1159–60), and 1721 (no. 1161), and the first Latin version of 1706 (nos. 1162–3), the third edition of his *Principia*, 1726 (no. 1172), as well as works by such authors as Apollonius (two of his titles were duplicated), 's Gravesande, David Gregory, Grew, and Christiaan Huygens.[1] Except for one instance, the entries for works of which Newton had two (or three) copies are not recorded consecutively in Huggins and often appear on different pages of the document. Though it is an obvious disadvantage that the '3 Dozen of small chymical books' and all the various 'Wast Books & Pamphletts' are not described individually, it is quite possible accurately to identify some of these items by their later separate reappearance in the Musgrave Catalogue and from the books themselves – particularly the volumes of tracts now in Trinity.

The six 'Books that has Notes of Sir Is. Newtons', as cited in the Huggins List, are, in order: 'Newtoni Principia interleavd imperf sewed 1687' (no. 1167); 'Optics lat. Interleavd 1706' (no. 1163); 'Descartes Geometria Tom I Lugd [Batav.] 1649' (no. 506: the few annotations in this, however, are not in Newton's hand); 'Bible with Service Dirty & leaf wanting Field [Cambridge] 1660' (no. 189);[2] 'Secrets Reveald or an Entrance to the Shutt Pallace 1669' (no. 1478); 'Ripleij Opera omnia chemica Casselis 1649' (no. 1406). I presume that Conduitt and his wife, as well as acquiring these six books with the corpus of Newton's manuscript writings, also retained all books given to them by Newton before he died, along with other items not regarded as legally published.[3] Three of these at least – Newton's interleaved

---

[1] Titles recorded more than once in Huggins are indicated at the separate entries for these works in the present catalogue.

[2] I cannot, however, ignore the possibility that the 'Bible with Service Dirty & leaf wanting Field 1660' (no. 189) may be the very same book (no. 188, in Trinity College) which Newton is reported to have given to the woman who nursed him in his last illness (see p. 19). If Newton himself handed over the volume, then it is not the same, but if we assume that it was Conduitt who actually presented it, then it could well have been recorded in the Huggins List before being subsequently withdrawn from the sale. The points of similarity between the Huggins listing and the surviving book are that both are annotated by Newton, both include a 'Service' or Prayer Book, and both were issued in 1660. On the other hand I would describe the Trinity volume as well-thumbed rather than 'dirty' and though it has a large part of a fly-leaf torn off I saw no trace of a complete 'leaf missing'. Finally the imprint on the Trinity Bible reads 'Printed in London by Henry Hills and [in smaller type] John Field' whereas the Huggins List offers merely 'Field' – and there were two editions of the Bible published in 1660 with imprint of John Field alone. In view of this uncertainty I have made two separate entries for these books in the catalogue below.

[3] For example, the set of sheets of the *Opticks*, 1704, corrected by Newton for its 2nd English 1717 ed. (no. 1158) (see pp. 22–3 above) and also five sheets (A–E4) of a preliminary proof of the opening section of vol. 2 of Des Maizeaux's *Recueil de diverses pièces...* 1720 (no. 1380) (see p. 24 above).

Principia, 1687; Schooten's first Latin edition of Descartes's *Geometria*, 1649; and Starkey's *Secrets reveal'd*, 1669 – passed by way of Conduitt into the possession of the Portsmouth family, ultimately to be listed in the 1888 catalogue of their Newton holdings,[1] along with a similarly interleaved (and likewise much-annotated) copy of the second edition of Newton's *Principia*, 1713 (no. 1169). All of these except for *Secrets reveal'd* were at that time made over to Cambridge University by the (fifth) Lord Portsmouth as part of his generous benefaction of Newton's scientific papers. This last emerged into public view at the 1936 Sotheby's sale of the Portsmouth family's remaining holdings of Newton's papers and related *Newtoniana*, where it was listed as Lot 121, 'Containing corrections & additions in the hand of Newton on almost every page, some completely filling the margins'. Bought by the dealer Francis Edwards, it was quickly resold to Denis I. Duveen and has now found a permanent home in the Library of the University of Wisconsin at Madison along with the other items of his manuscript and printed collections of works on chemistry and alchemy.[2]

Each separate entry in the Huggins List is preceded by a long dash and ends with a number ranging from one to nine, set in parentheses – as for instance '(6th)', '(2)', '(1st)', for the first three items – not entered in the hand of the original compiler. Instead of a bracketed number the sign 'x' appears at the end of the registration of thirty-two titles which include twelve of the duplicate copies described immediately above (seven of these by Newton), five other works by Newton, two large atlases, books by Bradley, Mead, Morgagni, and Raphson, and the rest a variety of items on differing subjects. The six books with 'Notes of Sir Is. Newtons' (which did not go to Huggins) have no mark at all opposite them. The purpose for which these additions were made is not known, but they look to have been supplied some time after the main list was compiled and may indicate its use as a basis for some sort of book-check or may possibly record an arrangement of the volumes into loads or containers in preparation for their removal from London to Chinnor. It is possible that the items marked 'x' may have been the subject of some further haggling by Conduitt which perhaps may have been in part successful, since nine of the thirty-two titles were not recorded in the Musgrave Catalogue some forty years later, and furthermore Musgrave lists only single copies of those works which had been entered twice or even three times in Huggins.

During his twenty-two years at Chinnor, Charles Huggins seems to have

---

[1] *Catalogue of the Portsmouth Collection of Books and Papers written by or belonging to Sir Isaac Newton* ...(Cambridge, 1888); see Section VII, 'Books'. The missing items do not appear to have been sold with Conduitt's own books when these were auctioned in about 1750 (see Appendix A).

[2] The book is described in Duveen's *Bibliotheca alchemica et chemica*...(London, 1949, repr. 1965), p. 470, with an illustration of its annotations on plate XII.

carried out his pastoral duties in a quiet and conscientious manner.[1] There is, however, no record that he did anything active with his Newton books; if any visitor was invited to look through them at this time I can find no clear contemporary mention of it,[2] and if he made his own private catalogue of their titles I can find no trace of its existence. There is no sign in any of the books that he had any system of marking their position on his shelves, nor have we any way of knowing how he maintained them, and it seems highly unlikely that he kept the Newton books separate from the others in his library which had also been published before 1727. Huggins was not entirely without books of his own when, as a still young man of twenty-eight, he arrived at the Rectory, and he certainly acquired others afterwards, some dating from before 1727 and some whose date of publication (after 1727) makes it impossible that they could have been Newton's. We can only speculate as to how many pre-1727 volumes he may have introduced into his library, though such indications as there are suggest that the total may not have exceeded a score or so. It is likely that his successor at Chinnor, an older man than Huggins had been on arrival there, and probably a more affluent one, added more. Viger's *De præcipuis Græcæ dictionis idiotismis...* 1647 (no. 1686) carries 'Ch Huggins' on its fly-leaf, and Rogers's *A cruising voyage round the world...* 1712 (no. 1414) has 'C Huggins' in the same position, while Barlow's *An answer to a Catholike English-man...* 1609 (no. 117) has 'Charles Huggins Rect[r] of Chinnor' on the end-paper of the book. The manner of these signatures might imply that Huggins had come to possess the first two works before he became Rector and that he obtained the last after reaching Chinnor, while, as none of the three books are shown in the Huggins List, they were probably not Newton's (though one or more might possibly have formed part of one of the blanket entries in it). Along with these

[1] To the questions of Thomas Secker, Bishop of Oxford, Huggins replied in 1738 that 'I reside constantly in ye Parsonage House', that 'all the rites and ceremonies of the Church are duly and regularly perform'd. The Sacramt administred five times in the year, at which there are a great number of Communicants', and that 'the Church-Wardens take the Offerings and divide it amongst the Poor People who attend at the Holy Sacrament' (H. A. L. Jukes (ed.), *Articles of enquiry addressed to the clergy of the Diocese of Oxford at the primary visitation of Dr. Thomas Secker, 1738*, Oxfordshire Record Soc., xxxviii (Oxford, 1957), p. 44). He also gave sixty poor people a sixpenny loaf every St Thomas's Day [21 Dec.] – a practice discontinued by his successor to the living and to the ownership of the library (*Victoria history of the counties of England: Oxfordshire*, viii, 72).

[2] Thomas Birch, the author of the article on Newton in *Biographia Britannica*, v (London, 1760), 3210–44, refers on 3212 n. [D] to the notes Newton made in some of his books, adding that 'the rest [of Newton's books] were purchased by the late Mr Huggens of St Martin's-Lane, London, whose son was a little while ago possessed of them'. For this, as for his earlier biographical account of Newton in *A general dictionary, historical and critical...* vii (London, 1738), 776–802, Birch drew much of his material from William Jones, and 'a little while ago' seems very likely to have been a repeat of information given to him by Jones in the early 1730s. I do not think that either Jones or Birch ever saw Newton's books after they were transferred to Chinnor.

books, all bearing Huggins's personal bookplate, Trinity Library has copies of John Dart's *Westmonasterium*. . .2 vols., 1742 and of *A candid and impartial account of the behaviour of Simon Lord Lovat*. . .1747 (bound with another 1747 item) similarly labelled.[1] Charles's bookplate, measuring 86 by 85 millimetres, bears the Huggins coat of arms (granted to John Huggins in 1725) and the legend 'Rev⁴ Carol⁵ Huggins, Rector Chinner in Com. Oxon.' (see plate 6), and was pasted (I presume) into all of his books. But the absence of this label from its front does not in itself rule out the possibility that a particular book once belonged to Newton: a number of those now located have clearly lost their original front boards and with them their bookplates, or have just had the latter since removed. On the other hand, the presence of the plate does not provide a watertight guarantee that the volume in which it appears was ever owned by Newton, for, as we have seen above, the books which once belonged to him and the 'intruders' among these were identically labelled.

Charles Huggins died in August 1750[2] a bachelor, and without making a will. His estate was administered by his elder brother William, who had inherited the patronage from the younger John Huggins in January 1736/7.[3] William, having on the death of John, the eldest brother, laid aside any idea of personal responsibility for the curacy,[4] presented the living at Chinnor to his future son-in-law, Dr James Musgrave from Kirby (or Kirkby) in Yorkshire.[5] The library as it now was – that is to say, the Newton collection together with any other books that Charles Huggins had introduced into it – was sold to the incoming Rector for £400, and in this way began the Musgrave connection with Newton's library. When William Huggins died in 1761, the property he owned at Chinnor passed to his daughter Jane, and Musgrave (probably moderately wealthy in his own right) became lord of the manor.

---

[1] Trinity College, NQ.17.1, 2 and NQ.8.105.

[2] At the age of fifty. He was buried at Chinnor on 12 Aug.; see the Chinnor Parish Register.

[3] See p. 30 n. 2.

[4] J. R. Bloxam, *A register of*. . .*Saint Mary Magdalen College*. . .*The Demies*, III (Oxford, 1879), 185.

[5] Musgrave came to Chinnor as Rector on 24 Dec. 1750 and married William Huggins's second daughter Jane there on 4 Feb. 1750/51 (see the Chinnor Parish Register). Both men had been at Oxford University, though not as contemporaries. Huggins was at Magdalen College from 1712 to 1722, when he resigned his Fellowship; see J. Foster, *Alumni Oxonienses*. . .*1500–1714*, II (Oxford, 1892), 760. Musgrave began his residence at St John's College in June 1727 and took the degrees of B.C.L. in 1734 and D.C.L. in 1738; see Foster, *Alumni Oxonienses*. . .*1715–1886*, III (Oxford, 1888), 1001. (But I can find no confirmation of R. de Villamil's assertion in his *Newton: the man*, p. 4, that Huggins had made Musgrave's acquaintance there.) By the time James Musgrave came to Chinnor a strong Nonconformist spirit was already established in the village. During his Rectorship (1750–78) Methodism came to alter the village's way of life, and a third of the parishioners attended Methodist meetings. Musgrave attributed the increasing dissent to 'love of novelty' as the number of his communicants dwindled, see *Victoria history of the counties of England: Oxfordshire*, VIII, 75.

Finding a suitable place to set the library in the twenty-two-room Rectory at Chinnor[1] would not have presented Huggins with any difficulty, and when, as seems probable, Musgrave undertook a rearrangement of the books there (if only to incorporate his own books with those from the Newton collection), he too would have had little trouble in finding any extra room needed. This old Rectory was pulled down in 1815 (thirty-five years after the Newton books were removed to Barnsley Park in Gloucestershire), to be replaced by a smaller building,[2] and so the room itself in which the volumes once stood no longer exists. As outlined below, however, we are able by means of Musgrave's Catalogue to reconstruct the sequence in which they were there shelved.

In contrast to his predecessor (who may have lacked funds for the purpose, or may just not have been interested) Musgrave was proudly aware of the Newton library in his Rectory; not only did he proceed to record his new possessions in a systematic way which I shall describe later, but he was also evidently prepared to display its treasures to visitors. The Swedish traveller J. J. Björnståhl wrote home on 24 October 1775 from Oxford giving this account of a visit he paid to see the library:

I shall report a circumstance which, I know for certain, has appeared in no book before. It is this, that we, along with others in this neighbourhood, made a journey for the explicit purpose of seeing the personal collection of books of the great and immortal Knight Newton. They are now in the possession of Dr Musgrave, Rector at Chinnor, eighteen miles from Oxford. It cost him about four hundred pounds sterling. Here are found all the editions of Newton's works, and – what is the most remarkable thing – in their margins they are filled with notes in his own hand, and occasionally yet other leafs at the end of the books are wholly written upon. I have no doubt that a Newtonian would meet here with a lot to delight and much to enlighten. . .For the rest, one may see that Newton had an excellent library. All the Greek and Latin classical writers are to be found in it.[3]

---

[1] Chinnor had been a wealthy parish, and one of its distinguished Rectors, Nathaniel Giles, the incumbent 1629–44, had built the large vicarage at a reputed cost of some £2000. The building suffered damage from the Parliamentary militia during the Civil War, but at the Restoration the Rector was given £100 towards its repair. Robert Plot, the seventeenth-century antiquary, after listing 'several structures of the minor Nobility' of Oxfordshire, continues 'To which add the Parsonage House at the Rectory at Chinner, little inferior to some of the aforementioned, either in greatness, commodiousness, or elegancy of Building' (*The natural history of Oxford-shire* (London, 1677), p. 266).

[2] At the instigation of the incoming parson Sir William Augustus Musgrave, a lifelong bachelor and Rector for nearly sixty years (1816–75). In 1962 this too was demolished and replaced by a new Rectory built on the same site.

[3] (My English translation from the original German of his) *Briefe aus seinen ausländischen Reisen*. . .III (Rostock and Leipzig, 1781), 288–9; cited by Edleston in p. lxxx n. 196 of the 'Synoptical view of Newton's life' which prefaces his edition of *The correspondence of Sir*

Even if his excess of enthusiasm may have led Björnståhl here to exaggerate the extent to which Newton's notes were crammed in the margins of the books and to overemphasize their coverage of the classical authors, his account does make it clear that he went to Chinnor 'along with others'; and we may infer that the location of Newton's books was known to the cognoscenti at Oxford and maybe elsewhere as well as that Musgrave was not averse to having visitors inspect his Newtonian holdings.

Musgrave gave the volumes in the library their own individual shelf-marks (the first time this is known to have been done to the collection) and compiled a catalogue incorporating them; furthermore, he undertook the urgently needed task of preserving Newton's numerous flimsy pamphlets by having them bound up into handily sized and durable volumes (usually five or six items at a time), and inserted his own personal printed plate in every volume on his shelves.[1] This consists (see plate 6) in a combination of the Huggins and Musgrave coats of arms together with the motto 'Philosophemur' and carries the name of its engraver, 'B. Green' (probably Benjamin Green, 1736–1800). Some of the books which appear at first glance to bear only Musgrave's label will be found on closer examination to have this plate (103 by 93 millimetres) on occasion pasted over the slightly smaller Huggins one, but less frequently so than has sometimes hitherto been supposed; much more often the books which I have seen have both bookplates, usually side by side (on facing pages of the front-papers). There are some volumes with the Musgrave label only, but these are either Newton items bound by Musgrave or 'alien' works which he himself added to his library.

The system of numbering introduced by Musgrave is straightforward, comprehensive, and of great help to us in determining if a book came from his (and so probably from Newton's) library. The shelf-marks, written in ink, usually at the top left-hand corner of the front paste-down or first fly-leaf of the book, consist of a letter (denoting the bookcase, it would appear), followed by a number (that of the shelf therein), then a dash and another number (that of the book itself) (see plate 6). For instance F5 – 15, in this case Morden's *Geography rectified*...4th ed., 1700 (no. 1109), would be [Bookcase] F, [Shelf] 5, [Book] 15 [on that shelf]. I have reconstituted a shelf-list based on the marks given in the Musgrave Catalogue, and checked this against the surviving books; from this it is clear, as would be expected, that the volumes on each shelf of the library were arranged uniformly by size. The letters used (denoting bookcases) go from A to J, the shelf-numbers of the cases

<hr>

Isaac Newton and Professor Cotes...(London, 1850), where Björnståhl is described – perhaps unjustly – as 'the simple-hearted Swede'.
[1] These volumes of tracts are all in the same style of marbled binding, and their end-papers contain a watermark which shows that they cannot have been bound before 1740. As they all bear the Musgrave bookplate only, they were almost certainly bound during his ownership of the library. See P. E. Spargo, 'Newton's library', *Endeavour*, xxxi (1971), 32.

range from 1–6 to 1–11 (doubtless because of the differing heights of the books they carried), and the individual running numbers go at one extreme 1–11 (these are very large tomes) and at the other 1–56 (for small books). The first shelf-mark is A1 – 1, the last J9 – 31, and the individual cases hold respectively:

A1 – 1 to A9 – 13 [175 vols.]   B1 – 1 to B6 – 26 [140 vols.]
C1 – 1 to C6 – 30 [156 vols.]   D1 – 1 to D8 – 18 [184 vols.]
E1 – 1 to E8 – 18 [304 vols.]   F1 – 1 to F8 – 19 [248 vols.]
G1 – 1 to G11 – 21 [444 vols.]   H1 – 1 to H9 – 23 [341 vols.]
J1 – 1 to J9 – 31 [393 vols.]

making an overall total of 2385 volumes.

Musgrave appears to have made no serious attempt to establish an arrange-ment of the books in strict subject-order, though in some instances (whether so planned by him or by Huggins) books on similar topics stood together and groupings of an author's works were made if they were of like size: thus seven volumes by Boyle were placed at E4 – 3 to E4 – 9, and eight works by Newton were E6 – 25 to E6 – 32. The modern value of this coding as an identifier of stray books from Newton's library is well illustrated by Cambridge University Library's copy of *Norwood's epitomie...* 1645 (no. 1189), a very rare navigational book of considerable mathematical interest presented by F. P. White to the library a few years ago. Though now without Huggins or Musgrave bookplate(s) it has the coding G9 – 12 on its front paste-down, exactly in Musgrave's manner, and is so listed in his catalogue. It was our noticing this mark which led us to examine this book in detail and locate the correction in a Newtonian hand which we have already cited (p. 22). There can therefore be no doubt that this is indeed Newton's copy. All Musgrave shelf-marks are reproduced in the present catalogue.

After the transfer of the Chinnor library to Barnsley Park in (or soon after) 1778, the Newton collection remained, as far as I know, unnoticed and unseen by visiting scholars for the next 140 years. The owner of the house and estate there had been Cassandra Perrot, whose aunt was the Rev. James Musgrave's mother. When she died a spinster in 1778, the terms of her will, recognizing this branch of the family as her heirs, appointed the Rev. James's son, also James – then a young man of twenty-seven – to succeed to her fortune and estate at Barnsley. As it happened, her cousin, James *père*, died within a month of Miss Perrot's demise, so that James *fils* not only became in quick succession the new owner of Barnsley but also in-herited the living at Chinnor and the contents of the Rectory along with it. This Musgrave it was who had the Newton library removed from Oxfordshire to Barnsley Park, where the rehousing of the books evidently necessitated a change in their shelf-marking. The new coding was added along the lower edge of the Musgrave bookplate (see plate 6) and reads, typically, 'Case

B.C.10 Barnsley' or 'Case EE.G.11 Barnsley' and so on. For example, Morden's *Geography rectified* cited above with the Musgrave mark F5 – 15, also carries on its bookplate 'Case AA.F.1 Barnsley' indicating the lettering of the bookcase, its shelf (here the letter F), and its individual number on that shelf. My examination of those books in the Trinity moiety, and of others elsewhere shows that all the letters of the alphabet (with the exception of D, Q, S, T, and X), together with the double letters AA, BB, and EE, appear to have been used to denote bookcases. To indicate the individual shelves of the cases one of the first few letters of the alphabet was normally used, though in some instances I have seen J and K, even O and Y. Judging solely from the markings in the books it would seem that fewer of them stood on individual shelves than in Musgrave's Chinnor ordering; the highest running shelf-number I have encountered is 23, that entered in La Hire's *Nouveaux élémens des sections coniques*. . . 12º, Paris, 1679 (no. 903) which carries 'Case H.A.23 Barnsley' (in addition to its earlier Musgrave coding G2 – 39). It is also clear from my reconstitution of the Barnsley Park layout, incomplete though this is, that the books were again grouped together according to their size, the Musgrave shelf-sequence being abandoned without substitution of any noticeably improved rearrangement by subject. By no means all the books were given the additional Barnsley codings when they went to their new home. Of the 862 volumes stemming from this source which are now in Trinity College, 696 (or 80 per cent) bear Barnsley Park shelf-marks, and 144 carry the Musgrave bookplates without the added coding,[1] while the remaining 22 have either lost or never had the Musgrave label and with it in consequence any Barnsley marking they might have borne. Books lacking these shelf-marks do not group themselves into any recognizable category by author, subject, language, or date and place of publication.

A more detailed analysis of how the volumes were located in their new quarters and how they were placed in relation to the books already at Barnsley Park cannot now be made, since there appears to exist no shelf-list or supplementary catalogue to which the Barnsley coding can be related. But let me note that the entries in Musgrave's Catalogue for items with the added shelf-marks have the letter B pencilled in after the original Musgrave coding. This may possibly indicate their onward transmission to Barnsley Park, though it more likely shows where new markings had been given to replace the Musgrave ones. In the volumes which I have studied, works listed in the Musgrave Catalogue without the added letter B almost invariably carry no Barnsley mark on (or near) their bookplate, yet they too are known to have been sent there along with those so favoured.

---

[1] Nor is this typical of the Trinity collection only. A similar proportion of Newton's other books elsewhere which were formerly at Barnsley Park do not bear that library's coding alongside their Musgrave bookplates.

From the time of the arrival of the young James Musgrave at Barnsley Park[1] until 1920, Newton's library remained there, so I believe, essentially intact.[2]

Occasionally, modern booksellers' catalogues mistakenly describe books as being with 'Newton's own press-mark' when they speak of Musgrave's shelving-code in volumes which had formerly belonged to Newton. However, a two-page manuscript headed 'Lib. Chem',[3] which is a listing by Newton of 119 chemical and alchemical books, does have what would seem to be a shelf-marking sequence. Here a series of numbers precedes the titles of the books cited; in Newton's order they run: 2.4.1 (for the first item) to 2.4.6, 2.4.8, 2.4.7, then 2.4.9 to 2.4.25; 2.5.1 to 2.5.13; 2.6.1 to 2.6.31; 2.8.1 to 2.8.29, 2.8.31, 2.8.30, 2.8.32; 2.7.1 to 2.7.18. I have positively identified 108 of the works so numbered and less affirmatively placed 9 more. The latest year of publication of any of the items recorded is 1696 (for Alipili, *Centrum naturæ concentratum: or The salt of nature regenerated*... (no. 25): this would suggest a rough dating of 1696–7 for the list. From a count of the chemical/alchemical books in Newton's library I find an overall total of 169 items (some of these published after 1696); the 'Lib. Chem' list therefore constituted something more than 70 per cent of Newton's final holding of titles in this category. Of 80 of these the present locations are known, and I have examined most of them; though they show many signs of Newton's ownership and usage, none has any numbering, either inside, or on its spine, which corresponds in any way with the 'Lib. Chem' sequence. It might well be, however, that the numbers thus added by Newton did indeed indicate a shelf-order among books whose precise location their owner knew and saw no reason to specify in the books them-

---

[1] James Musgrave otherwise left his stamp on the baroque mansion at Barnsley Park by extensive renovation and rebuilding; in particular the library was redesigned for him by John Nash in 1807. (See D. Verey, *Gloucestershire* (*The buildings of England*), 2nd ed., i (London, 1970), 98–100; also *Country Life*, xxiii (2 May 1908), 630–36 and cxvi (2 Sept. 1954), 720–23, (9 Sept. 1954), 806–9.) When the elder branch of the Musgrave family became extinct in 1812, James succeeded to the title of 8th Baronet and was followed in 1814 by his eldest son (yet another James). The 9th Baronet seldom resided at Barnsley Park and seems to have cared little for the house and its contents, preferring the hunting fields of Leicestershire. Never married, he died in 1858, and the family seat passed to his brother William Augustus Musgrave, the 10th and last Baronet, Rector of Chinnor for the previous forty-two years; and there he continued to dwell until his death in 1875. The estate at Barnsley passed to his sister Georgina, wife of Aubrey Wenman Wykeham of Swalcliffe Park, Oxfordshire. The following year Wykeham assumed the additional surname and arms of Musgrave by Royal Licence, thus beginning the double-barrelled Wykeham-Musgrave family name.

[2] Zeitlinger has stated that a number of books from Newton's library appeared on the market in the nineteenth century, notably at a sale at Leigh & Sotheby's in 1813 of the 'Collection of the Great Sir Isaac Newton' (*Newton's library*, p. 13). I can find no evidence for this. Certainly the books at the 1813 sale – whatever they were – did not come from Barnsley Park. See Appendix B.

[3] MS 418 in the Newton Collection at Babson College, Mass.

selves.[1] A pointer in this direction is that he was content, without qualm, when he thought it necessary to break the straight numerical sequence of his listing by writing...2.4.6, 2.4.8, 2.4.7...out of order (all three of them works by the same man, Michael Mayer) and by insisting that Johnson's *Lexicon chymicum*...1678, though numbered 2.8.30, should follow Cooper's *A catalogue of chymicall books*...1675 (2.8.31). It may well be that something more than a mere cataloguing of the chemical books on his shelves may have been in Newton's mind, though an examination of the list yields no obvious indication of what other purpose he might have had. Though the larger books tend to occur in the earlier divisions 2.4...and 2.5..., the variation in their height is not great, their arrangement is not by language, and there would appear to be no rigid classification by subject within the general field of alchemy.

There are two further kinds of coding in some of Newton's books; they are not, however, in his handwriting and do not relate to any system of shelf-marking known to me, if indeed this was their intended purpose. The first type has in some respects the appearance of a bookseller's private system of pricing rather than a device for arranging books on shelves. It consists of a series of letters (usually lower case, occasionally capitals), followed in most instances by the letter 'x' (sometimes one only, though at other times up to four are used, with some of them underlined) and, when appropriate, ending with a statement of the number of volumes: for example, 'bx' (in Simon's *Lettres choisies*...1700, no. 1517), or 'exxxx 2V (in the first volume of Nodot's *Nouveaux mémoires*...2 vols., 1706, no. 1177). The same combination of letters and underlinings is sometimes present in more than one work, thus 'bxx' is to be seen in five separate items which have no obvious connection with each other.[2] The coding, invariably written in the top left-hand corner of the front paste-down, is present in at least 54 of the 862 volumes now in Trinity, and though it may not have been added to all the books at the same time, it looks to have been the work of the same early-eighteenth-century hand. Of these books, only one (Lucianus, *Opera omnia quæ extant*...no. 988), published in 1615, has the appearance of having reached Newton second-hand, none was apparently a gift, and the rest came

---

[1] When the 'Lib. Chem' manuscript was offered for sale by Sotheby & Co. in 1936, their *Catalogue of the Newton Papers sold by Order of the Viscount Lymington*, Item 3, described it as 'A list...of books, with shelf marks (perhaps in Trinity Library)'. The suggestion is demonstrably incorrect: we would guess that the compiler (John Taylor) probably confused the marks on this listing with those present in Newton's Common Place Book (Item 235 in the same catalogue and now in the Keynes Collection, King's College, Cambridge), which were indeed Trinity shelf-marks (see p. 6 above).

[2] Arranged alphabetically the complete sequence reads: Ad, Add 7V, Agf, Ake, Ape, ax, axd.3, axx (used three times), bx, bxx (five times), bxxx, bxxxx, cb, ck 3 vol, cx, cxxxx (twice), dxx (twice), dxx (twice), dxxxx, dxxxx 2V, eb (twice), ex, exx 2 v (twice), exxxx 2V, fb 2V, gb, gc, hb, hb 2 Vol, hxx 2 Vol, Lg 2vol, lc, lxx (twice), mb, mc, pc3v, rbxx, rgc 2 Vol, Sx, ux, xgd 3 vol, xl.

out between 1652 and 1718; 40 of the total of 54 have title-page dates of from 1700 to 1718. They cover widely differing subjects and vary considerably in size, but it is significant that every one of them was published on the Continent, the text of 35 of them being in French and 18 in Latin, and the remaining work (the Lucianus) in Greek and Latin. Had this coding not been limited to 54 works, we might have suspected that it might have related in some way to box-numbers possibly used when the whole library was transferred from London to Chinnor in 1728; as it is, however, such evidence as there is leads me to suggest that the coding may be that of an English importer and dealer in foreign books patronized by Newton.[1]

The second kind of coding-marks found in Newton's books is even more puzzling and much more vague, and occurs so infrequently (only ten instances have been noticed) that I have been unable to trace any common characteristics or explicable pattern in them. They appear to be in a variety of hands, with some perhaps already present in certain of the books which Newton may have acquired second-hand; other markings may have been added to the rest later. I may repeat that Newton did not write them, but otherwise can provide only illustrations of their form and cite the titles of the books where they appear – none of which also carry the previous coding just now described. Boyle's *New experiments...touching the spring of the air...*1660 (no. 269) bears '$\frac{9n}{p2q}$' on the title-page, and the same author's *Some considerations touching the usefulnesse of experimental naturall philosophy...*2 vols., 1664–71 (no. 272) has '$\frac{2.7}{x.6}$' on the half-title of volume 1 as well as '$\frac{V-4}{x\ D}$' on the verso of its title-page. Descartes's *Geometria...operâ atque studio F. à Schooten*, 1649 (no. 506) carries '$\frac{6}{5}$qu' on the title-page while Starkey's *Secrets reveal'd...* 1669 (no. 1478) has 'A/nO' on the front paste-down.[2] The remaining six examples of this coding are: '$\frac{a}{\gamma}$' on the front paste-down of *Chronicon Carionis* ...1625 (no. 344), '$\frac{\delta}{\gamma}.2.\frac{\delta}{\gamma}$' on the title-page of *Pauli Iovii Historiarum sui temporis*, vol. 1, 1556 (no. 865), '$\frac{a\ H}{2}$' on the fly-leaf of Meursius's *Regnum Atticum...*1633 (no. 1077), '$\frac{2}{q3p}$' on the title-page of Ovid's *Opera omnia, ex recensione G. Bersmani...*1655–6 (no. 1221), 'G$\frac{\delta}{}$x' on the verso of the title-page of Schelstrate's *Ecclesia Africana...*1679 (no. 1462), and 'G$\frac{1}{5}$*' in a similar position in Selden's *De synedriis...*1679 (no. 1483). A few similar markings occur in other books, but they are too scrappy or unclear to make any reproduction here reliable, or even possible.

The Musgrave manuscript catalogue itself[3] is a leather-bound folio

---

[1] Mr J. C. T. Oates has drawn my attention to coding of a somewhat similar nature in accounts for books bought by Cambridge University Library from the booksellers Robert Scott, Edward Story, and Jonas Hart (all in 1679) and also Cornelius Crownfield (in 1709).

[2] The Descartes and Starkey volumes were two of the six that 'has Notes of Sir Is. Newtons'. D. I. Duveen describes the coding on this copy of Starkey, calling it 'Newton's press mark' in his *Bibliotheca alchemica et chemica*, p. 470; the work was entered as 2.6.1 in Newton's 'Lib. Chem' list.

[3] Trinity College NQ.17.36.

volume, lettered 'Catalogue of the Library of Dr James Musgrave, Rector of Chinnor, Oxon.' in gold on the front and back. It consists of ninety-one leaves, on which the contents of the library are set out in alphabetical order of authors or, failing that, titles. Its text is very clearly written (on the recto side of each leaf only) in what appears to be the hand of a single professional scribe[1] and was completed in 1767 (or soon after), since the latest year of publication of the books recorded is 1766. Each letter of the alphabet starts on a new sheet, and the pages have on average about twenty entries, each comprising author, title, number of volumes (where pertinent), date of publication, shelf-mark, and format, but without citing the place of publication. On occasion the compiler provides a little additional bibliographical information, somewhat similar in its content to that given in the Huggins List, though not using the exact wording of that document and so not copied from it. He tells us, for example, that Barrow's Latin epitome of Euclid's *Elements*...1655 (no. 581) is 'Liber notis MSS Isaac Newton' (see p. 19 above); that the second English edition, 1717, of Newton's *Opticks* (no. 1159) is 'with his Correction in MSS'; and that Oughtred's *Clavis mathematicæ*...Ed. 3ª...1652 (no. 1218) – a partially manuscript volume – has, we are mistakenly informed, 'MSS Addit. by Halley', while William Jones's *A new epitomy of the art of practical navigation*...1706 (no. 858) is a 'scarse Book' – the only one so described.[2] Other comments refer to the production of certain works: 'Black letter with Annot.', 'large paper', 'with fine Prints', 'with Cutts', 'rul'd with red lines', 'calf gilt leaves'; to their binding: 'stained blue Turkey'; or to other features: 'interleaved', 'the last leaf written', or simply 'a little stained'. The separate works are so evenly spaced on the page as firmly to suggest that the Catalogue was completed methodically and in a single operation, with no titles being added subsequently; this would perhaps indicate that it was to some appreciable extent based on – or even copied from – an earlier list, possibly one made by Charles Huggins, or simply made from a 'rough' copy.

The shelf-marks added to the books are in the hand of the Catalogue scribe. But whereas he was careful to write a separate shelf-mark in each volume of a multipartite work, he was content to enter in the Catalogue itself the coding of the first book only; thus Larrey's *Histoire de France*...1718 (no. 919), whose four separate volumes are individually marked G8 – 22 to

[1] A short entry in the Chinnor Parish Register 'by me James Musgrave' differs considerably from the handwriting in the Catalogue.
[2] Björnståhl, pp. 288–9, reports [here in English translation] that he 'saw that rare book by Mr Jones's father, of which ... even his son did not possess a copy. The title is: *Epitome of the art of practical navigation*...London, 1706. 8. Another very rare book by this same Jones (this is quite extraordinarily rare): *Synopsis palmariorum matheseos*...1706. 8...' (nos. 858 & 859; both copies are in Trinity). The words 'scarse Book' look to have been added to the Catalogue (in a different hand and written over a series of dots at the end of the title) after the latter's initial compilation, perhaps as a direct result of Björnståhl's estimate of the book's rarity.

G8 – 25, is there listed as G8 – 22 only. The document has altogether 1601 separate entries comprising 2385 volumes; of these 161 entries (356 volumes) relate to books published after Newton's death in March 1726/7 and are consequently excluded from the Catalogue printed below. This leaves 1440 titles (2029 volumes) which solely on the basis of their publication date might formerly have been the property of Newton. Between the books truly originating from Newton's possession and those which were later added to the Huggins–Musgrave collection the Catalogue makes no distinction. Many of the surviving books can of course be independently authenticated as Newton's by any of the usual signs of his ownership: his autograph signature or other authentic contemporary inscriptions, notes in his hand, his characteristic 'dog-earing', and so on. In the case of existing volumes which lack such signs or of those whose present location is unknown their registration in Huggins may fairly safely be taken as proof that they too were in Newton's library. A recording in Musgrave is good, but a deal less dependable, evidence of this. The entries in our present catalogue, therefore, specify if the titles were present in one or both of these listings. But we have always to remember that a not inconsiderable number of separate works which were unhelpfully lumped together in the 1727 list (see p. 32 above) under such blanket entries as '3 Dozen of small chymical books' are set out under their individual author and title in Musgrave, and that many of these prove on examination to be Newton's own copies. If we assume that the 1727 inventory figure of 1896 volumes[1] is reliable, then it appears that some 133 additional volumes issued before 1727 found their way into James Musgrave's library. This seems to be about right if we recall that Charles Huggins added some books of his own after 1727 to the library his father bought for him, and that very likely Musgrave both brought further pre-1727 books to Chinnor in 1750 and acquired yet others during his twenty-eight years' residence there. (Though his use of a personal bookplate from about 1760–65 would thereafter have made it unnecessary for Musgrave to write his name in a volume, let me add that I find it surprising that his signature is not present in any of the books I have examined.) Furthermore, his wife may well have contributed more than the single little prayer-book[2] inscribed with her maiden name 'Jane Huggins her Book' which is now in Trinity with other items in the Musgrave list stemming from Musgrave himself. We have to accept, then, that there are up to 130 or so volumes which are potential 'intruders' among the majority which do derive from Newton's shelves. They need all to be included in the present catalogue, however, simply because we have no sure means of deciding which ones in particular are to be omitted from it.

---

[1] See p. 29 above.

[2] Trinity College NQ.16.6: *La liturgie, ou Formulaire des prières publiques, selon l'usage de l'Eglise Anglicane. Nouvelle éd....* Amsterdam, 1731. This is in the Musgrave Catalogue as 'Liturgie in French, 1730 [*sic*]. G2–50. 8vo.', but is here excluded by reason of its publication date.

While Rectors of a country parish such as Chinnor are most unlikely to have collected advanced scientific or mathematical works of before 1727 – virtually none of the 356 volumes recorded in Musgrave with dates after 1726 are of this kind – the great majority of the doubtful books are theological (particularly sermons), and the rest comprise a few classical texts, works on modern history and biography, and works of English literature, together with a reference book or two. Where internal evidence from the book itself suggests that it may not have belonged to Newton this is shown below in the catalogue entry for that work.

Even in the case of its Newtonian items the Musgrave Catalogue presents certain difficulties and in some cases near-insurmountable obstacles to anyone who seeks precisely to identify its titles. The entries are brief, never running to more than one line, and there are various inaccuracies and miscopyings of titles and especially of dates (considerably more than in Huggins). These deficiencies, together with the omission of the place of publication and (when relevant) the number of the edition, would have caused many problems had it not been possible to check the shelf-marks in Musgrave with those in the surviving books and to compare these titles against those in the more accurate Huggins (where, I may again mention, every separately listed book has been traced). Even so there remain twenty-six entries where we have been unable to identify the title of a work, or its exact edition, and for these I have had to be content with reproducing the Musgrave description. Included in the twenty-six is the only manuscript in the list: '*Observationes on scripture, MSS by R. W.* J9 – 24. F°.' (no. 1197); but since none of his manuscripts would seem to have passed to Huggins, this probably never belonged to Newton. I have previously mentioned that the compiler of Huggins was often content to name but the first item only of a volume containing more than one distinct work; in Musgrave this happens even more frequently, and in such cases unless the actual volume is available the additional titles (whatever they might have been) cannot now be recovered. Much more serious, the compiler failed to provide detail adequate for the modern bibliographer when he 'catalogued' the numerous pamphlets and other small printed papers which Musgrave had had bound up together into manageable lots. Under the headings 'Tracts', 'Plays', and 'Sermons' the compiler noted these collections in short joint entries which in themselves make it impossible to determine the titles therein embraced. For example, we read of 'Tracts Mathematical by Collier, Martin, Moxon, &c., 1701. J3 – 24, 8°.' Fortunately, I can cite below, under their individual authors, the component tracts in this volume by referring to the book itself, now in Trinity, which is still lettered 'Tracts Mathematical' on its spine and on its front paste-down bears a Musgrave bookplate and the shelf-mark J3 – 24: it collects seven separate items, six (nos. 414, 424, 577, 1122, 1596, 1692) having publication dates ranging from 1697 to 1715, but the seventh printed as late as 1757. (The year

1701 furnished by the compiler is, I may add, the title-page date of the item appearing first in the bound volume.) Again, Raphson's *History of fluxions*... 1715[-1717] (no. 1371), though separately listed in Huggins, is concealed in Musgrave under 'Tracts Mathematical 2 Vol., by Grandus, Cotes, &c., 1720 D6 – 4 [to 5]. 4to'. Altogether, the Musgrave Catalogue records forty-six 'Tract' volumes, of which the twenty-three now in Trinity alone contain 129 individual items (118 pre-1727), many of them exceedingly rare. The majority of these latter are scientific or mathematical in nature, written by such British authors as Cheyne, Cotes, Craige, Gregory, Halley, Hooke, and Whiston, and by Continentals such as Jacob Bernoulli, 's Gravesande, and Huygens. A further collected volume of such scientific 'tracts' now at the University of Wisconsin at Madison contains three separate alchemical works (nos. 18, 1652, 1685).

At 'Plays' there are five entries (ten volumes); the three of these volumes in Trinity contain twelve separately published plays. Typical of the information offered under this heading is 'Plays by Mrs Centlivre, Owen &c., 1722. H8 – 19. 8vo.' The original volume[1] in fact comprises four separate pieces for the stage by, respectively, Centlivre (published in 1740), Owen, 1722 (no. 1227), Congreve, 1711 (no. 434), and Shakespeare (his *Tempest* as adapted by Davenant and Dryden in 1710, no. 1505). Further details of the works skimpily recorded jointly as 'Sermons' are even more elusive to track down; indeed, I have been unable here to make a single positive identification, having not yet succeeded in locating any of the volumes so described. The following excerpt typifies the scanty information given by Musgrave: 'Sermons by several Hands 3 Vol., Boards. 1720. B6 – 13 [to 15]. 4to'. The heading 'Sermons' is found in ten entries (comprising sixteen volumes) cloaking an indeterminate number of items not yet traced. The high rate (just over 50 per cent) of the 'Tracts' volumes whose whereabouts are now known as compared with those of 'Plays' and 'Sermons' which have been located is welcome both because they would seem more likely to have belonged to Newton (some indeed carry easily determinable signs of his ownership) and because of their greater intrinsic value to modern scholars. For volumes not yet traced I can only transcribe in the catalogue the imperfect Musgrave titles and give also a bare cross-reference at the authors quoted. However unsatisfactory this is, until the volume so described itself resurfaces, such vague entries cannot be made any more precise.

Books have an unfortunate tendency to disappear over the years from libraries both public and private, and this, we may guess, happened perhaps more than once in the case of the Huggins–Musgrave collection. I have no reason to think that the compiler of the Musgrave Catalogue was at fault by failing to record all the books then present at Chinnor – the reconstituted shelf-list shows that he was not – nor is it likely that they were given away or

---

[1] Trinity College NQ.10.92[1].

destroyed in any significant numbers. Yet a comparison of the Huggins and the Musgrave listings shows that sixty-two titles comprising 113 volumes (these totals include duplicate copies)[1] which appear in the former were not at hand to be included in the latter. We know that the six 'Books that has Notes of Sir Is. Newtons' registered in the 1727 document did not finally go to Huggins, but the remaining 56 works – by such authors as Thomas Burnet, Geber, 's Gravesande, Locke, and Whiston, as well as by Newton himself – are works of considerable interest and importance. It is possible that some of these items may have disappeared soon after Huggins acquired them, since nine of the missing titles are among the thirty-two marked 'x' on the 1727 list,[2] though it must be added that the remaining twenty-three so marked survive into Musgrave, if only as single copies where forty years earlier there had been two or even three. Others may have been bound into 'Tract' volumes – certainly in the case of some of the mathematical items – and thus became buried under this group of Musgrave headings, but this seems unlikely to have happened to more than a small number, since few were of a size to be suitable for binding in with other items. The twenty-three 'Tract' volumes in Trinity apparently yield only two of these missing works – Poleni's *De motu aquæ mixto libri II*...1717 (no. 1336) and the Raphson *History of fluxions* already mentioned, though even this may have been a variant issue and not that now in Trinity – and no other traces have been found of these early disappearances.[3]

The foregoing account of the deficiencies of Musgrave's Catalogue is presented to indicate how it may hinder as well as help our preparing a complete modern catalogue of Newton's library as it was at his death. By the standards of its own time, and considering the purpose for which it was made, it is an adequate and workmanlike document, and we are in many ways fortunate that it has survived. It is well known that, after being thus tucked away inaccessibly (but intact) for more than another century in the possession of Musgrave's descendants at Barnsley Park, the greater part of Newton's books were put up for auction in 1920, and individual volumes from that sale have since then appeared from time to time in the second-hand market. The man, Mr H. W. Wykeham-Musgrave, who put them up for auction when selling off his house at Thame Park in Oxfordshire, evidently sent over a quantity of the books at Barnsley Park which he also then owned to be bid for at the same time. The sale catalogue, compiled by Hampton & Sons of 3 Cockspur Street, London,[4] embraced 1013 lots, auctioned over 13–15 January 1920. Only the last 123 of them (Lots 891–1013) were made up of books under the general head 'Library of Miscellaneous Literature (of

---

[1] See p. 33.   [2] See p. 34.
[3] A list of the catalogue numbers of these books is given at Appendix D.
[4] *Thame Park, Thame, Oxon...The Greater Portion of the Contents of the Mansion, Early 19th Century English Furniture...A Large Collection of Porcelain & Glass. Pictures. Books...*

Etats de l'Afie. Ceux qui veulent Tirer
en *Suede*, en *Dannemarck*, en *Pologne*, à D.
tous les autres Etats du Nord, & en toute l
ont leur Correfpondance à *Amfterdam*, à A
n'y a que l'*Angleterre*, où la plufpart des Ban
des Villes de France, où il fe fait des Manuf
ce confiderable, font les Traites & Remifes d

Des Monnoyes Réelles de la France, o.

Efpeces d'or.

Des Loüis,	Qui valoient fur l'ancien pié 11L
Des Lys,	Qui valoient	7 L
Des Ecus fol,	Qui valoient	5 L

On ne voit plus de ces deux derniéres fortes d'E
été portées aux Monnoyes & refonduës pour faire des
Loüis & demi Loüis d'or.
   Par avis, les vieux Loüis d'or n'y valent que 12 livres 5 fols.

Efpeces d'argent.

Des Ecus, ou Loüis, Qui valoient fur l'ancien pié 3 Liv.& à prefent 3 Liv.
   12 f. tournois.
Des demi-Ecus, — Qui valoient —— 1 Liv.10 f. & à prefent 1 Liv.16 f.
Des Piéces, ou Reaux, Qui valoient —— 5 f. & à prefent —— 6 f.
Des Piéces, —— Qui valoient —— 3½f. & à prefent —— 4 f.
   Quoy que ledit Ecu de 60 fols, ou de 3 Livres Tournois, vaille à
prefent en France 3 livres 12 fols, ou 72 fols Tournois, néanmoins les
Changes fe font toûjours fur le pié d'un Ecu de 60 fols, étant à re-
marquer qu'en l'année 1690, que les fufdites Monnoyes augmenté-
rent de prix, & que lefdits Ecus furent mis à 62 fols, puis à 66 fols,
& finalement à 72 fols tournois, & les autres Efpeces à proportion,
les prix des Changes fur les Places Etrangéres diminuerent, mais Dieu
nous ayant donné la paix depuis le 20 Septembre de l'année 1697, il y
à apparence que les fufdites Monnoyes feront remifes fur l'ancien pié,
auquel cas les prix des Changes augmenteront pour les Païs Etrangers.

Efpeces de méchant aloy, ou de cuivre.

Des Sols marquez, qui valoient fur l'ancien pié 12 Deniers, & à prefent 15 ½
Des Deniers de cuivre, defquels les —— 12 font 1 fol.
   L'on voit à l'entour des nouveaux Loüis d'or & femblables efpeces, un
petit cordon, & à l'entour des Ecus & demi-Ecus des lettres, pour em-
pêcher qu'elles ne foient roignées, ce qu'on ne voit point aux vieilles efpeces.
   Les Monnoyes étrangeres ont cours en France, fuivant leur valeûr & à
proportion de l'augmentation ou diminution des fufdites efpeces.
                     A a 2   Det

Plate 4 'Dog-earing' by Newton in Ricard, *Traité général du commerce*...1700 (no. 1399).

XCVIII. Ayant accordé de payer pour le Laſt de Froment 122 ggl. combien de Laſts en pourray-je avoir pour 4270 Florins?

Réponſe, 25 Laſts.

XCIX. Si l'on peut avoir une aune de Ruban pour 3 ſtooters, combien d'aunes en pourrez-vous avoir pour 7 guldens 10 ſtuyvers?

Réponſe, 20 aunes.

C. Si le Galon vaut 1 blanq combien d'aunes en pourra-t-on avoir pour 8 ß?

Réponſe, 64 aunes.

CI. Si un homme achete de la Biere de Deventer pour 1346 Ducatons, en comptant la tonne renduë en Cave ſur le pied de 12 guldens 12 ſtuyvers combien en a-t-il achété de tonnes?

1346 Ducatons.

4——————

Rép. 336 tonnes & demie.

12——————

28 Laſts & demi-tonne.

CII. L'aune de galon à Fleur d'or coû-
·ins combien d'aunes dudit galon
172 Louis d'or & 6 ß, le
·ns 2 ſtooters?
aunes.

CIII.

Plate 5   'Dog-earing' by Newton in Desaguliers, *La science des nombres par rapport au commerce*...vol. 1, 1701 (no. 503).

Plate 6 Newton's copy of Calvin, *Institutio Christianæ religionis*, 1561 (no. 335), with his signature, the Huggins and Musgrave bookplates, and the Musgrave and Barnsley Park shelf-marks.

upwards of 3000 volumes), including early works on Alchemy, Astronomy, Mathematics, Agriculture, Classics, Theology, Etc.', but neither here nor on the title-page of the catalogue is there any mention of their provenance from Newton's library. I surmise that the books, their Newton association apparently either just forgotten or ignored as irrelevant, were regarded as a petty sideline to the real business in hand of auctioning the contents of Thame Park. For the most part each lot of books consisted of a number of single volumes bundled together: thus, Lot 935 was one of 'Theology (Old), *calf, etc.*, 173 [volumes]; Lot 980 comprised 'Books (various), Classics, Theology, etc., *about* 200'; Lot 1008 contained 'Books (various), *about* 100'. These are extreme cases, but only thirty lots consisted of a single work, twenty-two of these being multivolume sets such as, for example, the journals *Acta eruditorum*, the *Journal des sçavans* and the *Philosophical transactions*. In most other instances, to be fair, the catalogue does give individual titles of a number (varying from two to five) of the works in each lot, following this by 'various' and then '20' or whatever number made up the overall total of volumes in the bundle. By my count, a total of 2979 volumes were thus put up for auction; of these, 2156 were categorized simply as 'various', without date or other identification. Titles were provided for 92 of the remaining individual works (comprising 307 volumes) published before 1727, and for 119 works (516 volumes) of those issued after that year. In spite of these extremely meagre descriptions it is possible to identify with certainty 54 individual works (comprising 238 volumes) out of those which appear either in the Huggins List or in the Musgrave Catalogue, or in both. I can offer no exact estimate of the number of books from Newton's original collection which may have lurked hidden among the 'various' items, but even if no more than half of them received this blanket appellation their total would amount to a thousand or so. Nor, because of the relatively low number of identifiable titles, can I do other than offer my speculations about the criteria used for choosing those of the books at Barnsley which were taken away to be sold. No personal selection seems to have been made; maybe the intention was merely to clear out unwanted volumes from bookcases grown overfull. I have so far been able to trace fifty-six Barnsley shelf-marks present in volumes which are known to have been sold at Thame, and of these, thirty-one carry the same letters indicating the bookcase and shelf as books now in Trinity College which remained at Barnsley Park after 1920.[1] On the other hand, the remaining twenty-five bear letters listing shelves of which there is no equivalent in the items now in Trinity, which would suggest that in

---

[1] An illustration of haphazard selection of items to be sold is provided by the present location of Newton's copy of *Miscellanea curiosa*. . . [Ed. by E. Halley], vols. 1 & 2, 1705–6 (no. 1086). Vol. 1, with Barnsley shelf-mark AA.H.1, was sent to be sold at Thame and is now at the University of Chicago Library, whereas vol. 2 (AA.H.2) was left on the shelves at Barnsley for nearly ten years more, to be eventually acquired by Trinity in 1943.

most of such instances whole shelves were cleared out. It is likely that runs of periodicals such as those cited above (which in the 1920s would have fetched a good price at auction) would have been obvious choices for a quick thinning out of the contents of a library. Because in part, no doubt, of the all but complete failure of the auctioneer – through ignorance, it must have been – to dwell upon the association of the books with Newton, they fetched no more than about £170.[1]

Five lots at the sale contained eight books described as being 'With Isaac Newton's Autograph' or the equivalent. I know the present location of seven of these books and have authenticated the Newton signature in each case, viz. *Marci Antonini Imperatoris & philosophi, de vita sua libri XII*. . .12°, Lugduni, 1626 (no. 55); *The fame and confession of the Fraternity of R:C:*. . .8°, London, 1652 (no. 605); More, *A plain and continued exposition. . .of the Prophet Daniel*. . .4°, London, 1681 (no. 1115); *L. & M. Senecæ Tragædiæ*. . .12°, Amsterdami, 1645 (no. 1489); Stillingfleet, *A discourse concerning the idolatry practised in the Church of Rome*. . .8°, London, 1671 (no. 1562); Saint Vincent of Lérins, *Adversus profanas omnium novitates hæreticorum commonitorium*. . .12°, Cantabrigiæ, 1687 (no. 1688); and Walker, *A modest plea for infants baptism*. . . 12°, Cambridge, 1677 (no. 1705). Several of the other works sold at Thame (notably the heavily annotated set of Zetzner's *Theatrum chemicum*) are now known to carry notes by Newton on their margins and end-papers, but no hint of this is suggested in the catalogue. Heinrich Zeitlinger, who handled a number of the books from the Thame sale purchased at his urging by his employer Sotheran & Co., afterwards remarked upon the Huggins and Musgrave bookplates stuck in their fronts, pointing also to their Newton notes, their 'dog-earing' and their markedly good state of preservation, as well as noticing the works by Boyle which were there sold.[2] For many years afterwards books deriving from this 1920 auction of half or more of the Newton books at Barnsley Park continued to appear in the catalogues of many London second-hand booksellers, thence to be scattered into private and public ownership in various parts of the world, and the United States especially. Zeitlinger's Sotheran catalogues offered a choice variety of the Newton books 'as sold at Thame' along with a wealth of detail on their content.[3]

[1] Zeitlinger, *Newton's library*, p. 16.
[2] 'A Newton bibliography', in *Isaac Newton, 1642–1727: a memorial volume*, ed. W. J. Greenstreet (London, 1927), pp. 168–70.
[3] Sotheran, *Bibl.* Suppl. 2, I–II (1937): no. 1652 (ex Catalogue no. 783, 1923); nos. 3395, 3976 (ex Catalogue 786, 1923); nos. 5728–31 (ex Catalogue 789, 1924); no. 7520 (ex Catalogue 795, 1925); nos. 10087, 10375–6, 10608, 10868, 10892, 11461, 11953, 12098 (ex Catalogue 800, 1926); and no. 14128 (ex Catalogue 806, 1927). See also Suppl. 1 (London, 1932) of *Bibl.* whose p. 267 is headed 'A few Books [viz. nos. 3603–7] from Sir Isaac Newton's library, as sold at Thame'. All these books were sold fairly quickly, with the sole exception of no. 3603, the 1st edition of Barrow's *Euclid*, 1655. (For more on this book see p. 51.)

Similarly the items auctioned by Sotheby's on 20 April 1926[1] included a number of other works, perhaps as many as thirty, which must have come from the same source. The sale catalogue states in particular that Lot 563, Archimedes, *Opera*...Paris, 1615 (no. 75) bore 'Is. Newton' on the fly-leaf, while of the twenty titles by Robert Boyle there listed, Lot 579, *Essays of...effluviums*...London, 1673 (no. 259), is described as having the inscription 'For Mr. Isaac Newton from the Author'; and Lot 583, *Memoirs for the natural history of humane blood*...London, 168¾ (no. 266), is listed as possessing one 'in the handwriting of Isaac Newton, showing that the book was a gift to him from the author'.[2] The subsequent history of Lot 597, '*Euclidis Elementorum libri XV. breviter demonstrati, operâ Is. Barrow*, FIRST EDITION, diagrams, two or three leaves a little stained, calf, 16 mo: Cantabrigiæ, 1655' (no. 581), will be of special interest to students of the second-hand book trade in London between the wars. The auctioneer's marked-up sale catalogue records that this item was knocked down to Sotheran's for five shillings. The following year the same volume was offered for sale at a price of £500 as no. 3794 in Zeitlinger's Catalogue 804, with fulsome accompanying blurb, including (in Clarendon type) the information that it was 'with numerous MS. notes by Sir Isaac Newton'. It was re-advertised in 1931 as no. 3603 in Catalogue 828, again for £500 and with very similar detail as previously, though here adding that the book originated from the Thame Park sale. Zeitlinger recounted some years later that 'the work was mentioned in the auctioneer's [Sotheby's] catalogue without reference to notes...Only after its purchase was I able to discover to my great surprise that it contained important notes in Newton's handwriting'.[3] Apparently no customer was prepared to pay the inflated price asked for it by Sotheran's until finally in 1943 the item was bought by the Pilgrim Trust and presented to Trinity College, Cambridge along with 858 other volumes from Newton's library. I should add that the book[4] does indeed have both copious annotations by Newton, and also (as the Sotheby catalogue states) 'two or three leaves a little stained'.

The first published report of the remaining moiety of the Newton collection at Barnsley Park appeared in the *Morning Post* of 8 February 1928 under the caption 'Newton's Library Discovered'. The article, by the newspaper's science correspondent, gave credit to one Richard de Villamil for tracking it down, but did not reveal the exact location of the find, being content to place it as being 'in a private house in Gloucestershire' and speaking of 'at least 500 or 600 volumes (possibly more)'. On 7 January 1929, the same paper printed a fuller piece, headed 'More Newton Discoveries...A

---

[1] *Catalogue of Printed Books*...[sold on] *19–20 April 1926*, Lots 559–633: The Property of Hugh C. H. Candy, Esq., J.P., formerly Arnott Professor in Queen's College, London.
[2] See the entries in the catalogue below for these two Boyle items where their present location is given and the Sotheby descriptions are authenticated. (I do not know the whereabouts of the Archimedes volume.) [3] *Newton's library*, pp. 18–19.
[4] Now Trinity College NQ.16.201.

"Residue" of 860 Volumes', and here a description by de Villamil himself of the Barnsley moiety was directly reported at length. In his book *Newton: the man*[1] published three years later, de Villamil's brief account repeats almost verbatim part of the second *Morning Post* article where he tells of unearthing at Barnsley altogether 860 volumes of printed books along with the Musgrave manuscript catalogue. For a man of seventy-eight years he was remarkably energetic and persistent in his search; in thanking the owners of Barnsley Park, Mr and Mrs Wykeham-Musgrave, he records that they 'allowed me practically to "ransack" their mansion, cupboards and all other hiding-places for books, manuscripts, etc.' But whereas in the *Morning Post* de Villamil was quoted as saying that 'the discovery of the catalogue was not entirely an accident, since I suspected (from Edleston) that such a book had once existed', this admission was not repeated in his *Newton: the man*, where he was content merely to recall that 'by a series of lucky accidents' not specified by him he became aware in 1927 that a very substantial residue still remained at Barnsley. Who first provided him with the clue? According to Zeitlinger[2] it was he who did so when he had occasion to show him Newton's set of Zetzner's *Theatrum chemicum*, then on the shelves at Sotheran's. De Villamil apparently found no difficulty thereafter in tracing back to the Musgrave family home the 'Barnsley' part of the shelf-mark written in along the edge of the bookplates. Yet I find it hard to believe that Zeitlinger himself in 1927 was unaware of at least some of the circumstances of the 1920 Thame Park sale and that he remained totally ignorant that there were still Newton books at Barnsley Park. He may well have swapped in gossip with other members of the London second-hand book trade the possibility of there being others also unsold. But let us give credit where it is due: de Villamil was the first to arrive on the scene at Barnsley Park and to be invited into the house.

A substantial part of de Villamil's book[3] is devoted to his transcription *in extenso* of the Musgrave Catalogue. To this he appended the results of a preliminary collation of the Musgrave Catalogue made with the Huggins List, with items occurring in Huggins there starred, and also a supplementary

---

[1] Pp. 5–6. For a biographical sketch of de Villamil see the introduction by I. Bernard Cohen to the Johnson reprint of the book (New York, 1972), v–vii. In an earlier article, issued in March 1927, de Villamil had expressed the view that all the 'Philosophemur' [i.e. Musgrave] books had been included in the Thame Park sale. His own subsequent discoveries at Barnsley proved that this was not so. The title of the article, 'The tragedy of Sir Isaac Newton's Library', *Bookman*, LXXI (1926–7), 303–4, reflects its tone, and in it de Villamil wrote of the 1920 sale 'A bookseller [presumably Zeitlinger] assured me that had the books been properly catalogued and advertised, they should have fetched at least £10,000 ...Of the books that were sold by the purchasers (one a small furniture dealer), most, I gather, went to the United States. But – yet one more subject for lamentation – I understand that a great many were sent to the mills for pulping!' It is to be hoped that the information on which de Villamil based his final lament was ill-founded.
Zeitlinger, *Newton's library* (1944), p. 19.   [3] *Newton: the man*, pp. 62–110.

list of books appearing in Huggins only. Though his transcription has been of considerable (if limited) value to its variety of users, de Villamil there introduced several mistakes and miscopyings of his own over and above those already present in the original document.[1] The chief shortcoming of de Villamil's version for the modern bibliographer, however, is its omission of the shelf-marks which provide the most reliable means of determining the provenance of books from Musgrave's library and so possibly from Newton's own shelves.

One who, like myself, counts it a happy fortune, as well as a convenient aid, to find the personal library of a famous figure maintained substantially intact by its owner must welcome that owner's resistance of the temptation to sell off its items piecemeal – if sell them he has to – no matter what his personal motives are for doing so. In that way such a world-wide dispersal of the books as occurred after the 1920 Thame sale is prevented, and there remains some reasonable chance that the library will pass without further subdivision into public ownership. So firm was the policy to keep the remaining Newton books from Barnsley together – it is very hard to believe that no offers at all were made for some of the choicer individual items – that Sotheran's were empowered to retain them for twelve years until this remaining moiety of the library was finally purchased *en bloc* in 1943. Unfortunately, since the firm's correspondence dealing with the Newton library transactions was destroyed through bombing and flooding during the Second World War, we cannot know the details. My assumption that Sotheran's were throughout acting as agents for the Wykeham-Musgrave family, and not as principals, is grounded on a letter dated 17 September 1936,[2] in which the firm's Managing Director, J. H. Stonehouse, replied to an enquiry from J. M. Keynes about the library; 'the original price', he wrote, 'which the owners asked for this collection was £30,000 – they are now willing to sell it for £5,000 nett'.

The first public announcement that the collection was up for sale was made through a specially printed six-page brochure, undated, but put out by Sotheran's in 1929.[3] Headed 'The Newton Library, for Sale by Henry Sotheran Limited', this surveys the background to the library of 858 volumes and its discovery, and then briefly describes a score or so of the more outstanding books, giving pride of place to Newton's annotated copies of the first and second editions of his *Principia* (nos. 1168 & 1170) and laying

[1] The 'great help' de Villamil received from Mr Gordon D. Knox (who published his book), especially in the comparison of the two listings, is acknowledged in the author's preface, but ironically this particular part of the work has proved to be the most faulty and unreliable.

[2] King's College, Cambridge, Keynes Papers, Box 6. I am grateful to Sir Geoffrey Keynes for permitting me to examine his brother's correspondence and to quote excerpts from it.

[3] In a letter to Sotheran's dated 15 Oct. 1929 J. M. Keynes thanked the firm for sending him a copy (Keynes Papers, Box 6).

particular emphasis on their autograph additions, corrections, and cancellations. Two pages of the pamphlet are devoted to photo-facsimiles of the title-page and page 114 of the first edition, exhibiting their manuscript annotations. The last page ends by informing potential customers that 'the collection is at present deposited in London, in the possession of the advertisers, while the principal volumes of it may be seen at their West End House, 43, Piccadilly'. No asking price was named. The next advertisement of the collection came in 1931, when it appeared as no. 3602 in Sotheran's Catalogue 828 under the rubric 'For Sale: The Newton Library, including the portion (by far the larger and more important one), which was not included in the sale at Thame, in 1920, and consisting altogether of 858 volumes...' Taking up a complete page of the closely printed catalogue, it covers much the same ground as the earlier brochure; special mention is again made of the annotated editions of the *Principia* and information provided on the more important of the other titles. Once more, details of its London whereabouts are given, without indication of what sum purchasers might be expected to pay for the library. Prices were stated for all other items in the catalogue, among which are a few books from the Thame sale – notably the 1655 Barrow's *Euclid* mentioned above offered for £500 – as well as de Villamil's newly published *Newton: the man*.

Whatever enquiries or negotiations ensued as a result of these announcements, they evidently came to nothing, and we find the collection being advertised yet again in 1940, though in a much less prominent manner, in Sotheran's Catalogue 865 'Choice Books, Manuscripts and Engravings, including the Newton Library'. Much of the detail of the 1929 brochure is repeated, and we here read for the first time that 'Price [will be given] on Application'. Soon afterwards the books were despatched by Sotheran's away from the perils of wartime London to the comparative safety of the depository of Barnby, Bendall & Co., Ltd, Furniture Removers and Warehousemen, Cheltenham (where thirteen packing-cases of the books were stored) and of the National Provincial Bank at Banbury (which found space for the remainder in four more cases). In 1943, with no private buyer yet in sight, the Pilgrim Trust at length stepped in. (The Trustees had, I may note, already the previous November honoured the tercentenary of Newton's birth by purchasing his birth-place, the manor house at Woolsthorpe, in order to give it to the nation.) The historian G. M. Trevelyan, then Master of Trinity College, Cambridge, wrote to Lord Macmillan, Chairman of the Trust, urging the desirability of maintaining the collection together in England.[1] The speed of the Trust's subsequent positive response indicates their readiness to act forthwith on Trevelyan's suggestion, and the collection was swiftly purchased through, it is said, the timely benefaction of an American citizen.[2] Sadly, I might add, no one appears to have considered the

---

[1] *Newton's library*, pp. 5–6. [2] *Ibid.* p. 7.

overriding advantages accruing from housing the books in Cambridge University Library,[1] where they could to mutual profit be conjoined with the millions of words of Newton's autograph scientific manuscripts – the largest part by far of those surviving – and containing *inter alia* the drafts of his printed books.

The first public news of the purchase of the Newton library by the Pilgrim Trust appeared in an article in *The Observer* of 11 April 1943,[2] where, after the main features of the collection had been described, the guess was hazarded that the British Museum, the Royal Society, and Trinity College, Cambridge would be powerful rival claimants to house the books. *The Times* of 12 April published a similar piece,[3] in which it was also stated that the ultimate destination of the collection was yet to be decided. As we now know, the Pilgrim Trustees finally plumped for Trinity College, and early in August 1943 the College Librarian, Herbert M. Adams, was instructed to inspect the collection in its temporary places of storage and to arrange for its transfer (by rail) to Cambridge.[4] By the beginning of October all seventeen cases of the books had arrived safely in Trinity, and they were housed in their special

[1] On 9 April 1943 the Secretary of the Royal Society, A. V. Hill, aware of Lord Keynes's interest in Newton's manuscripts and books, wrote telling him of the acquisition of the library by the Pilgrim Trust and listed the possible future homes which had already been suggested for it: namely, Trinity College, the Royal Society, and the British Museum. He added that he and Sir Henry Dale (then P.R.S.) would prefer Trinity or the Royal Society, but concluded by inviting Keynes's views regarding the books' most fitting permanent place and asking him to discuss the matter with Lord Macmillan. In his ensuing letter of 13 April to Macmillan, Keynes, after expressing his delight that the library had been secured from dispersal abroad, made clear his feelings about its eventual destination in these words:

'I hope you will not pass it over to the British Museum, where it will be sunk and buried out of sight. The strong, obvious claimants, I should say, are Trinity and the Royal Society...My own inclination, if I may venture an opinion, would be slightly in favour of the Royal. Their personal, sentimental and historical association with Newton is very special. He is their tutelary god, their one god – there is a bit more polytheism in Trinity.'

He went on to recall that he first saw the library 'a few years ago...At that time they [Sotheran's] did not own it – I do not know whether you have now bought it from them as principals or whether they were simply acting for the previous owners'. Keynes also replied the same day to A. V. Hill in very similar, if somewhat more concise, terms, setting a 'certainly not' on the British Museum as possible future depository and reiterating that he was 'inclined to favour the Royal Society' since 'Their connection with Newton is so very extra special'. (See his correspondence preserved in King's College, Cambridge, Keynes Papers, Box 6.) For all that the University Librarian of the time and Lord Keynes were Fellows of the same college, King's, they were both apparently equally ignorant of the true extent and bulk of that Library's holding of Newton's manuscripts.

[2] P. 5, under the caption 'Pilgrim Trust buys Newton books lost for 200 years, by a Special Correspondent'.

[3] P. 6, 'Isaac Newton's library. Purchase by Pilgrim Trust'.

[4] Not the easiest of tasks to undertake during the war. For correspondence relating to the despatch of the books during August and September 1943, see the folder marked 'Notes, Papers etc. relating to the Newton Library', Trinity College, NQ.17.37.

bay in the Wren Library in time for the official presentation ceremony held there on the thirtieth of the month. Formally handed over by Lord Macmillan on behalf of the Pilgrim Trust and officially accepted for the College by the Master,[1] the gift consisted of 858 volumes discovered fifteen years before by de Villamil at Barnsley Park, together with Musgrave's manuscript catalogue and two additional volumes, viz. Barrow's 1655 edition of Euclid and a 1653 Greek translation of the Old Testament containing a sheet of notes by Newton (no. 195). (This last item, which was then valued at £75, was given to the College by Sotheran's.)[2] Not only did the Pilgrim Trustees find the £5500 asking price for the books, but they also offered to defray the cost of any repair or rebinding they might need. To commemorate the benefaction they commissioned a bookplate designed for the occasion by R. A. Maynard, and this is now to be found pasted inside the front cover of the individual volumes.

The published text of the ceremony in the Wren Library put the total number of printed books handed over to Trinity at 860, but there are, by my count, now there 864 volumes (all deriving from the Musgrave holdings) with the Pilgrim Trust bookplate pasted in. I have not been able to determine if this 864 is the actual number handed over in October 1943 or if the four volumes additional to the official 860 were subsequently sent to the College.[3] The 864 volumes received by Trinity comprise 907 separate titles, but eight at least of these volumes (one containing two titles) could not have formed part of Newton's library, because they were published after his death. These range over the years 1727 to 1747. Seventeen smaller items from 1727 to 1757 are bound up in volumes of tracts together with works issued before 1727. The number, therefore, of the volumes given to Trinity in 1943 which on the sole basis of their date of publication may once have stood on Newton's own shelves is at most 856. These comprise 882 individual titles, four of which are represented twice. Prior to the Trust's 1943 gift the College owned six volumes (containing eight individual items) which had been Newton's, two of which (nos. 55 & 1489) stemmed from the Thame Park sale and the rest from other, earlier, sources (the entries for nos. 188, 240, 507, 629, 1442, 1560 include what is known of their various provenances), so that Trinity's present Newton holding amounts to 862 volumes containing 890 separate works. Let me reissue the caution that not every book which has travelled

[1] Zeitlinger, *Newton's library* (1944), pp. 6–12, gives an account of the ceremony and reproduces the text of the speeches. A similar version was also included in *The Pilgrim Trust, 13th Annual Report* (1944), pp. 5–7.

[2] *Ibid.* p. 6 (part of the unsigned prefatory note). As Lot 1497 in Sotheran's Catalogue 843 of 1935 it had been priced at £100.

[3] Trinity College NQ.17.37 contains a typescript book-list compiled in 1929 by H.[einrich] Z.[eitlinger] itemizing 858 volumes, to which Mr Adams subsequently added (he cannot now recall when) five more works, including the *Euclid*; these, together with the Greek Old Testament, make up my total of 864 volumes.

the Huggins/Musgrave route to Trinity (or elsewhere) may automatically be assumed to have formerly belonged to Newton himself. I also admit to considerable doubt about the Newtonian provenance of a few of the other items in the Pilgrim Trust's benefaction.[1]

Books from Newton's library have appeared less and less frequently in booksellers' catalogues during recent years.[2] Yet there must be many such volumes still in private hands, or even on the second-hand market, whose Newtonian pedigree is not at present recognized by their owners. To cite just one instance in point, in August 1975 a colleague of mine spotted in a Cambridge bookshop Newton's copy (there unidentified) of Optatus, *Opera*...1631 (no. 1206) and bought it for £4: the book has Musgrave and Barnsley Park shelf-marks, and its fly-leaf bears 'pret 15ˢ.' in Newton's hand. How many more of Newton's books are yet to surface in a similar way?

To sum up, my present Newton catalogue is based very largely on the documents and books brought to light by de Villamil's persistence and ingenuity – the 1727 house 'inventory', the Huggins List and the Musgrave Catalogue, above all upon the substantial moiety of Newton's books themselves which has survived. Huggins and the somewhat less reliable Musgrave have provided a sound foundation on which to build the catalogue, but I have always, wherever this has been possible, dug down to the bedrock of the surviving books themselves. Many of the doubts and hesitancies arising from the various imperfections of these listings of Newton's books have been removed by the physical examination I have been able to make of the nine hundred of these in British libraries as well as in photocopy of yet others which are now found abroad. Not quite all of these passed through the hands of the Rectors of Chinnor and their descendants, having been parted from the main collection either during Newton's lifetime[3] or immediately after his death.[4] Those which still survive have a special importance, not least because of the autograph annotations which many of them contain, and it is fitting that they now (as may be seen from the catalogue) in great part stand on the shelves of the libraries of his College and University, having for so long been lost to public view.

[1] I am, for example, extremely suspicious of the authenticity of Beveridge's *The great necessity and advantage of public prayer*...4th ed., 1709 (no. 178). It has no markings by Newton, is not in Huggins or Musgrave, and has no Musgrave shelf-mark, and its only bookplate is the commemorative Pilgrim Trust one.

[2] When Sotheby's auctioned the collection of scientific books from the library of Professor E. N. Da C. Andrade on 12–13 July 1965, two items from Newton's library were among them: Lot 103, *Boyle's Works epitomiz'd by R. Boulton*, vols. 1–3, 1699–1700 (no. 276): and Lot 234, *Miscellanea curiosa*...[Ed. by E. Halley], vol. 1, 1705 (no. 1086), to which I have already referred (p. 49 n. 1 above). These originally came on to the market from the 1920 Thame Park sale.

[3] For example Descartes's *Geometria*...*operâ atque studio F. à Schooten*...Ed. 2ᵃ 2 pts. 1659–61 (no. 507).

[4] Notably Newton's *Principia*...1687 (no. 1167), Starkey's *Secrets reveal'd*...1669 (no. 1478), and the other four 'Books that has Notes of Sir Is. Newtons'; see pp. 33–4 above.

CHAPTER 3

THE COMPOSITION OF NEWTON'S LIBRARY

Newton's fame and reputation rest first and foremost on his mathematical and scientific talents and achievements. We may therefore be initially surprised to find that only 31 per cent of the books in his library at his death were explicitly scientific in content.[1] In table 1 his holdings are analysed by subject.

The totals given in this table are the aggregates of the individual titles set out in the present catalogue. I have already pointed out that some books (maybe as many as 130) which are included were not from Newton's own shelves, having been added to the collection by Huggins and Musgrave.[2] These dubious items which cannot separately be identified and so excluded are almost entirely of a theological, literary, or historical nature. Their inclusion somewhat inflates the totals given for these particular categories, and consequently the correct percentages for scientific titles may be a little higher than those here listed. A further unknown factor which may also detract from the accuracy of my figures is the number and breakdown of items contained in the twenty-two volumes of tracts so far untraced – comprising perhaps as many as 120 individual titles on varying subjects.[3] The classification of a book into one or other subject category must often be arbitrary and imprecise. The division between theology and philosophy, for example, cannot always be readily defined or consistently observed, and the like is true of that between alchemy and chemistry, astronomy and mathematics, or mathematics and physics. To underline the point, one might well ask under what head should one pigeon-hole Newton's *Principia*?[4] Add to this the many further hazards which ensue on applying twentieth-century subject-divisions

---

[1] An analysis of the books in his library published before 1696 when he left Cambridge, in all 1001, shows the virtually identical proportion of 32 per cent of purely scientific books. We may assume that Newton acquired a few of these after 1696, but this would apply equally to pre-1696 books on non-scientific subjects, so that the overall percentage is likely to be reasonably accurate. It is worth noting that 156 of the final total of 169 (al)chemical books were dated before 1696, in astronomy 24 of the eventual 33, in mathematics 62 out of 126.

[2] See p. 45.

[3] See pp. 46–7. Trinity's twenty-three 'Tract' volumes contain 129 separate items.

[4] For the record I here set it under 'mathematics'.

TABLE 1. *Newton's library: subject analysis of the 1752 titles*[1]

	Titles	Proportion (%)
A. Scientific books		
1. Alchemy, 138, and chemistry, 31	169	9½
2. Mathematics	126	7
3. Medicine and anatomy	57	3⅓
4. Physics (including optics, 15)	52	3
5. Astronomy	33	1⅔
6. Other scientific subjects (including general works, 28; natural history, 18; zoology, 7; botany, 6; mineralogy, 3)	101	5⅔
B. Non-scientific books		
1. Theology (including general works, 205; Bibles, Testaments, and Biblical studies, 99; Church Fathers, 61; Church history, 28; religious controversy, 28; Jewish rites and customs, 24)	477	27½
2. Classical literature, Greek and Latin	149	8⅔
3. History, 114 (general, 5; ancient, 19; modern, 90), together with chronology, 22, and biography, 7	143	8⅓
4. Reference and periodicals (including dictionaries, 43; grammars, 11; periodicals, 18)	90	5⅓
5. Voyages and travel, 46; geography, 30	76	4½
6. Modern literature (including English, 40; Latin, 10)	58	3⅓
7. Philosophy (ancient, 9; modern, 24); logic, 6	39	2
8. Law, 22; politics, 15	37	2
9. Economics (including currency, 10)	31	1⅔
10. Other subjects (including antiquities, 18; numismatics, 10; medals, 6)	114	6½

(and their inbuilt preconceptions) to topics of the sixteenth to eighteenth centuries. But none of these reservations bear significantly on the general problem of deciding whether a book is scientific or non-scientific in theme; I do not, however, claim that the respective percentages for the component parts of these two main divisions are rigorously exact.

It is worth stressing again that Newton's was a working library, with its books there for use, not mere decoration. This means that the works there (totalling 69 per cent in all) which deal with theology and church history, the classics, history and chronology, geography and travel, and the rest, accurately reflect in their several proportions his broad range of interests in

[1] In arriving at this total for this and the following tables, I have disregarded Newton's ownership of more than one copy of identical editions of the same book. In the catalogue, however, it is convenient for reference purposes to give each surviving duplicate copy a separate number; thus the entries therein run to 1763.

these widely differing areas, and it could indicate that over the course of his long life these subjects, together with alchemy, might well have occupied as much (if not more) of Newton's time than his work on mathematics and other scientific topics. But this should be taken only as an indication, nothing more. It would be misleading, not to say ridiculous, to proceed further and suggest that it might be possible narrowly to relate the total of the books Newton held on any one subject to the amount of time he may have spent in that field of study.[1] I cannot believe (and it goes against all else we know of his life to suppose) that he spent more time on classical literature (149 titles) than on mathematics (126), or more on voyages and travels (46) than on astronomy (33). To assert as much certainly fails to take any account of the much higher 'density' of content of the scientific and especially the mathematical books of his time, as compared with that of works dealing with, for example, ephemeral theological matters. Newton's books on mathematics and physics have in the main the appearance and feel of having been new when he obtained them: few were likely, indeed, to have been on the shelves of the Cambridge booksellers, and he must have had to wait several months for them to arrive from the Continent. On the other hand, many of the works in other categories such as theology and the classics, and particularly alchemy, were second-hand purchases, often bearing the names, notes, and other markings of their earlier owners, and may even have been bought in job-lot bundles at book-sales.[2]

In considering what conclusions we may draw from the relative splay of scientific as opposed to non-scientific titles in the library, we must pay heed to a variety of weighting factors, other than that of a work's 'density'. Not least to be considered is that there were ready to hand for purchase in Cambridge fewer works in science than in most of the other fields of learning. Another factor which may have had some bearing on the composition of Newton's library is the ease of access which he had to Cambridge institutional libraries and to Isaac Barrow's bookshelves until the latter's death in 1677.[3]

Fortunately, not much more than a quarter of Newton's mathematical books were auctioned at the Thame Park sale of 1920, and the remainder, containing many highly important items, stayed together, to be housed eventually in their present bay in Trinity Library. That they have not been split up has made it conveniently possible for detailed study to be made of

---

[1] In his contribution to *The Royal Society: Newton tercentenary celebrations, 15–19 July 1946* (Cambridge, 1947), p. 18, Professor E. N. Da C. Andrade stated (in my view mistakenly) that 'Newton devoted probably as much time and effort to alchemy and chemistry...as he did to the physical sciences.' Yet it must normally have taken Newton hours, days, even weeks and months, to digest and react to a single page of mathematics, whereas an alchemical pamphlet may (on occasion) have been absorbed at a single sitting, even if his chemical experiments took up a deal of his time.

[2] A fate which was to lie in store for at least some of the very same volumes at the Thame Park sale in 1920. [3] See pp. 6–7, 61–4.

them. Some account of the more significant volumes has already been published. D. T. Whiteside in the first volume of his edition of Newton's *Mathematical papers* has discussed the several early works on trigonometry present – by Gunter (no. 728), Norwood (no. 1190), Oughtred (no. 1220), and Ward (no. 1711), and has drawn attention to Newton's bound set of James Gregory's mathematical tracts (nos. 711, 712, 714) as well as to his copy of Schooten's *Exercitationum mathematicarum libri V...* 1657 (no. 1471); a leaf of paper now stuck (but once loose) in the last bearing one of Newton's earliest mathematical annotations is printed by him.[1] Among Newton's other mathematical books, the classical Greek mathematicians – Apollonius, Archimedes, Diophantus, Euclid, Proclus – are strongly represented, most of them in finely printed folios and several of them presentation copies to him from their editors. We should not entirely overlook the possibility that Newton may have lost or discarded or even worn out a few books over the long span of his life. Nevertheless, we have to accept that a number of highly important mathematical and astronomical items which one might have expected to find in his library seem never to have been there. But we must not overlook the books then in Barrow's library which Newton could well have consulted, even if he was unable to avail himself of the freedom as often as he might have wished.[2] The absence of books from his own shelves need not be quite so crucial if he might readily have had access to Barrow's copies. The catalogue of Barrow's library made at his death in 1677 exists.[3] What books are listed in it? In all 1099 volumes are recorded, all of these substantial works. If Barrow owned any pamphlets, single sermons, or other small items, they were either despatched by Newton to the bookseller 'S. S.', who commissioned the catalogue of the library, without being listed individually therein, or they were simply not sent at all. (It is, I may add, hard to believe that Barrow possessed no material of this nature. Perhaps it was thought that the minor items would fetch such a small amount of money in London that they were not worth the time and effort of cataloguing them separately, or maybe they were simply offloaded in Cambridge for cheapness and ease.) A comparison of the 992 titles in the 1677 document with the 1752

---

[1] Newton, *Math. papers*, I (1967), 46.

[2] It is possible that too much is made here (and by others) of the intimacy of contact between Barrow and Newton, especially in the 1660s. The earliest known reference to Newton made by Barrow in the course of his correspondence with John Collins over some five years from 1664 occurs in a letter of 20 July 1669 which begins: 'A friend of mine [not named by Barrow] here, that hath a very excellent genius to those [mathematical] things, brought me the other day some papers.' On 31 July following, Newton is still unidentified in a further letter of Barrow's accompanying 'the papers of my friend I promised'. Finally, on 20 Aug. 1669 the 'friend' is identified: 'I am glad my friends paper giveth you so much satisfaction. his name is Mr Newton, a fellow of our College, & very young (being but the second yeere Master of Arts) but of an extraordinary genius & proficiency in these things' (Newton, *Correspondence*, ed. H. W. Turnbull, I (1959), 13–15).

[3] Bodleian MS.Rawl.D.878, fols. 39r–59v.

items in Newton's library reveals that, other than in the fields of classical literature and of religion (patristic works especially), the works that the two men held in common were quite small in number. Barrow devoted nearly all his mental attention to religious subjects after resigning his Chair of Mathematics in 1669,[1] so it is no surprise to find that theology singly makes up 31 per cent of his holdings, while classical literature forms a further 21 per cent (he had been Regius Professor of Greek from 1660 before becoming Lucasian Professor of Mathematics in 1664). The combined percentages of books on mathematics (151 titles) and on astronomy (24 titles) reach the fairly high total of $17\frac{1}{2}$, just about double the combined percentage holdings of Newton in these subjects. All other scientific holdings together amount to $7\frac{3}{8}$ per cent (76 works).

Heinrich Zeitlinger has laid special stress on works by several well-known foreign mathematicians and astronomers which are not recorded in Huggins and Musgrave[2] – in this he is followed by H. A. Feisenberger[3] – but some of these were certainly present in Barrow's library. In the 1677 list we may find, for example, works by Tycho Brahe (including *Opera*, 2 vols.), Cardano (*Opera omnia*, though only vol. 2), Cavalieri (*Specchio ustorio overo trattato delle settioni coniche, Exercitationes geometricae* and *Trigonometria plana*), Kepler (*Dioptrice* and *Epitome Astronomiae Copernicanae*), Stevin (*Opera*, 2 vols.), and Viète (Schooten's 1646 edition of his collected *Opera mathematica*): none of these authors were represented in Newton's library except for Kepler's *Dioptrice* – and that in the 1682 reissue of the 1653 edition, where it was published with Gassendi's *Institutio astronomica* (no. 651).[4] In addition Barrow owned scientific books by Boulliau, Lansberg, Mersenne, Riccioli, Tacquet, Witelo, and Zucchi – none of these authors are to be found in Newton's collection – along with certain works by Galileo, Hobbes, Hooke, and Leibniz which Newton did not have (though he owned other books by them). Also with Barrow were no fewer than nine separate works by John Wallis, the Oxford

---

[1] However convenient and sentimentally attractive it may seem, the modern version that Barrow relinquished the professorship in favour of Newton is not proven by reliable existing evidence. See Newton, *Math. papers*, III (1969), xiv n. 14.

[2] *Newton's library* (Cambridge, 1944), pp. 22–3.

[3] H. A. Feisenberger, 'The libraries of Newton, Hooke and Boyle', *Notes and Records of the Royal Society of London*, XXI (1966), 42–5.

[4] Newton therefore acquired this edition of the *Dioptrice* well after Barrow's library had gone from Trinity. He had certainly studied the 1653 1st edition of the joint compendium, which was then required student reading, as an undergraduate in the early 1660s, but had at that time, it would seem, paid most attention to the Gassendi item (from which he drew a short MS summary of the 'Systema mundanum secundū Copernicum'). See D. T. Whiteside, 'Before the *Principia*: the maturing of Newton's thoughts on dynamical astronomy, 1664–84', *Journal of the History of Astronomy*, I (1970), 5–18, esp. 16 n. 11. Galileo's *Sidereus nuncius* was also published in the same 1682 compendium, but neither this nor Kepler's *Dioptrice* is mentioned in de Villamil's reproduction of the Musgrave Catalogue, and Zeitlinger was thereby misled into stating that there was no work by Kepler in the Newton collection, even though he probably handled the very volume here at issue.

Savilian Professor, whose influence on Newton's mathematics in his earlier years was second only to Descartes's; in Newton's final library list, however, Wallis is represented only by the *Mechanica*...1669–71 (no. 1709) and *Opera mathematica*, 3 vols., 1693–9 (no. 1710). It is important to remind ourselves again, however, that Barrow's collection was not the only one available for Newton to consult. There was a far more extensive range of astronomical, mathematical, and scientific books in the Public [University] Library, which at this period contained works by all the authors just now named and many more besides;[1] and Trinity College Library's 3000–4000 volumes also included additional copies of most of them.[2]

Newton had some small interest in and knowledge of anatomy and medicine, but even so the total of his fifty-seven works in this area is rather more than might have been anticipated. He had a mere half-dozen of straight anatomical texts, but two of these were the illustrated folios by John Browne (no. 303) and William Cowper (no. 451). The medical books ranged, if somewhat scrappily, over aspects of the history, principles, and practice of the art. Prominent among these are ones dealing with the maintenance of good health and the prevention of disease, and with remedies and cures – notably Luigi Cornaro's *Sure and certain methods of attaining a long and healthful life*... 2nd ed., 1704 (no. 443) and George Cheyne's *Essay of health and long life*, 3rd ed., 1725 (no. 368), though Newton could only have acquired the latter (if indeed the book was his) during the last few months of his long life. Richard Mead, Newton's doctor, presumably gave his patient the two editions of his book on plague and its prevention (nos. 1055–6). The physician

---

[1] The late-seventeenth-century author catalogue of Cambridge University's 'Old Library' (now in the 'Star' classes) of the University Library, taken together with the contemporary shelf-lists, demonstrates that these authors were amply available there in Newton's Cambridge days. The original bookplates still present in a number of the volumes provide confirmation of this – and in several instances precise dating of their arrival in the Library. One such plate, dated 1656, commemorates Richard Foxton (1590–1648), a member of Emmanuel College and sometime Mayor of Cambridge, from whose legacy of £40 to the Library a number of scientific works were bought; a second, dated 1657, shows that a bequest of £50 from the benefactor Alexander Ross (1591–1654) of Aberdeen was put to similar use; a third plate labels books as gifts from Henry Lucas of St John's College, who after his death in 1663 founded the Lucasian Professorship of Mathematics by the terms of his will, and there are yet other books bearing Lucas's signature. The University Library has a set of the twin volumes of Wallis's 1656–7 *Opera*, but since, unfortunately, the 'Old Library' author catalogue now lacks all its leaves from the letter 'V' onwards we cannot be sure if this copy was already there in the 1660s when Newton read the book (see his notes on the work reproduced in Newton, *Math. papers*, I (1967), 96–142).

[2] The Trinity Catalogue Add.MS.a.101, by author and subject, of the College Library in 1667 has on fol. 35 a specially added entry (undated) of 'Libri Mathematici' which lists thirty-nine volumes presented by Barrow when he was but a Fellow of the College (that is, before 1673, when he became Master); and on fol. 64 further records twenty-four volumes, mainly on scientific subjects 'Ex dono D^{ris} Barrow Collegij Mag^{ri}'. Trinity's holding at this time of works by Wallis was very slight, and the College does not appear thus to have possessed the *Opera* of 1656–7.

William Cockburn was much concerned with the cure of dysentery and scurvy, and four of his publications on these and related matters found their way into Newton's library (nos. 404–7). Virtually all these books of remedies and cures were smallish items (some mere tracts), issued for the most part well after 1700, and I suspect that many of them, like Mead's book, for example, were gifts.

If, as has lately been conjectured,[1] Barrow had any deep interest in chemistry and alchemy it certainly is not reflected in the 1677 list of his library. Whereas books dealing with other scientific subjects are there in substantial numbers, we find five titles only on chemistry, and yet more significantly, just a single alchemical work. Similarly my examination of the contemporary catalogues of the University and Trinity libraries leads me to assess their (al)chemical holdings as comparatively small.[2] These deficiencies in the sources from which Newton might otherwise have readily borrowed what he needed to read may, therefore, have been one of the reasons which led Newton to build up his own collection of (al)chemical books and texts to the very high total of 169 works. Taking account, indeed, of the poverty of such works in the 1677 list, I cannot altogether resist the speculation that Newton, who undertook the tabling of Barrow's library, may have abstracted some alchemical items for his own library,[3] though I hasten to add that there is nothing – inscriptions or notes – in the surviving books themselves which begins to substantiate this conjecture. We do know that Newton had already started to buy (al)chemical books in the late 1660s[4] and that he continued to acquire many more without break after Barrow's death.[5]

Almost every survey of Newton's own library published since de Villamil's recounting of his discoveries at Barnsley Park has laid special stress on the numerous alchemical books which it contained.[6] E. N. Da C. Andrade

---

[1] See B. J. T. Dobbs, *The foundations of Newton's alchemy, or 'The Hunting of the Greene Lyon'* (Cambridge, 1975), pp. 95–102.

[2] Trinity College Add.MS.a.101 includes their (al)chemical books under the general head 'Medici &c', with a few more at 'Philosophi Recentiores'.

[3] Newton's library did contain one work from Barrow's library, Cosin's *Historia transubstantiationis papalis*...1675 (no. 444), which is recorded in the 1677 list; see p. 11 n. 2 Furthermore, Cambridge University Library's copy of Zucchi's *Nova de machinis philosophia*...1649, the first item in a bound volume (M.5.42) containing three other works, has the inscription 'G. Atkinson. Trin: Coll: Cant. E Biblioth. D^{ris} B. 1677. 4º. Id. Junij' [= 10 June]. Since Barrow had died on 4 May and the 1677 list (which tells us that the books were sent off to 'S. S.' on 14 July) records Zucchi's work followed immediately by the other three, it is very possible that some degree of private trading took place within the College between the completion of the list and the departure of the books for London, or indeed in other instances before the list was made.

[4] He purchased Zetzner's *Theatrum chemicum*...6 vols., 1659–61 (no. 1608) in April 1669. See p. 8.

[5] Newton's list of books which is now Bodleian MS New College 361, vol. II, fol. 78 (which I have dated tentatively 1702 to mid-1705), contains twelve alchemical items. See p. 9.

[6] It is strange that de Villamil himself, who handled 860 volumes when going through the library, apparently did not think them worthy of special comment (*Newton: the man* (London

claimed that the library was 'well stocked with the standard alchemical and mystical books', adding that alchemy and chemistry were one study in Newton's time.[1] When I myself distinguished the items into two separate categories comprising 138 alchemical and 31 chemical titles I was aware that the dividing-line between the two was by no means clear-cut. F. E. Manuel's interpretation of the boundary between the two aspects of the 'one study' was very different to mine when he had earlier come to review the contents of Newton's library.[2] His given total for alchemical works was about a hundred with seventy-five more for chemistry/mineralogy, but his conclusion that overall they formed about a tenth of the library comes reassuringly near to my own calculation. Zetzner's six-volume *Theatrum chemicum*, with its reputed mass of annotations in Newton's hand, formed the centre-piece of his alchemical collection, and such other similar compendia as Ashmole's *Theatrum chemicum Britannicum*...1652 (no. 93) supplemented it. Of foreign authors, nine separate works by the German alchemist Michael Mayer are identifiable, as are eight by the Catalan polymath Raymund Lull and four by the reputed Benedictine monk Basil Valentine, and several French alchemists are also represented. English writers include George Starkey with nine items by (or attributable to) him: one published under his own name, one anonymously, and the rest under one or other variant on the pseudonym Eirenæus Philalethes. Thomas Vaughan, under his supposed pen-name Eugenius Philalethes, is present with four items, and Samuel Norton's tracts supply eight titles. It is strange that in spite of all Newton's wide reading in the alchemical books he owned – and, I may add, of the mass of his surviving autograph papers and notes on the subject – he remained ever unwilling to make detailed publication of his ideas or conclusions on this enduring interest of his Cambridge years.[3]

[1931]), pp. 6, 9–11, 39). Nor was J. J. Björnståhl on his visit in 1775 to Chinnor (where he was able to inspect the entire library) sufficiently impressed to give them a mention (*Briefe aus seinen ausländischen Reisen* (Rostock and Leipzig, 1781), III, 288–9).

[1] In *The Royal Society: Newton tercentenary celebrations*, p. 18.
[2] *A portrait of Isaac Newton* (Cambridge, Mass., 1968), p. 163. In this book and in his *Isaac Newton, historian* (Cambridge, 1963), esp. pp. 37–47, Professor Manuel offers a valuable survey of the strengths and weaknesses of Newton's holdings, though one based on an examination of the Huggins List only. An interesting and salutary illustration of the potential pitfalls in making an assessment of Newton's coverage of any one subject is provided by contrasting the verdict of Manuel, who found that the books on astronomy were 'plentiful' (*Isaac Newton, historian*, p. 43), with Zeitlinger's judgement (based on the Musgrave Catalogue and on the books that went to Trinity) that there were 'only relatively few' (*Newton's library*, p. 22). The catalogue of Flamsteed's private library prior to 1685 (see E. G. Forbes, 'The library of the Rev. John Flamsteed, F.R.S., first Astronomer Royal', *Notes and Records of the Royal Society of London*, XXVIII (1973–4), 119–43, lists 64 items on astronomy out of an overall total of 231. When I compare this with Newton's holding of 33 such books from a total of 1752, I am inclined to go along with Zeitlinger in this instance.
[3] For a further survey of Newton's alchemical holdings, especially those in the Trinity College collection, see Dobbs, pp. 49–51. Mrs Dobbs, at Appendix A, pp. 235–48, gives

A glance at table 1 will reveal the large number of works of Biblical texts and commentaries (several of the latter on prophecy) which Newton owned and the extent of his holdings of the available printings of the doctrines of the Church Fathers (mostly in folio editions), several of which may well have come from the library of his stepfather Barnabas Smith.[1] Newton's interest in Jewish rites and customs is clearly manifested. His theological books also include many relating to religious controversies in the England of his day, while others contain sermons, and there are a few on Socinianism. The Huggins List records that Newton had two copies of *Oratio Dominica in diversas omnium fere gentium linguas versa*. . .1715 (no. 984), a book which contains in parallel text and transliteration well over a hundred different versions of the Lord's Prayer in a wide range of Asian, African, European, and American languages and dialects. This work, issued 'Typis Guilielmi & Davidis Goerei, Amstelædami', and in its way a minor masterpiece of the printer's art, was edited by John Chamberlayne, Fellow of the Royal Society, who had some acquaintance with Newton; one of the two copies (not that which survives)[2] which is described in Huggins as 'Spanish calf, gilt leaves' was presumably a presentation. Another religious work which we would not perhaps expect to find in Newton's possession is *The Massachuset Psalter: or, Psalms of David with the Gospel according to John, in columns of Indian and English*. . .8°, Boston, N.E., 1709 (no. 1039).[3] Newton's need of background information for his *The chronology of ancient kingdoms amended* published posthumously in 1728 would account for the considerable number of works on world history and chronology. Books on geography and travel are well represented,[4] and there

a detailed finding-list of Newton's alchemical manuscripts that are available for study, most of them in the Keynes Collection of King's College, Cambridge.

[1] William Stukeley in his *Memoirs of Sir Isaac Newton's life*. . .ed. A. Hastings White (London, 1936), p. 16, recounts that in the library of Barnabas Smith 'There were some years ago 2 or 300 books . . . chiefly of divinity and old editions of the fathers. . .These books Sir Isaac gave to his relation Dr. Newton of Grantham, who gave some of them to me, when I went to live there.' Stukeley cannot be taken to imply that Newton left the collection intact at Woolsthorpe. Indeed, the most famous tome from that source is not a printed book at all, but the 'Waste Book' (now Cambridge University Library Add.MS.4004), originally a theological commonplace book of Smith's which Newton took to Cambridge and used for entering up many of his early (and some later) mathematical and mechanical research notes. At least one of the printed books can be identified with confidence: Newton's copy of Livy's *Historiæ Romanæ principis libri omnes superstites*. . .1609 (no. 964) is heavily annotated in a hand which, when compared with Smith's autograph in the 'Waste Book', leaves virtually no doubt that this volume also came to Newton from the same source.

[2] Trinity College, NQ.8.24.

[3] Trinity College, NQ.16.160. There must, however, be considerable doubt about Newton's ownership of this. It is not recorded in the Huggins List and so may well have been added to the collection by Huggins or Musgrave, and on the fly-leaf it carries the name of a former owner, 'Jonathan Yate Gifford'. (These 'Red Indian Bibles' and the like were often given as New Year or other anniversary presents.)

[4] Though perhaps not to the extent suggested in J. Harrison and P. Laslett, *The Library of John Locke* (Oxford, 1965; 2nd ed. 1971), pp. 12–13, where we compared the contents of

is a miscellany on the history of races and peoples in Europe and elsewhere.

Periodical publications and reference material constitute in their bulk a larger and more significant part of the collection than the mere 5⅓ per cent holding of titles might suggest. This proportion compares very favourably with Locke's 2⅔ per cent for similar items, and in terms of number of volumes, rather than of titles, they form about 16 per cent of his library. It was his wont, it would appear, to acquire long, unbroken runs of the foremost learned journals of his time. Chief among these are *Philosophical transactions of the Royal Society*, nos. 1–380 in 16 vols., 1665–1723 (no. 1304), *Acta eruditorum*, 55 vols., 1682–1726 (no. 7) and *Journal des sçavans*, 75 vols., 1665–1722 (no. 863); the whereabouts of these sets is at present unknown to me, but let me hope that they will surface again before long.[1]

Lest I misleadingly convey any insistent impression that Newton was an assiduous purchaser of foreign journals I should state that many of them came to him as periodic gifts, usually unsolicited, often perhaps even unwanted. Thus the *Acta eruditorum* came to Newton as a present from Otto Mencke,[2] its co-founder with Christian Pfautz, and the *Journal des sçavans* doubtless reached him in a similar fashion. The *Histoire de l'Académie Royale des Sciences, Année M.DCCX – Année M.DCCXXII. Avec les Mémoires de mathématique & de physique*...13 vols., 4°, Paris, 1712–24 (no. 1250) together with the official French astronomical almanack *Connoissance des temps pour l'année MDCCXIV, MDCCXVI–MDCCXXII*...(no. 435) – the latter eight volumes had disappeared from the library by the time the Musgrave Catalogue was made – were received by Newton as of right as an *associé étranger* of the Paris Académie through the good offices of Pierre Varignon.[3] We know likewise that the English stationer Thomas Johnson, who published at the Hague, sent Newton a run of his *Journal litéraire*, vols. 1–9, 1713–17 (no. 864),[4] and

Locke's and Newton's libraries, offering the estimate that roughly 15 per cent of Locke's titles in all subjects were the same as Newton's and 20 per cent of Newton's were the same as Locke's – though not necessarily in identical editions. My recent research leads me to think that we exaggerated the similarity of the two libraries in the field of travel literature.

[1] The three sets can readily be authenticated by their Musgrave shelf-marks: D2-1 to D2-16, D2-17 to D2-71, G5-31 to G5-105(?). They were all auctioned off at the Thame Park Sale, January 1920.

[2] See Newton, *Correspondence*, ed. H. W. Turnbull, III (1961), 270. Mencke died in 1707, but it is likely that his son, who took over his father's role in the editorship of the *Acta*, continued regularly to send the journal to Newton.

[3] See Varignon's letters to Newton on 17 Nov. (N.S.) 1718 and 18 Sept. (N.S.) 1721 and Newton's acknowledgements of the gifts in letters to Varignon on 29 Aug. (O.S.) 1718, 19 Jan. (O.S.) 1720/21 and 26 Sept. (O.S.) 1721, printed respectively in Newton, *Correspondence*, ed. A. R. Hall and L. Tilling, VII (1977), 14–16, 152–6, and 2–4, 119–23, 160–65 (this last was first published in D. Brewster, *Memoirs of the life, writings, and discoveries of Sir Isaac Newton* (2 vols., Edinburgh, 1855), II, 497–501).

[4] Johnson was the Royal Society's foreign bookseller for items like the 1712 *Commercium Epistolicum* (The Royal Society Journal Book records on 17 June 1714 a parcel of fifty

Johnson may well have presented the set of *L'Europe savante*, 12 vols., 1718–20 (no. 587), also published at The Hague, as well as being responsible for other Dutch serials finding their way on to Newton's shelves, such as Jean Le Clerc's *Bibliothèque ancienne et moderne*...vols. 1–18, 1714–22 (no. 928) and his *Bibliothèque choisie, pour servir de suite à la Bibliothèque universelle*, vols. 16–27, 1708–13 (no. 929), both issued at Amsterdam, though Newton did not own the parent *Bibliothèque universelle et historique* itself. Surprisingly, among the books of one not usually noted for any active participation in English literary life and manners is a complete set of both the *Tatler* (no. 1594) and its successor the *Spectator* (no. 1542). These two publications might have been sent to Newton by Addison or Steele, or possibly come from their friend and patron Charles Montague, Earl of Halifax, but it would seem more likely that they were introduced into the library by Newton's niece Catherine Barton when she lived with Newton.[1] Catherine had, to some imprecisely known extent, enjoyed a close intimacy with Halifax over a number of years and was also a long-time friend of Swift, and through him was acquainted with other leading literary figures of early-eighteenth-century London. Examination of the runs of Newton's periodicals in Trinity reveals no signs of use other than a very infrequent 'dog-earing' of their pages (and that might have been accidental); but of course his yet unlocated contemporary copies of the *Philosophical transactions*, the *Acta eruditorum* and the *Journal des sçavans* might just possibly contain significant autograph annotations or other markings by him, against the pattern of what we know already.

When Newton needed to read certain books written in languages in which he was not fluent he often arranged to have pertinent passages in them just translated for him, by Abraham De Moivre and others; but for his own immediate use he had to hand a wide range of dictionaries. Chief among his lexicons were six in Greek, another six Greek–Latin ones, five polyglot (two of these comprising Oriental languages), three Hebrew, two Latin–English, and a Latin one. He also possessed (doubtless needed for his Biblical studies) Chaldee and Syriac dictionaries, two French, and one French–English, as might be expected; but the occurrence of a French–Italian (no. 611) and a

---

copies delivered to him at a considerable discount), and it was in his *Journal litéraire* that a series of pieces on the fluxion priority dispute written by Newton (under John Keill's name) during 1713–19 appeared. Johnson's English representative was a 'Mr Darby' (probably the London stationer John Darby) through whom gift copies of the *Journal* were disbursed in England. Newton certainly received his copy of the *Journal* for Nov./Dec. 1713 that way (see his letter to Keill on 2 April 1714, Trinity College, Cambridge, MS.R. 16.38B(428), printed in J. Edleston (ed.), *Correspondence*...(London, 1850), pp. 169–70, and in Newton, *Correspondence*, ed. A. R. Hall and L. Tilling, VI (1976), 79–80. The close ties between Newton and Johnson are further shown in a letter of Halley to Keill, 3 Oct. 1715, reproduced in Edleston, pp. 184–5, and in *Correspondence*, VI, 242–3: 'We have printed a French translation of y⁰ account of the Commercium given in the Transactions, in order to send it abroad: S⁰ Isaac is desirous that it should be publisht in the Journal Literaire.'
[1] See also p. 29 n. 2.

Dutch–Latin wordbook (no. 514) with them is strange (if they were not gifts or Catherine Barton's). The dictionaries were supplemented by a variety of grammars: five Greek, two Hebrew, one English, one Latin, and even Anton's *Grammatica española*...1711 (no. 53) and *The Hig Dutch Minerva...or A perfect Grammar...whereby the English may...learne the neatest dialect of the German mother-language*...[1670?] (no. 763) – these last two items also probably presentations. Further philological works included books on various aspects of the English, Greek, Latin, and Hebrew languages, as well as a volume on ancient Egyptian hieroglyphics (no. 1658). Newton's dictionaries were by no means all of a linguistic nature: several of them were concerned more narrowly with other subjects such as agriculture (no. 373), the Bible (no. 609), chemistry (no. 855), geography (no. 614), history (no. 417), law (no. 449), navigational terms (no. 1529), Oriental peoples (no. 755), personal names (no. 1559, this signed by Newton and dated 1661), and place-names of England and Wales (no. 8, a remarkably comprehensive and well-compiled gazetteer). The dictionaries of this nature which I have seen usually bear signs of frequent handling and in several instances some of their pages are still 'dog-eared'.

Chief among the few strictly bibliographical aids was Hyde's *Catalogus impressorum librorum Bibliothecæ Bodlejanæ in Academia Oxoniensi*, 2 vols., 1674 (no. 824) – the most extensive record of any English library's contents published up to then. John Locke also owned a copy of this work in which he inserted interleaves, thereafter adopting it as the basic catalogue for his own library.[1] (This practice was also followed by a number of Oxford and Cambridge college libraries at the time, among them Trinity;[2] but whether this Trinity copy was consulted by Newton is not known.) When Newton visited Locke at Oates in the autumn of 1702, he was shown his host's *Essay on the Corinthians*[3] and the two men worked together on interpreting the Scriptures.[4] Locke, proud of his library, may well have invited his visitor to look over the books he had so precisely arranged, pointing out the use he had

[1] Harrison and Laslett, pp. 30–31.
[2] Now Trinity College Add.MS.a.103a, b, it was presented to the College in 1675/6 by Robert Scot, a London bookseller, was marked up to show the College's holdings, and was finally superseded when the revised and enlarged edition of Hyde's *Catalogus*...of 1738 was acquired.
[3] Peter (Lord) King, *The Life of John Locke*, new ed. (2 vols., London, 1830), II, 37–8.
[4] As Locke's annotations in a Bible from his library demonstrate (Harrison and Laslett, pp. 9–10). The two evidently got on together well enough, and in a following letter of 30 April 1703 Locke told his nephew Peter King that
  'Mr Newton, in Autumn last, made me a visit here...I have several reasons to think him truly my friend, but he is a nice [i.e. pernickety] man to deal with, and a little apt to raise in himself suspicions where there is no ground...[and after these reservations continues] Mr Newton is really a very valuable man, not only for his wonderful skill in mathematics; but in divinity too, and his great knowledge in the Scriptures wherein I know few his equals' (King, II, 37–9).

made of Hyde's *Catalogus*, and mentioning his other library methods. But Newton, too careless even to put his name or other mark of ownership on most of his books, was hardly likely to adopt any of Locke's bookish habits, and in fact, as we have seen, he did not. His copy of Hyde's book has not so far been located, but it seems practically certain that Newton never used it as a basis for cataloguing his own collection of books.[1] The fact that the compiler of the Huggins List made no special comment about the book would seem evidence that Newton had not done so.

The large number of works of classical literature must reflect Newton's conventional formal education in his youth – it will be recalled that he inscribed an Ovid and a Pindar at the age of sixteen[2] – and his enduring interest in the ancient world. The figure of $8\frac{2}{3}$ per cent falls far short of the comparable estimate of 21 per cent in the case of Barrow's library (not surprising for a former Professor of Greek), but is fairly close to Locke's 10 per cent and would seem to be about the proportion to be expected in the library of a cultured man of that period. Writings of most of the better-known classical authors were on Newton's shelves, and especially noteworthy are works on the Argonauts and their voyages; to these he subsequently had recourse for various datings he required in preparing his chronology of the ancient kingdoms. We meet with no medieval literary works of major significance as we turn from the classics towards more modern literature. Those brought together under this latter heading as 'Latin' relate to the writings of some of the later Humanists and of others such as Theodorus Beza, Justus Lipsius, Francis Bacon, George Buchanan, and John Owen.

In the field of English literature the tendency of reviewers of Newton's library has been to dismiss his holdings as virtually negligible. This attitude is in part misguided; for it is my impression that large private libraries of Newton's age normally contained only a very little material of an English literary nature. As a comparison, Barrow at his death apparently had just two titles of such works in all, and the proportion in the case of Locke's library was 2 per cent (73 titles out of 3641), whereas Newton's share was $2\frac{1}{3}$ per cent (40 titles). I suggest, therefore, that in this respect Newton's library conformed to the pattern of his time, and while I do not imply that he was engrossed in literary activities, he was far from unaware of what was then going on – Catherine Barton would have seen to that. I have already suggested that some of the books on Newton's shelves in 1727 may have originally been the property of his niece, and she may well have been responsible for other works being sent there by her friends as gifts to her uncle. In fact the works of English literature in the library may well reflect her taste (as well as her influence), rather than Newton's own. It is also worth re-

---

[1] It will be seen in the catalogue below that thirteen items by Locke were in Newton's library, including three on money, but not Locke on education or government, and somewhat surprisingly not his *Reasonableness of Christianity*.   [2] See p. **2**.

iterating here my earlier warning[1] that Newton may not in fact have owned all the works which were published before 1727, such as those listed by Musgrave but not present in the Huggins recording of 1727, and that Charles Huggins and James Musgrave may each have added to the collection certain works of the sort now under discussion.[2]

In the library one may notice works of English poetry,[3] drama, and prose, most of them publications from the late seventeenth or early eighteenth century; while Chaucer is not among them, two of Shakespeare's plays are. I have referred in another context[4] to his copy of Halifax's *Works and life...* 1715, and among other major items are two volumes of Sir Richard Blackmore's *Essays*, Abraham Cowley's *Works*, Milton's *Poetical works* in two volumes, Matthew Prior's *Poems* in the 1718 folio edition, and two separate issues of Pope's six-volume translation of Homer's *Iliad*. To these we can add Swift's *Tale of a tub* and editions of Samuel Butler's *Hudibras*, Garth's *Dispensary*, and *Poetical recreations* by Mrs Jane Barker and others. Leading the dramatists is Congreve with five separate plays (and a volume of his poems); Shakespeare's two items I have mentioned; and there are further pieces by less prolific playwrights. Even the field of literary history and criticism is not entirely bare, since here the library offers Langbaine's *Lives and characters of the English dramatick poets* and Addison's *Notes upon the twelve books of Paradise lost*, along with the *Tatler* and the *Spectator*.

Modern European literature took up very little space on Newton's shelves. He found room for a three-volume edition of Montaigne's *Essais*, Richard Simon's *Lettres choisies, où l'on trouve un grand nombre de faits anecdotes de littérature*, Nicolas Barat's *Nouvelle bibliothèque choisie...* in two volumes, poems and letters of René Le Pays in his *Amitiez, amours, et amourettes*, 1678. The only signs of possible interest Newton may have had in any other European literature are furnished by two English translations from the Spanish – Alvarado's *Spanish and English dialogues...* 1718, which in any case came as a gift,[5] and Guevara's *Spanish letters...* 1697. Should Newton have ever looked for any modern literature in Barrow's library, he would have encountered works by

---

[1] P. 45.

[2] Reference to the pertinent entries in our catalogue for literary items (as for all others) will show if they appear in the Huggins List or the Musgrave Catalogue, or (as would be the case for any major works stemming from Catherine Barton) both.

[3] The conclusion of de Villamil was that 'on the balance of evidence, Newton did not care for poetry' (*Newton: the man*, p. 10). He proceeds to quote from Joseph Spence's collection of *Anecdotes, observations, and characters, of books and men...* (London, 1820), p. 368, where Lord Radnor is credited with this reminiscence: 'A friend once said to him [Newton], "Sir Isaac, what is your opinion of poetry?" – His answer was; "I'll tell you that of Barrow; he said, that poetry was ingenious nonsense."' [4] See p. 15.

[5] An undated letter (?1718) from Alvarado to Newton printed in Newton, *Correspondence*, ed. A. R. Hall and L. Tilling, VII (1977), 377–8, refers to 'this small present' (no. 35) and to 'my Benefactors [of whom Newton was evidently one], who did me ye favor, to accept my Translation, ye Booke of Common Prayer into Spanish' (no. 244).

Ariosto, Boccaccio, and Tasso, but very little else. Whether his copies of Stark's Latin translation of the *Kalīlah Wa-Dimnah*...1697 (no. 874) or Pococke's Latin version of *Lamioto'l Ajam, carmen Tograi*...1661 (no. 1648) were acquired in connection with his study of ancient kingdoms or whether they evidence any special interest in Arabic literature on Newton's part is open to question, even if the latter volume does show signs of his characteristic page-marking.

The books on economics are few in number but not without their importance if we remember Newton's position at the Mint. Works on trade and commerce by some of the foremost writers of the day were in his library: Francis Brewster, Josiah Child, Charles Davenant, William Petyt, and John Pollexfen among the English; those from the Continent included Pierre Daniel Huet and Samuel Ricard. On fiscal currency and other topics which related narrowly to the field of Newton's official appointment I may here notice items by Locke, William Lowndes, Rice Vaughan, and (in French) Pierre La Court and François Le Blanc, together with two separate editions of Jean Boizard's *Traité des monoyes*. The most noteworthy volume, however, is a collection of 180 official documents issued by the French *Cour des Monnoyes* over the years 1689–1701 (no. 632), all carefully preserved and bound together in vellum,[1] with many of the pages having signs of earmarking in Newton's characteristic manner. The numismatic books in the library would have assisted Newton when he was called upon for his opinion on coinage problems, and they may have been of some use in his researches relating to world chronology, comprising as they do studies of ancient Greek, Roman, and Hebrew coins, by Oiselius, Foy-Vaillant, and Reland, together with John Evelyn's *Numismata*...1697 (no. 593). Two of the medal volumes illustrate the principal events of the life and reign of Louis XIV (nos. 1069 & 1248), while the rest were of a general historical nature.[2] Printed editions of several of the ancient Greek and Latin philosophers are to be seen listed in the catalogue, along with a Latin version of Confucius, but a search will turn up nothing from the medieval philosophers. Stanley's *History of philosophy*... 3rd ed., 1701, and his *Historia philosophiæ Orientalis*...1690 were in the library. Of contemporary English philosophers we find works by Glanvill, Locke, and Power, and from abroad, Bayle, Descartes, Leibniz, and Malebranche, but the number of titles in this area is not great. The major works on political thought are absent from his library, the items grouped together under 'politics' in table 1 being mainly smallish books on contemporary British affairs and policy. If, however, Newton had ever wished to consult the works of some of the more important theoreticists he could have used Barrow's

[1] Trinity College, NQ.10.43.
[2] A number of Newton's medals are recorded in the inventory made of his house after his death, viz. 'Item thirty nine silver medals' (de Villamil, *Newton: the man*, p. 16). See also his annotation in *The works and life of...Charles, late Earl of Halifax*...reproduced on p. 15 above.

copies of Bodin's *De republica*, Hobbes's *De cive* and *De homine*, Pascal's *Pensées* and *Lettres provinciales*, Spinoza's *Tractatus theologico-politicus*, and three books by Machiavelli.

The pair of incunables in Newton's library (Rolewinck's *Fasciculus temporum*, c. 1485 and Breydenbach's *Peregrinatio in Terram Sanctam*, 1486, bound up together)[1] are likely to have found their way on to his shelves more because of their content than for their rarity. Other works there present provide examples of Newton's interest in topics other than those already touched upon here: for instance Thomas Shelton's shorthand manual (no. 1506), which recalls his preoccupation with 'short-writing' as a youth; James Weir on accountancy (no. 1720), which brings to mind his life-long concern with tallying accounts; John Perry's account of the navigational experiments in the Thames at the 'Daggenham Breach' (no. 1280), manifesting his interest in navigation; two works of Nicolas Gauger on building chimneys so as not to smoke offensively (nos. 654 & 1058) – his curiosity about mechanical and other devices might account for his having this, though perhaps it was just that his opinion on the two books was sought and they stayed with him when their immediate purpose was fulfilled. A variety of books, some of them unwanted, were, we know, sent to Newton for his expert judgement, for 'approval', and often simply out of courtesy. We may also guess that for every gift copy he sent out of his own published works as they appeared off the press he received at least one book in return. There are, however, a few volumes in the library – always supposing that they were Newton's and not Catherine Barton's – whose presence is somewhat unexpected. These include books on painting and sculpture (nos. 1617, 647, 925), the art of cooking (no. 56),[2] floristry (no. 669), and gardening, with works by Richard Bradley, John Lawrence, and others (nos. 279, 920–22, 917, 1057); and the catalogue shows other items whose claim to a place on Newton's shelves would have seemed to be weak, judging from their subject-matter. At first sight, perhaps the most curious volume of all is the rare first edition of Sir Thomas Parkyns's *The Inn-play, or Cornish-hugg wrestler...* 1713 (no. 1258), which despite its title does not relate to tavern-wrestling in Cornwall but tells of the methods taught by the wrestling-master (a Mr Cornish) at Gray's Inn of the Law Court. The author had been Fellow-Commoner at Trinity College during 1680–82, and a statement in his preface[3] that Newton had shown some

[1] Trinity College, NQ.11.32. This volume, which had first come to the College in 1546–9 as a gift from William Filey, had gone from its Library by 1667, but was presented to Trinity a second time, by the Pilgrim Trust in 1943 (P. Gaskell and R. Robson, *The library of Trinity College, Cambridge: a short history* (Cambridge, 1971), plate 2).

[2] Newton was among the eighteen subscribers to this book.

[3] The author's preface (A1v, A2r) relates that 'his useful hydrawlicks, and the use and application of the mathematicks here in wrestling he owes to Dr Bathurst his Tutor, and Sr. Isaac Newton Mathematick Professor, both of Trinity College in Cambridge, the latter seeing his inclinations that way, invited him to his publick lectures, for which he thanks him'. (Parkyns does not tell us if he actually went to the lectures, the deposited ones of which in 1680–82 were on algebra.)

TABLE 2. *Newton's library: language analysis*

	Titles	Proportion (%)
1. English	722	41¼
2. Latin	707	40¼
3. Greek–Latin	113	6½
4. Greek	25	1¼
5. French	152	8¼
6. Others (polyglot, 6; Arabic–Latin, 4; Chaldaic, 1; Hebrew, 5; Hebrew–Chaldaic, 1; Hebrew–Latin, 1; Syriac, 1; Dutch, 1; Dutch–Latin, 1; French–Italian, 1; Greek–French, 1; Italian, 5; Latin–English, 2; Spanish, 1; Spanish–English, 1; American Indian–English, 1)	33	2

interest in his 'inclinations' at that time goes some way towards explaining why a copy of the book found its way to Newton. While Parkyns's title-page enthuses that the manual had been 'digested in a method which teacheth to break all holds, and throw most falls mathematically', it is, even so, very hard to picture Newton spending much time with the book, which has all the appearance of a presentation copy.[1]

The analysis of Newton's books by language is more straightforward, and I can be precise in grouping the 1752 titles under the headings of table 2.

Latin in the later seventeenth century still maintained its pride of place as the medium of the learned. In addition to being the language of classical, medieval, and theological texts it continued widely at this time its role as the *lingua franca* used by scholars in their philosophical and scientific books. It may be argued that the Royal Society in publishing its *Transactions* in English had some effect in reducing the dominance of Latin in the field of science, and several other scientific books by such authors as Dee, Recorde, Gilbert, Wilkins had earlier been written in English, but the advance towards wide use of the vernacular was only steadily gaining momentum. Though the titles in Newton's library which are in English just outnumber those in Latin, the former works often tend to be less substantial and important than the latter. In addition, when comparing the proportion of books in English in Newton's library with that of works written in other languages, we should take account of the fact that 45 per cent of the overall total were published abroad (see table 3), well over two-thirds of these being in Latin. Some loose connection may be made between the restricted range of languages of Newton's library books and the limitation of his own (far from wide) linguistic talents and abilities, adequate as these were for most of his working needs. His early education at Grantham Grammar School gave him a sound

[1] Trinity College, NQ.9.61.

grounding in Latin and some acquaintance with both Greek and Hebrew. His fluency in Latin was further polished at Cambridge where he was of course required not merely to read Latin but to speak and dispute in it. After 1667-8 almost till the end of his life he made all but universal use of it in his private papers, and we need not stress that statute required Latin to be employed for all University business, and in particular in the delivery of his Lucasian lectures. Newton's thorough grasp of Latin weakened a little during the last years of his life, and he was no doubt relieved in 1706 to be able to publish Samuel Clarke's Latin translation of his 1704 *Opticks* rather than adapt his own (yet unpublished) initial Latin version of it of some dozen years before; even so he took care to correct not a few of Clarke's vaguenesses and errors. With Greek, Newton was never entirely happy, though short quotations in it from various sources are to be found in a number of his manuscripts. Books in his library written in Greek alone are small in number, and it is clear that he manifested a preference for works with Greek and Latin in texts parallel.

In Locke's library the predominance of books in French over those in other Continental tongues was very marked; his proportion of items in that language was more than twice that of Newton. But, unlike Locke, Newton never travelled abroad – indeed till his seventy-eighth year he possibly never journeyed outside the narrow triangle formed by Lincolnshire, Cambridgeshire, and London,[1] and so had no personal contacts with foreign booksellers (though there were those, like Collins, ever willing to supply him with any publication he required). The texts grouped together as 'Others' in table 2 show plainly how little Newton ventured into European languages other than French, and of this he achieved no more than a stumbling knowledge. He tells us of his lack of fluency there in a letter written to John Collins on 20 May 1673: 'I received y^r two last letters with Heuret's *Optiques* w^{ch} (not being so ready in ye French tongue my selfe as to reade it wthout the continuall use of a Dictionary) I committed it to ye perusal of another who gives me this account of it...'[2] Sixteen years later Newton's command of the language had scarcely improved, and we hear of Fatio de Duillier not only offering to help Newton with the reading of Huygens's *Traité de la lumière*,[3] but also writing to Huygens on 24 February 1689/90: 'Il [Newton] a quelque peine à entendre le François mais il s'en tire pourtant avec un Dictionaire.'[4] Since he had 152 French titles in his library, many of them presentations, no doubt, but several showing signs of use – with the proportion of these which are of a scientific nature higher than that occurring

[1] See Stukeley, *Memoirs*, p. 13: 'In August that year [1720] Sir Isaac went to Oxford in company of Dr John Kiel; he having not been there before.'
[2] Newton, *Correspondence*, ed. H. W. Turnbull, I (1959), 281. 'Heuret' is Grégoire Huret, and the book referred to is his *Optique de portraiture et peinture, en deux parties*...F^o, Paris, 1670 (*Correspondence*, ed. A. R. Hall and L. Tilling, VII (1977), 387). The work was not present in Newton's library at his death.   [3] *Ibid.* III (1961), 390.   [4] *Ibid.* p. 69.

TABLE 3. *Newton's library: analysis of the 1752 titles by place of publication*

	Titles	Proportion (%)
1. Britain	962	55
London, 826; Oxford, 82; Cambridge, 48; others, 6		
2. The Netherlands	278	16
Amsterdam, 134; Leyden, 69; Utrecht, 19; The Hague, 17; Antwerp, 14; others, 25		
3. German states, etc.	187	10¼
Frankfurt, 66; Cologne, 36; Strasburg, 12; Hamburg, 11; others, 62		
4. France	182	10¼
Paris, 149; Lyons, 23; others, 10		
5. Switzerland	73	4⅛
Basle, 35; Geneva, 31; others, 7		
6. Italy	35	2
Venice, 10; Rome, 7; Florence, 5; others, 13		
7. Other countries		
Denmark (Copenhagen, 4); Spain (Cadiz, 1); America (Boston, 1)	6	⅓
8. Not known	28	1½
9. Manuscript transcript	1	–

in other language groups – Newton's widely accepted inability to understand written French may be exaggerated.

The researches Newton pursued into Jewish rites and customs will partly explain why he owned Hebrew books. He knew the elements of Hebrew well enough from an early age, and with the aid of his lexicons he would have been easily able to deal with the Hebraic references and quotations as they came up in his study of Jewish history. Another, simpler, explanation of the presence of some of these volumes is that they came from the library of Newton's stepfather.

A breakdown by place of publication of the books in Newton's library brings out the increasing concentration of the English book trade in London (where Newton himself of course lived after 1696). Works issued there – the large majority of them published after 1700 – form nearly half of the entire collection.

Since most of the editions of the classics in Newton's library were published in the Netherlands, we will not be surprised to find that the number of works issued in Amsterdam far exceeds that of those published at Oxford, and that the Leyden total is greater than the Cambridge one. Except for the books published in France (in their markedly greatest part at Paris), which are mostly in the vernacular, those carrying the imprint of other European

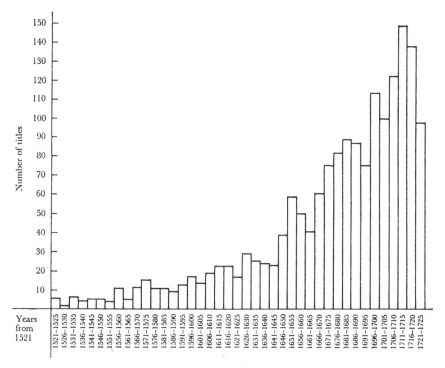

Newton's library: breakdown of its 1719 titles issued over the period 1521–1725 by date of publication. (For multi-volume works the date of the earliest published volume has been used.)

towns and cities are, as I have shown in table 2, nearly all in Latin and not in the native tongue.

The accompanying histogram, in which 1719 titles of Newton's books separately distinguishable as being issued over the period 1521–1725[1] are broken down into their individual totals by five-year units of their dates of publication, will show at a glance the reasonably uniform way in which he built up his holdings over his adult life.

This histogram can only be relied upon as a general pointer to the growth of Newton's library in respect of the works which he received off the press in their year of publication, and this would not allow for the considerable number of new foreign works which (with rare exceptions) must have taken several months at least to reach him. Furthermore, we have very little solid evidence, other than that provided by previous owners' or donors' inscriptions, which permits us to distinguish between the second-hand acquisition and one bought brand-new. If we look upon 1661 as the year when

[1] The remaining 33 titles here omitted which make up the overall total of 1752 consist of 5 items published over the period 1485–1517, 20 dated 1726, 3 dated 1727 (but in the Huggins List), 4 with date of issue unknown, and 1 manuscript transcript.

he began to form a personal library (I have traced ten books which have that year included in Newton's inscription) we can directly allow neither for the 488 items of before that date which he acquired second-hand at some time in his long life not precisely known, nor for the works published from 1661 onwards which he obtained in a similar manner. But even while we are unable narrowly to relate the growth of his library to the imprint dates of the individual books, at least the distortion must predominantly lie in one direction only and decrease in its degree the nearer one comes to 1727, the year of Newton's death.

Of the titles in his library, 632 (or 36 per cent of the collection) were published in the eighteenth century; if to these we add the 101 works dated between 1697 and 1700, the inference here is that after Newton left Cambridge for London in 1696 his library continued to expand (but at a decelerating rate of growth), this due no doubt in part to a number of presentations after he grew to fame and became one to be revered. The largest annual totals are 39 books issued in 1720, 38 in 1713, 30 in 1700 and again in 1710, 27 in 1697, 28 in 1684, 26 in 1678, and 22 in 1668. Works which on the basis of their date of imprint alone might possibly have been with Newton before Barrow died in 1677 amount to 717 (41 per cent of the final total); by the time he left Cambridge the number had grown (as I have indicated[1]) to 1001 (57 per cent).

A final reservation. The foregoing survey of the nature and content of the books which Newton once owned is presented with the realization that his library is and can only be just one manifestation, intricately complex though it be, of his wide-ranging intellectual knowledge and activity, and as such has to be taken on a par with all the others. It is tempting to pass facile, ready-made judgements based on an examination of the volumes he had on his shelves, but this way could well lead us in particular instances to too easily distorted or even broadly false conclusions should we ever forget that very little is straightforward about Newton and that he remained a law unto himself. However outstandingly significant in their several ways some of the individual items in it may be, his library as a whole should not necessarily of itself and without the supporting external evidence of his mass of surviving autograph writings be taken as a precise index of the quality and range of his mind. Nevertheless, it is an important touchstone of the multifaceted variety of his genius, one that has been too long neglected, and a sustained analysis along the lines which I have here suggested should have much more still to teach us about the bookish habits of one for whom the printed word was only a stepping-stone to a long train of intellectual and scientific speculation of his own.

[1] See p. 58, n. 1.

THE CATALOGUE: ITS COMPILATION
AND CONTENT

———

The catalogue which follows is based wherever possible on the Newton books which still survive. For the remainder their recording in the Huggins List and/or the Musgrave Catalogue has been expanded either by examining copies of the works in the University Library, Cambridge (whose current forms of author or other relevant headings I have adopted) or by consulting the published catalogues of other libraries. While trying to avoid an unnecessary clutter of cross-references, I have attempted to anticipate the places at which the user of the catalogue might reasonably search for a book. Headings for editors are used very sparingly, being confined chiefly to collections of documents or texts which have come to be known by the name of their compilers. The catalogue descriptions normally follow the title-page usage of the books themselves in respect of ligatured letters and of punctuation. Similarly diacritics are reproduced, though for French accents I have also in some instances introduced those present in the text of a book, as well as at times modernizing others.

The entries in the catalogue are numbered consecutively both for ease of reference in the foregoing introduction and also for precise pin-pointing from cross-references within the catalogue itself to the books' main listings, where the full available information is printed. When two identical copies of a title are known to exist, a separate number has been allocated to each, but if there now remains but a single survival (or none at all) of a work known to have been duplicated in Newton's library, then one numbered entry has been made to suffice.

Information which I have been able to gather relating to the individual books composing the collection is presented at the end of their catalogue description in the following sequence. First comes an indication of the work's presence in (or absence from) the Huggins List, thus: H (or Not H), followed by a record of its appearance in the Musgrave Catalogue together with its Musgrave shelf-mark, thus: M/F4 – 18, or where the work consists of more than one volume the first and last 'running' number: M/D5 – 16 to 24. Items not in Musgrave are shown as 'Not M'. Those stemming from the Portsmouth Collection and described in its catalogue published in 1888 are shown as 'Portsmouth (1888)', and any other known provenance is stated. Then follow particulars of the book's present location together with, where appropriate, the holding institution's class-mark or collection number, so that a work in Trinity College is shown as Tr/NQ.16.201 or one at Babson College as Babson 405. The negative information that the whereabouts is not known is indicated by a simple '?'

Special characteristics of the book itself are recorded. They include transcriptions of Newton's signature, with his price-notes, date of acquistion, and presentation inscriptions whether by Newton or by the donor (or his agent). The occasional serious physical imperfection is noted, and if a volume contains more than one separate work and appears likely to have been in that state when Newton owned it, then the catalogue makes this clear. When, however, the volume consists of a number of tracts bound up together during Musgrave's ownership, then I saw little value in giving a string of numbers linking the items which happen to be contained within the same covers. The visible evidence of Newton's method of page-marking is described; where pages still have their corners turned down (or up) I have cited their numbers, except for the very few cases (for example nos. 231 and 1697) where the total is unduly large. Signs of anterior 'dog-earing' and its extent are likewise given.

The history and descent of a book are related to its appearance in booklists and sale catalogues. In addition to the very few recordings by Newton in the Trinity, Fitzwilliam, and Cambridge University Library Notebooks of purchases he made, the catalogue entries contain the prices paid for the 'Books for Mr Newton' set out in the Bodleian Library MS New College 361 (Ekins Papers), thus: Ekins 3 – 6. The Thame Park sale catalogue of January 1920 is cited with its lot number as, for example, 'Thame/979', and the Sotheran catalogues of the 1920s and 1930s are covered either by reference to Sotheran's *Bibliotheca chemico-mathematica*...2 vols. and Suppl. 1-2, London, 1921-37, or to the numbered catalogues themselves. I have distinguished these by 'Sotheran, *Bibl.* [volume] 2 (1921) [item] 12535' for the former and by 'Sotheran [Catalogue] **828** (1931) [item] 3602' for the latter. Appearances in other sale or library catalogues are self-explanatory, thus: 'Sotheby, July 1936/121' or 'Duveen [*Bibliotheca alchemica et chemica*...1949], pl. xii'. Occasionally, information about Newton's copy of a book has appeared in print already. This is largely confined to D. T. Whiteside's edition of Newton's *Mathematical papers* and to the recently completed publication of the *Correspondence*; the relevant details have been added to the catalogue entries.

Finally, and most important, a record of the annotations in Newton's hand in eight-four of the books is set out in italics, beginning on a separate line at the end of the catalogue entry. His briefer notes are reproduced in full, but the majority are usually summarized with the intention of offering an indication of their nature and extent.

SUMMARY OF ABBREVIATIONS
USED IN THE CATALOGUE

━━━

Babson	*A Descriptive Catalogue of the Grace K. Babson Collection of the Works of Sir Isaac Newton...Babson Park, Mass.* New York, 1950. *Supplement...*1955.
Correspondence	Newton, *Correspondence*. Vols. 1–3, ed. H. W. Turnbull; vol. 4, ed. J. F. Scott, vols. 5–7, ed. A. R. Hall and L. Tilling. Cambridge, 1959–77.
Duveen	*Bibliotheca alchemica et chemica: an annotated catalogue of printed books...in the library of Denis I. Duveen.* London, 1949 (repr. 1965).
Ekins	'Books for Mr Newton'. Bodleian Library, Oxford, MS New College 361 (Ekins Papers), vol. II, fol. 78.
H	Huggins List, 1727.
M	Musgrave Catalogue, *c.* 1767.
Math. papers	Newton, *Mathematical papers*, ed. D. T. Whiteside. Vols. 1– . Cambridge, 1967– .
Portsmouth (1888)	*A Catalogue of the Portsmouth Collection of Books and Papers written by or belonging to Sir Isaac Newton...* Cambridge, 1888.
Sotheran, *Bibl.*	*Bibliotheca chemico-mathematica...*, ed. H. Zeitlinger and H. C. Sotheran. 2 vols. and Suppl. 1–2. London, 1921–37.
Thame	*Thame Park, Thame, Oxon. The Greater Portion of the Contents of the Mansion...Hampton & Sons...are... to sell...January 13th 1920, and two following Days* (sale catalogue).
Tr	Trinity College, Cambridge.
ULC	University Library, Cambridge.

CATALOGUE OF THE LIBRARY

ABENDANA, Isaac. *See* ISAAC ABENDANA.

1 ABULFEDA. De vita, et rebus gestis Mohammedis. . . . Ex Codice MS^{to} Pocockiano Bibliothecæ Bodleianæ textum Arabicum primus edidit, Latinè vertit, præfatione, & notis illustravit J. Gagnier. F°, Oxoniæ, 1723.
H (2 copies); M (1 only) C3 – 10; Tr/NQ.11.23.
ACADÉMIE ROYALE DES INSCRIPTIONS ET BELLES LET-TRES. *See* PARIS. *Académie Royale des Inscriptions et Belles Lettres.*
ACADÉMIE ROYALE DES MÉDAILLES ET DES INSCRIPTIONS. *See* PARIS. *Académie Royale des Médailles et des Inscriptions.*
ACADÉMIE ROYALE DES SCIENCES. *See* PARIS. *Académie Royale des Sciences.*
ACARETE DU BISCAY. Voyages and discoveries in South-America. . . 1698. *See* 1700.

2 ACCOUNT (An) of a Roman temple, and other antiquities, near Graham's Dike in Scotland. [By W. Stukeley.] 4°, [London, 1720].
Not H; M (Tracts) E6 – 35; Tr/NQ.10.34³.

3 — An account of Switzerland. Written in the year 1714 [by T. Stanyan]. 8°, London, 1714.
H; M/J4 – 23; Tr/NQ.9.39.

4 — An account of the affairs of Scotland, relating to the Revolution in 1688. As sent to the late King James II. when in France. By the Right Honourable the Earl of B— [Balcarres]. 4°, London, 1714.
Not H; M (Tracts) J3 – 14; Tr/NQ.9.20².

5 — An account of the Court of Portugal, under the reign of the present King Dom Pedro II. . . . [By J. Colbatch.] 8°, London, 1700.
H; M/J4 – 44; Tr/NQ.9.80 [a few signs of dog-earing].

6 ACOSTA, José de. De natura Novi Orbis libri II. . . 8°, Coloniæ Agrippinæ, 1596.
Not H; M/F1 – 13; Parke-Bernet, New York, June 1973/211; ?

7   ACTA eruditorum anno MDCLXXXII–MDCCXXVI publicata. . . 55 vols. 4°,
     Lipsiæ, 1682–1726.
     H; M/D2 – 17 to 71; Thame/984; ?; *see also* 894.
     ACTS of Parliament. 22 vols. ('Begun 7ᵐᵒ Gulielmi & end with 9°
     Georgii'.) [1697–1722?] *See* 1259.
     — The Acts of Parliament relating to the building fifty new churches
     in and about the cities of London and Westminster. . . 1721. *See*
     1260.
     ACUÑA, Christoval de. Voyages and discoveries in South-America. . .
     1698. *See* 1700.

8   ADAMS, John. Index villaris: or, An alphabetical table of all the cities,
     market-towns, parishes, villages, and private seats, in England and
     Wales. F°, London, 1680.
     H; M/G11 – 1; ?

9   ADAMS, Thomas. The happines of the Church. Or, A description of
     those spirituall prerogatives wherewith Christ hath endowed her. . .
     (2 pts.) 8°, London, 1619.
     Not H; M/H2 – 28; Tr/NQ.9.81 [a few signs of dog-earing].

10 ADDISON, Joseph. Notes upon the twelve books of Paradise lost.
     Collected from the Spectator. 12°, London, 1719.
     Not H; M/G9 – 36; ?
     — The Spectator. [By J. Addison, Sir R. Steele and others.] 8 vols.
     1712–15. *See* 1542.

11 ADDISON, Lancelot. The present state of the Jews. . . 8°, London,
     1675.
     H; M/H7 – 30; ?

12 — The primitive institution, or A seasonable discourse of catechizing. . .
     2nd ed. 12°, London, 1690.
     Not H; M/G9 – 24; ?

13 ADO, *Abp of Vienne.* Adonis Viennensis Archiepiscopi, Breviarium
     chronicorum ab origine mundi ad sua usque tempora. . . 1353. 8°,
     Basileæ, 1568.
     H; M/F1 – 16; Tr/NQ.16.9² [bound with 1712; a few signs of
     dog-earing].

14 AELIANUS, Claudius. Varia historia. T. Faber emendavit. (*Greek &*
     Latin.) 2 pts. 8°, Salmurii, 1668.
     H; M/G2 – 3 & 4; ?

15 AESCHINES, *Socraticus.* Dialogi tres Græce et Latine vertit et notis
     illustravit J. Clericus. . . 8°, Amstelodami, 1711.
     H; M/F4 – 19; ?

16 AESCHYLUS. Αἰσχύλου Προμηθεὺς Δεσμώτης, Ἕπτα ἐπὶ Θήβαις,
     Πέρσαι, 'Αγαμέμνων, Εὐμενίδες, 'Ικετίδες. 8°, Parisiis, 1552.
     H; M/F3 – 32; Tr/NQ.16.58.

17 — Tragoediæ septem: cum scholiis Græcis omnibus; deperditorum dramatum fragmentis, versione & commentario T. Stanleii. F°, Londini, 1663.

H; M/C2 – 7; Tr/NQ.11.30.

18 AGRICOLA, Georg. De animantibus subterraneis liber...editus...a J. Sigfrido. 8°, Witebergæ, 1614.

Not H; M (Tracts) H8 – 18; Memorial Library, University of Wisconsin – Madison [bound with 1685]; Thame/974.

19 — De ortu causis subterraneorum lib. v, De natura eorum quæ effluunt ex terra lib. iiii, De natura fossilium lib. x...F°, Basileæ, 1546.

Not H; M/J9 – 17; Thame/969; ?

20 — De re metallica libri xii...Ejusdem De animantibus subterraneis liber...F°, Basileæ, 1621.

H; M/D8 – 10; Tr/NQ.11.4 [a few signs of dog-earing].

21 AGRIPPA, Heinrich Cornelius. De occulta philosophia libri iii. F°, [Cologne], (1533).

Not H; M/J9 – 22; ?

22 AICHER, Otto. Epitome chronologica historiæ sacræ et profanæ... 4°, Coloniæ, 1706.

Not H; M/C1 – 15; ?

ALBINEUS, Nathan. Bibliotheca chemica contracta ex delectu & emendatione N. Albinei in gratiam & commodum artis chemicæ studiosorum. 1673. *See* 220.

23 ALEXANDRO, Alexander ab. Genialium dierum libri vi. Ed. ultima ...8°, Lugduni, 1616.

H; M/F2 – 15; ?

24 ALIMARI, Doroteo. Longitudinis aut terra aut mari investigandæ methodus...[Ed. by S. Ricci.] 8°, Londini, 1715.

H; M/E5 – 6; Tr/NQ.16.185.

25 ALIPILI. Centrum naturæ concentratum: or The salt of nature regenerated...Published in Low Dutch, and now done into English ...12°, London, 1696.

Not H; M/G9 – 17; Thame/973; ?

ALLESTREE, Richard. The art of contentment. By the author of The whole duty of man, &c. [R. Allestree]. 2nd impr. 1675. *See* 87.

— The lively oracles given to us...By the author of The whole duty of man, &c. [R. Allestree]. 1678. *See* 963.

26 ALLINGHAM, William. A short account of the nature and use of maps ...8°, London, 1703.

Not H; M/E2 – 43; ?

27 ALLIX, Peter. Reflexions upon the books of the Holy Scriptures to establish the truth of the Christian religion. 2 vols. 8°, London, 1688.

H; M/H2 – 23 & 24; ?

28 — Remarks upon the ecclesiastical history of the antient churches of the Albigenses. 4°, London, 1692.
Not H; M/F5 - 16; Tr/NQ.8.21² [bound with 29; a few signs of dog-earing].

29 — Some remarks upon the ecclesiastical history of the ancient churches of Piedmont. 4°, London, 1690.
H; M/F5 - 16; Tr/NQ.8.21¹ [bound with 28; a few signs of dog-earing].

30 — Two treatises: I. A confutation of the hopes of the Jews concerning the last Redemption. II. An answer to Mr. Whiston's Late treatise on the Revelations...8°, London, 1707.
H; M/H7 - 33; ?

31 ALMELOVEEN, Theodor Jansson van. Fastorum Romanorum con-sularium libri II...Accedunt Praefecti urbis Romae et Con-stantinopolis. (2 pts.) 8°, Amstelædami, 1705.
H; M/F1 - 17; Tr/NQ.8.88.

32 ALSTED, Johann Heinrich. Clavis artis Lullianæ, et veræ logices, duos in libellos distributa...accessit Novum speculum logices minimè vulgaris. 12°, Argentorati, 1609.
Not H; M/E5 - 42; bound with 999; Sotheran **800** (1926) 11461, **804** (1927) 3806, **828** (1931) 3606, **843** (1935) 1498; ?

33 — Thesaurus chronologiæ...Ed. 4ª limatior & auctior. 8°, Herbornæ Nassoviorum, 1650.
Not H; M/F4 - 30; ?
ALTESERRA, Antonius Dadinus. *See* DADINUS ALTESERRA, Antonius.

34 ALTHAM, Roger. Five charges to the Clergy of the Archdeaconry of Middlesex. (5 pts.) 8°, London, 1717–21.
Not H; M/H3 - 1; ?

35 ALVARADO, Felix Antonio de. Spanish and English dialogues. With many proverbs...12°, London, 1718.
H; M/J8 - 47; Tr/NQ.10.74; presented to Newton by the author ?1718 (*Correspondence*, VII, 377–8).

36 AMALTHEUS, *Astronomicus*. [*A1 begins:*] Amalthei Astronomici pars prior. Cap. I. De obliquitate eclipticæ telluris. 4°, [*c.* 1690]. *Not identified.*
Not H; M (Tracts) D5 - 25; Tr/NQ.8.51⁶.

37 AMATUS, *Lusitanus*. Curationum medicinalium centuria prima...12°, Parisiis, 1552.
Not H; M/G1 - 36; ?

38 AMBROSE, *St, Bp of Milan*. Omnia quotquot extant opera, primum per D. Erasmum...diligenter castigata...5 vols. in 2. F°, Basileæ, 1567.
H; M/G11 – 4 & 5; ?

39 AMES, William. De conscientia et eius iure vel casibus, libri v. Ed. nova. 12°, Amstelodami, 1643.
Not H; M/E1 – 31; ?

40 ANACREON. Anacreontis et Sapphonis carmina. Notas & animadversiones addidit T. Faber...(*Greek & Latin*.) 8°, Salmurii, 1680.
Not H; M/G2 – 9; ?

41 — Les poësies d'Anacreon et de Sapho, traduites de Grec en François, avec des remarques par Mademoiselle Le Fevre. (*Greek & French*.) 12°, Paris, 1681.
H; M/G7 – 14; Tr/NQ.16.64.

42 ANALYSE des infiniment petits, pour l'intelligence des lignes courbes. [By Le Marquis de L'Hôpital.] 4°, Paris, 1696.
H; M/E6 – 10; ?

43 ANCIENT (The) and modern history of the Balearick Islands; or of the Kingdom of Majorca...Transl. from the original Spanish [of J. B. Dameto]. 8°, London, 1716.
H; M/J4 – 5; Tr/NQ.9.69.

44 ANDERSON, Robert. The making of rockets...experimentally and mathematically demonstrated. 8°, London, 1696.
H; M/E2 – 45; Thame/979; ?

45 — To cut the rigging: and proposals for the improvement of great artillery. 4°, London, 1691.
Not H; M (Tracts) J3 – 41; Tr/NQ.8.111[7].

ANDREAE, Johann Valentin. *See* ROSENCREUTZ, Christian, *pseud.*

46 ANDREWES, Lancelot, *Bp of Winchester*. The pattern of catechistical doctrine at large: or A learned and pious exposition of the Ten Commandments...F°, London, 1650.
Not H; M/J9 – 10; ?

ANGLESEY, Arthur Annesley, *1st Earl of*. The privileges of the House of Lords and Commons argued and stated...1671. To which is added a discourse, wherein the rights of the House of Lords are truly asserted...Written by Arthur [Annesley], Earl of Anglesey. 1702. *See* 1350.

47 ANIMA magica abscondita: or A discourse of the universall spirit of nature...By Eugenius Philalethes [i.e. T. Vaughan]. 8°, London, 1650.
Not H; M/G9 – 30; Tr/NQ.16.134[2] [bound with 50].

48 ANSELME DE SAINTE-MARIE, Pierre de Guibours, *Père*. Histoire généalogique et chronologique de la maison royale de France...3e éd. F°, Amsterdam, 1713.
    H; M/B5 – 20; ?

49 ANSWER (An) to Several remarks upon Dr Henry More his Expositions of the Apocalypse and Daniel, as also upon his Apology. Written by S. E. Mennonite...4°, London, 1684.
    Not H; M/F5 – 10; Tr/NQ.9.173 ['Isaac Newton Donum R^{ndi} Authoris' in Newton's hand on fly-leaf].

50 ANTHROPOSOPHIA theomagica: or A discourse of the nature of man and his state after death...By Eugenius Philalethes [i.e. T. Vaughan]. 8°, London, 1650.
    Not H; M/G9 – 30; Tr/NQ.16.134^1 [bound with 47 & 1199].

51 ANTIDOTUM Britannicum: or, A counter-pest against the destructive principles of Plato redivivus [of H. Neville]...8°, London, 1681.
    Not H; M/G9 – 19; ?

52 ANTIQUÆ historiæ ex xxvii. authoribus contextæ libri vi....D. Gothofredi operâ...(2 pts.) 12°, Lugduni, 1590–91.
    H; M/G1 – 26 & 27; ?
    — Antiquæ musicæ auctores septem. Græce et Latine. M. Meibomius restituit ac notis explicavit. 2 vols. 1652. *See* 1060.

53 ANTON, Pasqual Joseph. Grammatica española...A Spanish grammar ...8°, London, 1711.
    H; M/J8 – 32; Tr/NQ.8.108.

54 ANTONINUS, *Augustus*. Itinerarium Provinciarum Vibius Sequester de Fluminū...P. Victor de regionibus urbis Romæ. Dionysius Afer de situ orbis Prisciano interprete. 8°, Lugduni, [*c.* 1545].
    H; M/E3 – 33; Tr/NQ.16.139.

55 ANTONINUS, Marcus Aurelius. Marci Antonini Imperatoris & philosophi, de vita sua libri xii. Græcè & Latinè...Accessit Marini Proclus item Græcè & Latinè. 12°, Lugduni, 1626.
    Not H; M/F1 – 23; Tr/Adv.e.1.13 ['Is. Newton pret 1^s' on fly-leaf]; Thame/972; not part of the Pilgrim Trust gift.

56 APICIUS, Caelius. De opsoniis et condimentis, sive Arte coquinaria, libri x. Cum annotationibus M. Lister... et notis selectioribus... 8°, Londini, 1705.
    H; M/F4 – 20; Tr/NQ.9.127; Newton was one of the 18 subscribers to this book.

57 APOCRYPHA. Codex Apocryphus Novi Testamenti, collectus, castigatus testimoniisque, censuris & animadversionibus illustratus à J. A. Fabricio. 8°, Hamburgi, 1703.
    Not H; M/A3 – 5; ?

THE LIBRARY OF ISAAC NEWTON

58 — Codex Pseudepigraphus Veteris Testamenti, collectus, castigatus testimoniisque, censuris et animadversionibus, illustratus a J. A. Fabricio. 2 vols. 8°, Hamburgi, 1713–23.
H; M/F3 – 19 & 20; ?

59 — Evangelium Infantiæ. Vel Liber apocryphus de infantia Servatoris. Ex manuscripto edidit, ac Latina versione & notis illustravit H. Sike. (2 pts.) (*Arabic & Latin.*) 8°, Trajecti ad Rhenum, 1697.
H; M/A1 – 2; Tr/NQ.10.75.

60 — Apocrypha. With marginal notes. 4°. *Not identified.*
Not H; M/F6 – 34; ?

61 APOLLODORUS, *Atheniensis.* Bibliotheces, siue De dijs libri III, T. Faber recensuit, & notulas addidit. (*Greek & Latin.*) 8°, Salmurii, 1661.
H; M/F2 – 24; ?

62 APOLLONIUS, *Pergaeus.* Apollonii Pergæi Conicorum libri VIII, et Sereni Antissensis De sectione cylindri & coni libri II. [Ed. by D. Gregory and E. Halley.] (*Greek & Latin.*) (3 pts.) F°, Oxoniæ, 1710.
H (2 copies, 1 of which is described as 'damaged'); M (1 only) B4 – 15; Tr/NQ.17.32 [not damaged; manuscript alteration at foot of separate title-page of pt 1 to make the editorial statement read 'Ex Codd. MSS. Græcis ediderunt David Gregorius & Edmundus Halleius...Professores Saviliani', followed by a 9-line Latin explanatory note signed 'Carolus Oliphant M.D.'].
— Apollonii Pergæi Conicorum libri IIII...1675. *See* 76.

63 — De sectione rationis libri II ex Arabico MS^to. Latine versi. Accedunt ejusdem de sectione spatii libri II restituti...Opera & studio E. Halley. 8°, Oxonii, 1706.
H (2 copies, the entry for 1 ending 'caret Tit.'); M (1 only) E5 – 9; Tr/NQ.9.126 [the complete copy].

64 APOLLONIUS, *Rhodius.* Argonauticωn libri IIII. Scholia vetusta. . . Cum annotationibus H. Stephani...(*Greek.*) 4°, [Paris], 1574.
H; M/F6 – 12; Tr/NQ.10.68¹ [bound with 1398].

65 — Argonauticorum libri IV. Ab J. Hoelzlino in Latinum conversi; commentario & notis illustrati, emaculati...8°, Lugd. Batavorum, 1641.
H; M/A4 – 15; ?

66 — Interpretatio antiqua et perutilis in Apollonii Rhodii Argonautica. (*Greek & Latin.*) 8°, [Paris], 1541.
H; M/E1 – 15; ?

67 APOSTOLIC FATHERS. The genuine Epistles of the Apostolical Fathers...Translated and publish'd...by William [Wake], Lord Archbishop of Canterbury. 3rd ed. 8°, London, 1719.
H; M/H3 – 20; ?

88

68 — SS. Patrum qui temporibus Apostolicis floruerunt. Barnabæ, Clementis, Hermæ, Ignatii, Polycarpi opera edita et inedita, vera, & supposititia; unà cum Clementis, Ignatii, Polycarpi, actis atque martyriis. J. B. Cotelerius ex MSS. codicibus eruit, ac correxit, versionibusque & notis illustravit...Nova ed....Recensuit...J. Clericus. (*Greek & Latin.*) 2 vols. F°, Antwerpiæ, 1700.

H; M/C4 – 7 & 8; Tr/NQ.18.11 & 12 [vol. 1 has p. 238 turned up, pp. 262, 322, 586–7 down, vol. 2 pp. 18, 20–21, 24, 34, 37, 373 down, and a few other signs in both vols.].

69 APPARITION (The). A poem [by A. Evans]. 8°, London, 1710.

Not H; M (where it is described as being bound 'with several others') J3 – 31; ?

70 APPIANUS. Romanorum historiarum pars prior (& altera)...A. Tollius utrumque textis multis in locis emendavit, correxit... (*Greek & Latin.*) 2 vols. 8°, Amstelodami, 1670.

H; M/F4 – 11 & 12; ?

71 ARABIA, seu Arabum vicinarumꝗ gentium Orientalium leges, ritus, sacri et profani mores, instituta et historia...12°, Amsterdami, 1635.

Not H; M/G1 – 37; ?

ARBUTHNOT, John. An examination of Dr. Woodward's Account of the deluge, &c. By J. A[rbuthnot]. M.D....1697. *See* 596.

72 — Tables of ancient coins, weights and measures, explain'd and exemplify'd in several dissertations. 4°, London, 1727.

H; M/C2 – 20; ?

73 ARCANA Gallica: or, The secret history of France, for the last century ...By the author of the Secret history of Europe [J. Oldmixon]. 8°, London, 1714.

H; M/J4 – 11; Tr/NQ.9.123.

74 ARCHER, John. Every man his own doctor, compleated with an herbal ...2nd ed., with additions...8°, London, 1673.

Not H; M/J8 – 46; ?

75 ARCHIMEDES. Opera quae extant. Novis demonstrationibus commentariisque illustrata. Per D. Rivaltum...(*Greek & Latin.*) F°, Parisiis, 1615.

H; M/C3 – 15; Sotheby, Apr. 1926/563 where it is described as having 'Is. Newton' on fly-leaf; ?

76 — Archimedis opera: Apollonii Pergæi Conicorum libri IIII. Theodosii Sphærica: methodo novo illustrata, & succinctè demonstrata. Per I. Barrow. 4°, Londini, 1675.

H; M/B1 – 19; Tr/NQ.16.193 [on fly-leaf 'pret. 3ˢ. 6ᵈ Is. Newton' beside the scored-through inscription (in Barrow's hand) 'Edm: Matthews Sid: Coll: Ex dono Reverendi Auctoris'; see pp. 10–11 above].

ARCHIMEDES, *pseud.* *See* PITCAIRNE, Archibald.

ARIAS MONTANUS, Benedictus. *See* MONTANUS, Benedictus Arias.

77 ARISTARCHUS, *Samius*. De magnitudinibus & distantiis solis & lunæ, liber nunc primum Græce editus cum F. Commandini versione Latina...notisque illustravit J. Wallis. 8°, Oxoniæ, 1688.

H; M/F2 – 2; ?

78 — Aristarchi Samii [or rather, G. P. de Roberval's] De mundi systemate, partibus, & motibus eiusdem, libellus. Adiectæ sunt Æ. P. de Roberval...notæ...12°, Parisiis, 1644.

Not H; M/G2 – 5; ?

79 ARISTOPHANES. Comœdiæ xi, Græcè & Latinè...emendationibus virorum doctorum, præcipuè I. Scaligeri...12°, Lugduni Batavorum, 1624.

H; M/G2 – 7; ?

ARNAULD, Antoine. Historia et concordia Evangelica...Opera & studio Theologi Parisiensis [i.e. A. Arnauld]. Ed. 2ª...1660. *See* 769.

— Logica, sive Ars cogitandi...[By A. Arnauld and P. Nicole]. E tertia apud Gallos ed. recognita & aucta in Latinum versa. 1687. *See* 980.

80 ARNOBIUS, *Afer*. Adversus gentes libri vii. Cum recensione viri celeberrimi, & integris omnium commentariis. Ed. novissima atque omnium accuratissima. (4 pts.) 4°, Lugduni Batavorum, 1651.

H; M/C1 – 27; Tr/NQ.7.49 [pt 3, p. 104 turned down and several other signs of dog-earing esp. in pt 1].

81 ARRIANUS, Flavius. De expedit. Alex. Magni, historiarum libri viii. Ex B. Vulcanii nova interpretatione...(*Greek & Latin.*) F°, [Geneva], 1575.

H; M/C3 – 6; ?

82 — De rebus gestis Alexandri Magni regis Macedonum libri viii... B. Facio interprete. 12°, Lugduni, 1552.

Not H; M/G1 – 24; ?

83 ARROWSMITH, John. Tactica sacra, sive De milite spirituali pugnante, vincente, & triumphante dissertatio, tribus libris comprehensa...4°, Cantabrigiæ, 1657.

Not H; M/B1 – 11; Tr/NQ.8.130.

84 ARS chemica, quod sit licita recte exercentibus, probationes doctissimorum iurisconsultorum. Septem tractatus seu capitula Hermetis Trismegisti, aurei. Eiusdem Tabula Smaragdina, in ipsius sepulchro inuenta, cum commento Hortulani Philosophi. Studium Consilii conjugij de massa solis & lunæ...8°, (Argentorati, 1566).

Not H; M/E3 – 28; Tr/NQ.10.145.

Short notes by Newton in margins and texts of pp. 9–13, 15, 21, 22, 29–31, 48, 92, 128, 129, mostly references to alchemical books; many signs of dog-earing.

85 — — [Another copy.]
H; M/E3 – 21; Tr/NQ.16.137¹ [bound with 168, 536, & 649; a few signs of dog-earing].

86 — Ars magica, sive Magia naturalis et artificiosa. . .effectus, virtutes, & secreta in elementis, gemmis, lapidibus, herbis, & animalibus secundum certas astrorum. . .figuras. . .horasque planetarias exhibens. . .12°, Francofurti, 1631.
Not H; M/E1 – 27; ?

87 ART (The) of contentment. By the author of The whole duty of man, &c. [R. Allestree]. 2nd impr. 8°, Oxford, 1675.
Not H; M/H3 – 37; Tr/NQ.10.97.

88 — The art of heraldry. [By R. Blome.] 2nd ed. 2 pts. 8°, London, 1693.
Not H; M/J8 – 35; ?

89 ARTEMIDORUS, *Daldianus*. De somniorum interpretatione libri v. A I. Cornario Latina lingua conscripti. 8°, Lugduni, 1546.
Not H; M/F1 – 12; ?

ARTEMONIUS, Lucas Mellierus, *pseud. See* CRELLIUS, Samuel.

ARTEPHIUS. Philosophie naturelle de trois anciens philosophes renommez: Artephius, Flamel, & Synesius. . .Dernière éd.... 1682. *See* 1309 & 1310.

90 ARTIS auriferae, quam chemiam vocant, volumina duo, quæ continent Turbam philosophorum, aliosᷓ antiquiss. auctores. . .Accessit noviter volumen tertium. . .8°, Basileæ, 1610.
Not H; M/D1 – 23; Tr/NQ.16.121.
Notes and references by Newton mainly relating to other alchemical works on verso of title-page of vol. 1, in vol. 1, pp. 20, 21, 68, on verso of title of vol. 2, in vol. 2, pp. 6, 23, 24, 99, 111, 158, 190, 245, 275; many signs of dog-earing.

91 — — [Another copy.]
Not H; M/E3 – 45; ?

ARVIEUX, Laurent d'. Voyage [du Chevalier d'Arvieux] dans la Palestine, vers le Grand Emir. . .Par M. de la Roque. 1718. *See* 918.

92 ASGILL, John. The succession of the House of Hannover vindicated, against the Pretender's Second declaration in folio, intitled, The hereditary right of the Crown of England asserted, &c. 8°, London, 1714.
Not H; M (Tracts) J3 – 13; Tr/NQ.9.153⁵.

93 ASHMOLE, Elias. Theatrum chemicum Britannicum. Containing severall poeticall pieces of our famous English philosophers, who have written the Hermetique mysteries in their owne ancient language. . .4°, London, 1652.
H; M/J5 – 30; Van Pelt Library, University of Pennsylvania, Philadelphia.

Numerous corrections, additions and references by Newton; several signs of
dog-earing; Sotheran **800** (1926) 10087.

94 — The way to bliss. In 3 books. 4°, London, 1658.
H; M/H2 – 19; ?

95 ATHANASIUS, *St, Abp of Alexandria*. Opera quæ reperiuntur omnia...
(*Greek & Latin*.) 2 vols. F°, Parisiis, 1627.
H; M/D7 – 3 & 4; ?

96 — Opera omnia quæ extant...Opera & studio Monachorum Ordinis
S. Benedicti. (*Greek & Latin*.) 3 vols. F°, Parisiis, 1698.
H; M/E8 – 1 to 3; ?

97 ATHENAEUS, *Naucratita*. Deipnosophistarum libri xv. Cura & studio
I. Casauboni...F°, [Heidelberg], 1597.
H; M/C3 – 9; ?

98 ATHENAGORAS. Apologia pro Christianis...Uterque Græcè &
Latinè. 8°, [Geneva], 1557.
H; M/E3 – 20; ?

99 ATLAS geographus: or, A compleat system of geography...Vols. 1–4.
Europe; Asia, Africa. 4°, London, 1711–14.
H; M/D5 – 17 to 20; ?

ATREMONT, d'. Le tombeau de la pauvreté...Par un philosophe
inconnu [i.e. d'Atremont]...1673. *See* 1619.

100 AUGUSTINE, *St, Bp of Hippo*. Meditationes, Soliloquia et Manuale
...Omnia...studio H. Sommalii. 16°, Coloniæ Agrippinæ, 1649.
Not H; M/G1 – 38; ?

101 — Omnium operum primus(— decimus) tomus, ad fidem exemplariũ
...repurgatorum...10 vols. in 5. F°, (Parisiis, 1531–2).
H; M/E8 – 7 to 11; Thame/990; ?

102 — Pious breathings. Being the Meditations of St. Augustine, his
Treatise of the Love of God, Soliloquies and Manual...Made
English by G. Stanhope. 5th ed. 8°, London, 1720.
Not H; M/H6 – 30; ?

103 AURIFONTINA chymica: or, A collection of fourteen small treatises
concerning the first matter of philosophers, for the discovery of their,
hitherto so much concealed, mercury...24°, London, 1680.
Not H; M/G9 – 56; Tr/NQ.16.119.
*Notes and references by Newton relating to other alchemical works in margins
of pp. 21, 93, 123, 124, 131, 147*; a few signs of dog-earing.

104 AUSONIUS, Decius Magnus. Opera, J. Tollius recensuit...aliorum-
que notis accuratissime digestis...8°, Amstelodami, 1671.
H; M/F4 – 13; ?

105 AYMON, Jean. Métamorphoses de la religion romaine...12°, La
Haye, 1700.
Not H; M/G2 – 48; ?

106 BACCHINI, Benedetto. De ecclesiasticæ hierarchiæ originibus dissertatio. 4°, Mutinæ, 1703.
Not H; M/F6 – 20; Tr/NQ.10.46.

107 BACCI, Andrea. De gemmis et lapidibus pretiosis...nunc vero in Latinum sermonem conversus. A W. Gabelchovero. 8°, Francofurti, 1643.
Not H; M/E3 – 37; Stanford University Library, Stanford, Calif.
BACHET, Claude Gaspard. Diophanti Alexandrini Arithmeticorum libri VI. Et de numeris multangulis liber I. Nunc primùm Græcè & Latinè editi...Auctore C. G. Bacheto. 1621. *See* 524.

108 BACON, Francis, *Viscount St Albans*. Essays, or, Councils, civil and moral. With a table of the colours of good and evil. And a discourse of the wisdom of the ancients...(3 pts.) 8°, London, 1706.
Not H; M/E4 – 25; Tr/NQ.9.24.

109 — Opuscula varia posthuma, philosophica, civilia, et theologica, nunc primum edita. Cura & fide G. Rawley. Una cum nobilissimi auctoris vita. 8°, Londini, 1658.
Not H; M/A3 – 11; Tr/NQ.10.88 [p. 17 turned down and a few other signs of dog-earing].

110 BACON, Roger. De arte chymiæ scripta. Cui accesserunt opuscula alia eiusdem authoris. 12°, Francofurti, 1603.
Not H; M/E1 – 47; Tr/NQ.16.120.
Notes and page references by Newton relating to other alchemical works on fly-leaf and pp. 43–4.
BAKER, Thomas, *B.D.*, †1740. Reflections upon learning...By a gentleman [T. Baker]. 4th ed. 1708. *See* 1384.

111 BAKER, Thomas, *Rector of Bishops Nympton*. The geometrical key: or The gate of equations unlock'd ... (*Latin & English*.) 4°, London, 1684.
H; M/J3 – 42; Tr/NQ.9.25.

112 BALAM, Richard. Algebra: or, The doctrine of composing, inferring, and resolving an equation...12°, London, 1653.
Not H; M/G9 – 54; Babson 400 ['Isaac Newton' on fly-leaf, written beneath the signature 'Tim. Burrage, Clare. Cantabr. 1659'].
BALCARRES, Colin Lindsay, *3rd Earl of*. An account of the affairs of Scotland, relating to the Revolution in 1688...By the Right Honourable the Earl of B—[Balcarres]. 1714. *See* 4.

113 BALL, John. An answer to two treatises of Mr. Iohn Can, the leader of the English Brownists in Amsterdam...4°, London, 1642.
Not H; M/H2 – 29; Tr/NQ.10.103[2] [bound with 114].

114 — A friendly triall of the grounds tending to separation; in a plain and modest dispute...4°, [London], 1640.
Not H; M/H2 – 29; Tr/NQ.10.103[1] [bound with 113; a few signs of dog-earing].

BARAT, Nicolas. Nouvelle bibliothèque choisie. . .[By N. Barat.] 2 vols. 1714. *See* 1191.

115 BARBA, Alvaro Alonso. The first book of the Art of mettals, in which is declared the manner of their generation; and the concomitants of them. Written in Spanish. Transl. into English. . .8°, London, 1670.

Not H; M (Tracts) H8 – 20; Tr/NQ. 10. 143² [a few signs of dog-earing].

116 — The first (& second) book of the Art of mettals. . .Transl. . . .by Edward Earl of Sandwich. 2 pts. 8°, London, 1674.

H; M/J8 – 45; ?

BARKER, Jane. Poetical recreations. . .In 2 parts. Part I. Occasionally written by Mrs. J. Barker. Part II. By several gentlemen of the Universities, and others. 1688. *See* 1334.

BARKSDALE, Charles. Doctorum virorum Elogia Thuanea. Operâ C. B[arksdale]. 1671. *See* 1614.

117 BARLOW, William, *Bp of Lincoln*. An answer to a Catholike English-man so by himselfe entituled, who, without a name, passed his censure upon the apology, made by. . .Prince Iames. . .for the Oath of Allegiance. . .4°, London, 1609.

Not H; M/H2 – 14; Tr/NQ. 10. 109 [on end-paper 'Charles Huggins Rect^r of Chinnor '; probably not from Newton's library].

118 BARNABAS, *St*. Epistola catholica. . .Hanc primum è tenebris eruit, notisque & observationibus illustravit H. Menardus. . .(*Greek & Latin*.) 4°, Parisiis, 1645.

H; M/F6 – 30 [bound with 399]; ?

119 BARNES, John. Catholico-Romanus pacificus. 8°, Oxoniæ, 1680.

Not H; M/A1 – 17; ?

120 BARONIUS, Caesar. Annales ecclesiastici. Ed. novissima. . .12 vols. in 6. F°, Coloniae Agrippinae, 1609–13.

H; M/E7 – 1 to 6; ?

121 BAROZZI, Francesco. Cosmographia in quatuor libros distributa, summo ordine, miraque facilitate. . .8°, Venetijs, 1585.

Not H; M/E4 – 41; Tr/NQ. 16. 50.

122 BARROW, Isaac. Lectiones XVIII, Cantabrigiæ in Scholis publicis habitæ; in quibus opticorum phænomenωn genuinæ rationes investigantur, ac exponuntur. Annexæ sunt Lectiones aliquot geometricæ. [Ed. by Newton.] (2 pts.) 4°, Londini, 1669(–1670).

H; M/B1 – 20; Tr/NQ. 16. 181.

On front paste-down Newton wrote the names 'Christianus Melder Pofess.[sic] Lugd. Math. Burcherus de Volder. Jacobus Vander Esch'; on fly-leaf 'Isaaco Newtono Reverendus Author hunc dono dedit July 7ᵗʰ. 1670.' also in Newton's hand; a few signs of dog-earing; Sotheran **828** *(1931) 3602.*

123 — Lectiones mathematicæ XXIII; in quibus principia matheseôs generalia exponuntur: habitæ Cantabrigiæ A.D. 1664, 1665, 1666. Accesserunt ejusdem Lectiones IV. . .(2 pts.) 8°, Londini, (1684–) 1685.
H; M/E3 – 40; ?

124 — A treatise of the Pope's supremacy. To which is added A discourse concerning the unity of the Church. 4°, London, 1680.
H; M/F5 – 14; Tr/NQ.9.113 [several signs of dog-earing].
See also 76, 581-3, 585.

125 BARWICK, Peter. The life of the Reverend Dr. John Barwick, D.D. sometime Fellow of St. John's College in Cambridge. . .Written in Latin. . .Transl. into English. . .8°, London, 1724.
Not H; M/G10 – 16; Tr/NQ.8.64.

126 BASIL, *St, Abp of Caesarea, the Great.* Opera omnia, quæ reperiri potuerunt. Nunc primum Græcè & Latinè coniunctim edita. . .3 vols. F°, Parisiis, 1618.
H; M/D7 – 11 to 13; ?

127 BASILIUS VALENTINUS. Azoth, ou Le moyen de faire l'or caché des philosophes. Reveu, corrigé & augmenté par Mr. L'Agneau Medecin. 8°, Paris, 1659.
H; M/G7 – 8; Tr/NQ.16.124^1 [bound with 130 & 169].

128 — Basilius Valentinus Friar of the Order of St. Benedict his last will and testament. . .(6 pts.) 8°, London, (1656–)1657.
H; M/E1 – 17; Tr/NQ.16.135.
Pt 2 ('XII. Keyes') has correction by Newton in text of p. 7 and references by him in margins of p. 13 to 'Maier. Embl. 24, p. 70' and p. 16 to 'Maier. Embl. 54, p. 150'; many signs of dog-earing.

129 — Basil Valentine his Triumphant chariot of antimony, with annotations of T. Kirkringius. . .8°, London, 1678.
H; M/E5 – 37; Tr/NQ.16.97 [many signs of dog-earing].

130 — Les douze clefs de philosophie. Plus l'Azoth, ou Le moyen de faire l'or caché des philosophes. Traduction françoise. 8°, Paris, 1660.
Not H; M/G7 – 8; Tr/NQ.16.124^3 [bound with 127].

131 BASNAGE, Samuel. Annales politico-ecclesiastici annorum DCXLV. a Cæsare Augusto ad Phocam usque. . .3 vols. F°, Roterodami, 1706.
H; M/B4 – 2 to 4; ?

132 BASNAGE DE BEAUVAL, Jacques. Antiquitez judaïques, ou Remarques critiques sur la République des Hébreux. 2 vols. 8°, Amsterdam, 1713.
H; M/G8 – 10 & 11; Tr/NQ.9.109 & 110 [vol. 1 has a few signs of dog-earing].

133 — The history of the Jews, from Jesus Christ to the present time...
   Transl....by T. Taylor. F°, London, 1708.
   H; M/F8 – 15; ?

134 BAXTER, Richard. The cure of church-divisions: or, Directions for
   weak Christians...8°, London, 1670.
   Not H; M/H1 – 36; ?

135 — Gildas Salvianus; the reformed pastor. Shewing the nature of the
   pastoral work...8°, London, 1656.
   Not H; M/H1 – 34; Tr/NQ.16.165 [a few signs of dog-earing].

136 — The Nonconformists plea for peace...8°, London, 1679.
   Not H; M/H1 – 33; Cornell University Library, Ithaca, N.Y.

137 — The right method for a settled peace of conscience and spiritual
   comfort. 8°, London, 1653.
   Not H; M/H1 – 37; ?

138 — A treatise of self-denyall. 4°, London, 1660.
   Not H; M/H2 – 35; Tr/NQ.10.105 [a few signs of dog-earing].

139 — The unreasonableness of infidelity: manifested in four discourses...
   8°, London, 1655.
   Not H; M/H1 – 35; ?

140 BAXTER, William. Glossarium antiquitatum Britannicarum, sive
   Syllabus etymologicus antiquitatum veteris Britanniæ atque Iberniæ
   temporibus Romanorum...8°, Londini, 1719.
   H; M/F4 – 23; ?

141 BAYER, Johann. Explicatio characterum aeneis urano-metrias
   imaginum, tabulis, insculptorum, addita, & commodiore hac forma
   tertiùm redintegrata. 4°, Ulmæ Suevorum, 1640.
   H; M/E4 – 33; Tr/NQ.16.15.

142 — Uranometria, omnium asterismorum continens schemata, nova
   methodo delineata, aereis laminis expressa. F°, Ulmæ, 1655.
   H; M/B2 – 26; Babson 401.
   Names of constellations written by Newton on every map.

143 BAYLE, Pierre. Lettres choisies, avec des remarques [by P. Marchand].
   3 vols. 12°, Rotterdam, 1714.
   H; M/G6 – 19 to 21; Tr/NQ.9.137 to 139.

144 BECAN, Martin. Disputatio theologica de Antichristo reformato...
   8°, Coloniæ Aggrippinæ [*sic*], 1608.
   Not H; M/A2 – 11; Tr/NQ.9.161³ [bound with 349].

   BECHAMEL, François Jean. Voyages and discoveries in South-
   America...1698. *See* 1700.

145 BECHER, Johann Joachim. Actorum Laboratorii Chymici Monacen-
   sis, seu Physicæ subterraneæ libri II...8°, Francofurti, 1681.
   H; M/E2 – 19; Tr/NQ.16.96 [pp. 475, 506, 526, 780 turned up and
   a few other signs of dog-earing].

146 — Magnalia naturæ: or, The philosophers-stone lately expos'd to publick sight and sale...4°, London, 1680.
Not H; M (Tracts) D5 – 24; Tr/NQ.16.79³.

147 BEDA. Ecclesiasticæ historiæ gentis Anglorum, libri v...12°, Coloniæ Agrippinæ, 1601.
Not H; M/E1 – 42; ?

148 — Historiae ecclesiasticae gentis Anglorum libri v...Cura et studio J. Smith. F°, Cantabrigiæ, 1722.
H; M/C4 – 3; ?

149 BEGER, Laurentius. Spicilegium antiquitatis sive Variarum ex antiquitate elegantiarum vel novis luminibus illustratarum. F°, Coloniæ Brandenburgicæ, 1692.
H; M/C3 – 20; Tr/NQ.17.33 [a few signs of dog-earing].

150 BÉGUIN, Jean. Tyrocinium chymicum, commentario illustratum. A G. Blasio. Ed. 2ª, priori locupletior & emendatior. 12°, Amstelodami, 1669.
Not H; M/E1 – 21; ?

151 BELLARMINUS, Robertus, *Cardinal*. De æterna felicitate Sanctorum libri v...8°, Coloniæ, 1626.
Not H; M/G1 – 41; ?

152 — De ascensione mentis in Deum per scalas rerum creatarum opusculum. 8°, Coloniæ Agrippinæ, 1626.
Not H; M/G1 – 40; ?

153 — De scriptoribus ecclesiasticis liber unus...8°, Coloniæ Agrippinæ, 1613.
Not H; M/F1 – 11; Tr/NQ.9.3.

154 — The notes of the Church, as laid down by Cardinal Bellarmin; examined and confuted [by W. Sherlock, S. Freeman, S. Patrick, etc.]. 4°, London, 1688.
H; M/F5 – 21; Tr/NQ.8.137 [a few signs of dog-earing].

155 BELLERS, John. An essay towards the improvement of physick... With an essay for imploying the able poor; by which the riches of the Kingdom may be greatly increased...4°, London, 1714.
Not H; M (Tracts) J3 – 41; Tr/NQ.8.111¹.

156 — Proposals for raising a colledge of industry of all useful trades and husbandry...4°, London, 1696.
Not H; M (Tracts) J3 – 41; Tr/NQ.8.111².

157 BELLINI, Lorenzo. De urinis et pulsibus, de missione sanguinis, de febribus, de morbis capitis, et pectoris. 4°, Francofurti, 1685.
H; M/C1 – 20; Tr/NQ.8.53 [p. 165 turned up and a few other signs of dog-earing].

158 BELOT, Jean. Œuvres. Contentant la chiromence, physionomie, l'art de mémoire de Raymond Lulle...Dernière éd., reveuë, corrigée & augmentée de divers traictez. 8°, Rouen, 1662.
Not H; M/G6 – 1; Tr/NQ.16.69.

159 BENNET, John. Collectio sententiarum, exemplorum, testimoniorum, nec non et similitudinum, in usum scholasticæ juventutis. 8°, Londini, 1707.
Not H; M/A4 – 16; ?

160 BENNET, Thomas. An answer to the Dissenters pleas for separation, or An abridgement of the London cases...5th ed. 8°, London, 1711.
Not H; M/H4 – 5; ?

161 — A discourse of the everblessed Trinity in unity, with an examination of Dr. Clarke's Scripture doctrine of the Trinity. 8°, London, 1718.
Not H; M/H4 – 4; ?

162 — An essay on the Thirty Nine Articles of religion, agreed on in 1562, and revised in 1571...8°, London, 1715.
Not H; M/H4 – 3; ?

BENTLEY, Richard. A dissertation upon the Epistles of Phalaris [etc.]. 1697. *See* 1757.

163 — A dissertation upon the Epistles of Phalaris. With an answer to the objections of the Hon. Charles Boyle, Esquire. 8°, London, 1699.
Not H; M/E4 – 29; ?
— Emendationes in Menandri et Philemonis reliquias, ex nupera editione J. Clerici...auctore Phileleuthero Lipsiensi [R. Bentley]. 1710. *See* 553.

164 — The present state of Trinity College in Cambridg, in a letter from Dr. Bentley, Master of the said College, to the Right Reverend John [Moore], Lord Bishop of Ely. 8°, London, 1710.
Not H; M (Tracts) J3 – 7; Tr/NQ.7.1¹.
— A short account of Dʳ Bentley's humanity and justice...1699. *See* 1509.

165 BERCHET, Toussaint. Στοιχείωσις τῆς χριστιανῶν πίστεως, ἢ κατηχισμός. Elementaria traditio Christianorum fidei, aut Catechismus...(*Greek & Latin.*) 8°, Londini, 1648.
Not H; M/A3 – 7; Tr/NQ.9.2.

BERLIN. *Societas Regia Scientiarum.* Miscellanea Berolinensia ad incrementum scientiarum, ex scriptis Societati Regiæ Scientiarum exhibitis edita. Vol. 1. 1710. *See* 1085.

BERNARD, Edward. Catalogi librorum manuscriptorum Angliæ et Hiberniæ in unum collecti...[By E. Bernard.] 1697. *See* 352.

166 — De mensuris et ponderibus antiquis libri III. Ed. 2ª, purior & duplo locupletior. 8°, Oxoniæ, 1688.

H; M/D1 – 15; Sjögren Library, Royal Swedish Academy of Engineering Sciences, Stockholm [inscribed 'Isaaco Newtono Mathematico præstantissimo & Philosophiæ instauratori perfelici E. Bernardus ex merito'].

167 BERNARD, Richard. Thesaurus Biblicus, seu Promptuarium sacrum ...(2 pts.) F°, London, 1642-4.
Not H; M/J9 – 25; ?

168 BERNARDUS, *Trevisanus*. Περὶ χημείας: opus historicum & dog-maticum, ex Gallico in Latinum simpliciter versum...8°, Argen-torati, 1567.
Not H; M/E3 – 21; Tr/NQ.16.137⁴ [bound with 85].

169 — Traicté de la nature de l'oeuf des philosophes. 8°, Paris, 1659.
Not H; M/G7 – 8; Tr/NQ.16.124² [bound with 127].

170 BERNOULLI, Daniel. Discours sur le mouvement des clepsidres ou sabliers. (Pièce qui a remporté le prix de l'Académie Royale des Sciences, 1725.) 4°, Paris, 1725.
Not H; M (Tracts) D6 – 3; Tr/NQ.8.1³.

171 BERNOULLI, Jacob. Positiones arithmeticæ de seriebus infinitis earum� summa finita. (4 pts.) 4°, Basileæ, [1689–98].
Not H; M (Tracts) D5 – 25; Tr/NQ.8.51².

172 BERNOULLI, Johann I. De motu musculorum, de effervescentia, & fermentatione dissertationes physico-mechanicæ. Ed. 2ª prior emendatior...4°, Venetiis, 1721.
Not H; M (Tracts) D6 – 5; Tr/NQ.10.28¹.

173 — Essay d'une nouvelle théorie de la manœuvre des vaisseaux...8°, Basle, 1714.
H; M/G10 – 27; Turner Collection, University of Keele Library [inscribed on fly-leaf 'Illustr. Newtono Mathematico D.D. Auctor'].

174 — Problema mechanico-geometricum de linea celerrimi descensûs. Problema alterum purè geometricum, quod priori subnectimus & strenæ loco eruditis proponimus. (Programma.) [*A broadsheet.*] Groningæ, 1697.
Not H; not M; Royal Society Library.
Annotated by Newton; for Newton's solution of the problems see his letter to Montague 30 Jan. 1696/7 (*Correspondence*, IV, 220–29).

— Tracts, Mathematical by Bernoulli, Stewart &c. 2 vols. 1713. *See* 1630.

175 BEROSUS, *the Chaldean*. Antiquitatum libri v, cum commentariis J. Annij...8°, Wittebergæ, 1612.
Not H; M/E2 – 34; Tr/NQ.8.119 [pp. 26, 111, 115 turned up, p. 37 down].

BERRIMAN, John. Sermons. *See* 1492.

BERTRAM, *priest of Corbie*. *See* RATRAMNUS, *Corbiensis*.

176 BETULEIUS, Xystus. In M. T. Ciceronis libros de officiis, de amicitia, de senectute, commentaria...4°, Basileæ (1544).
Not H; M/J6 – 5; ?

177 BEVERIDGE, William, *Bp of St Asaph*. The Church-catechism explained: for the use of the Diocese of St. Asaph. 2nd ed. 8°, London, 1705.
Not H; not M; Tr/NQ.10.108; part of the Pilgrim Trust presentation 1943 but without Huggins or Musgrave bookplates, probably not from Newton's library.

178 — The great necessity and advantage of publick prayer and frequent communion...4th ed. 12°, London, 1709.
Not H; not M; Tr/NQ.10.94; probably not from Newton's library, see note at 177 above.

179 — Institutionum chronologicarum libri II. Unà cum totidem arithmetices chronologicæ libellis. Ed. 2ª, priori emendatior. 4°, Londini, 1705.
H; M/C1 – 17; Tr/NQ.8.96 [p. 179 turned up].

180 — Thesaurus theologicus: or, A complete system of divinity: summ'd up in brief notes upon select places of the Old and New Testament ...4 vols. (Vols. 1 & 2, 2nd ed.) 8°, London, 1711.
Not H; M/H4 – 6 to 9; Tr/NQ.16.43 to 46.

181 BEZA, Theodorus. Annotationes maiores in Novum Dn. Nostri Iesu Christi Testamentum...(2 pts.) 8°, [Geneva], 1594.
Not H; M/F4 – 29; Tr/NQ.7.53 ['Isaac Newton Trin: Coll: Cant 1661.' in Newton's hand on fly-leaf].

182 — Epistolarum theologicarum Theodori Bezæ Vezelij, liber unus. 2ª ed., ab ipso auctore recognita. 8°, Genevæ, 1575.
Not H; M/D1 – 21; ?

183 — Poemata. Psalmi Davidici xxx. Sylvæ. Elegiæ. Epigrammata... Omnia, in hac 3ª editione, partim recognita, partim locupletata... 8°, [Geneva, *c.* 1576].
Not H; M/A3 – 2; Tr/NQ.10.96.
— Pseaumes de David, mise en rime françoise par C. Marot et T. de Bèze...1613. *See* 1356.

184 BIAEUS, Jacobus. Regum et imperatorum Romanorum numismata aurea, argentea, ærea...subjectis L. Begeri annotationibus. F°, Coloniæ Brandenburgicæ, 1700.

H; M/D7 – 2; Tr/NQ.11.9.

185 BIANCHINI, Francesco. De kalendario et cyclo Caesaris ac De Paschali canone S. Hippolyti Martyris dissertationes II...(2 pts.) F°, Romæ, 1703.

H; M/B2 – 28; Tr/NQ.11.33.

— Solutio problematis Paschalis. [By F. Bianchini.] (1703.) *See* 1535.

186 BIBLE, *English*. The Bible, that is, the Holy Scriptures...With most profitable annotations...4°, London, 1599.

Not H; M/C6 – 28; ?

187 — The Holy Bible & the Book of Common Prayer. (*Black letter.*) 4°, London, 1641.

Not H; M/J6 – 10; Thame/993; ?

188 — The Holy Bible containing the Old Testament and the New...8°, London, 1660.

Not H; not M; Tr/Adv.d.1.10² [bound with 240 & 1560; many signs of dog-earing]; presented to Trinity by John Cox, Fellow, July 1878 and said in a note attached to the volume, signed Joseph Cox, to have been given by Newton in his last illness to the woman who nursed him.

Numerous notes and biblical references by Newton throughout and esp. at Daniel and Revelation.

189 — The Holy Bible...(With Prayer Book.) 12°, Cambridge, 1660.

H (where it was listed as one of the six 'Books that has Notes of Sir Is. Newtons' and described as 'Dirty & leaf wanting'); not subsequently recorded; ?

190 — The Holy Bible. [With Apocrypha, Book of Common Prayer and Psalms.] 4°, Cambridge, 1683.

H; not M; ?

191 — The Holy Bible, containing the Old Testament and the New... 2 vols. [The 'Vinegar Bible'.] F°, Oxford, (1716–)1717.

Not H; M/B5 – 1 & 2; ?

192 — *French*. La Saincte Bible...4°, Genève, 1637.

Not H; M/F6 – 14; ?

193 — *Greek & Latin, New Testament*. Novum Iesu Christi Testamentum, Græcè & Latinè: T. Beza interprete...(2 pts.) 8°, [Geneva], 1611.

Not H; M/A4 – 13; Tr/NQ.16.180.

194 — Novum Testamentum Græce, cum vulgata interpretatione Latina Græci contextus lineis inserta...Ed. postrema. Cum præfatione... B. Ariæ Montani...8°, [Geneva], 1627.

H; M/E2 – 28; ?

195 — *Greek, Old Testament.* Ἡ Παλαιὰ Διαθήκη κατὰ τοὺς Ἑβδομήκοντα. Vetus Testamentum Græcum ex versione Septuaginta interpretum, juxta exemplar Vaticanum Romæ editum. (*With* In Sacra Biblia Græca ex versione LXX. interpretum scholia...) (2 pts.) 8°, Londini, 1653.

H; M [shelf-mark not known]; Tr/NQ.7.79; Sotheran **843** (1935) 1497; see p. 56 above.

A closely written sheet of notes in Newton's hand pasted on fly-leaf, on Biblical studies, including a list of contents of various Bible codices, chiefly those at Cambridge.

196 —— [Another copy of Pt 1.]

Not H; M/F2 – 18; Tr/NQ.7.80¹ [bound with 200; 'Isaac Barrow' on fly-leaf with 2-line Greek quotation also in his hand].

197 — Septuaginta interpretum tomus I (& tomus ultimus)...summa cura edidit J. E. Grabe. F°, Oxonii, 1707–20.

H (lists 'Vols 1 & 4 in quires'); M/C4 – 12 & 13; ?

198 — *New Testament.* Τῆς Καινῆς Διαθήκης ἅπαντα. 8°, [Antwerp, 1573].

H; M/A1 – 11; Tr/NQ.16.171¹ [bound with 1355].

199 — Ἡ Καινὴ Διαθήκη. Novum Testamentum. Huic editioni omnia difficiliorum vocabulorum themata, quæ in G. Pasoris Lexico grammaticè resolvuntur, in margine apposuit C. Hoole. 12°, Londini, 1653.

H; M/A1 – 10; Tr/NQ.16.169 ['Isaac Newton hujus libri verus est

£ s d

possessor. Pretium – 0 – 3 – 0, Aprilis 3 die Anⁿᵒ Dni 1661.' on fly-leaf].

200 — Τῆς Καινῆς Διαθήκης ἅπαντα. Novi Testamenti libri omnes. Ed. nova accurata. 8°, Londini, 1653.

Not H; M/F2 – 18; Tr/NQ.7.80² [bound with 196; with marginal notes by Barrow].

201 — Novum Testamentum. Ed. nova, denuo recusa...Studio & labore S. Curcellæi. 8°, Amstelodami, 1675.

H; M/A1 – 8; ?

202 — Τῆς Καινῆς Διαθήκης ἅπαντα. Novi Testamenti libri omnes. Accesserunt parallela scripturæ loca, nec non variantes lectiones... [Ed. by J. Fell.] 8°, Oxonii, 1675.

H; M/E2 – 29; ?

203 — Ἰησου Χριστου του Κυριου ἡμων ἡ Καινὴ Διαθήκη...12°, Londini, 1701.

Not H; M/A1 – 9; ?

204 — Novum Testamentum. Cum lectionibus variantibus MSS exemplarium...Studio et labore J. Millii. F°, Oxonii, 1707.

H; M/B4 – 18; ?

205 — Τῆς Καινῆς Διαθήκης ἅπαντα. Novum Testamentum. [Ed. by M. Maittaire.] 12°, Londini, 1714.
Not H; M/A1 – 7; ?

206 — *Hebrew.* [Hebrew Bible.] 4°, [Venice, 1517].
H; M/B1 – 2; Tr/NQ.8.22 [wants title-page].

— *Proverbs.* Proverbia Salomonis, iam recens iuxta Hebraicā veritatē translata, & annotatiōibus illustrata, autore S. Munstero. [1524.]
See 1128.

207 — *Latin.* Biblia Latina. Edit. Vulgata. 8°, *Not identified.*
Not H; M/A1 – 4; ?

208 — Testamentum Latinum per Hieronymum. 8°, 1500. *Edition not identified.*
Not H; M/G1 – 43; ?

209 — Biblia Latina Hieronymi. 8°, 1523. *Edition not identified.*
Not H; M/A1 – 6; ?

210 — Biblia Sacra, ad veritatem Hebraicam, & probatissimorum exemplarium fidem summa diligentia castigata...8°, Lugduni, 1568.
Not H; M/F2 – 9; Tr/NQ.8.87.

211 — Biblia Sacra, sive Testamentum Vetus, ab I. Tremellio et F. Iunio ex Hebraeo Latinè redditum. Et Testamentum Novum, à T. Beza è Graeco in Latinum versum. 8°, Amstelodami, 1648.
H; M/A1 – 5; ?

212 — *Genesis.* Genesis sive Mosis Prophetæ liber I. Ex translatione J. Clerici ...Ed. 2ª auctior et emendatior. F°, Amstelodami, 1710.
H; M/B2 – 3; Tr/NQ.18.15[1] [bound with 213].

213 — *Pentateuch.* Mosis Prophetæ libri IV; Exodus, Leviticus, Numeri, et Deuteronomium, ex translatione J. Clerici...Ed. nova auctior et emendatior. F°, Amstelodami, 1710.
H; M/B2 – 3; Tr/NQ.18.15[2] [bound with 212].

214 — *Historical Books.* Veteris Testamenti libri historici...ex translatione J. Clerici; cum ejusdem commentario philologico...F°, Amstelodami, 1708.
H; M/B2 – 4; Tr/NQ.18.16.

215 — *New Testament.* Iesu Christi D.N. Novum Testamentum, T. Beza interprete...8°, Londini, 1579.
Not H; M/A1 – 12; Tr/NQ.16.36.

216 — *Polyglot.* Biblia Sacra polyglotta, complectens textus originales... edidit B. Waltonus. 6 vols. F°, Londini, 1655–7.
H; M/C5 – 2 to 7; Thame/970; ?

217 — *Syriac & Latin, New Testament.* Novum Domini Nostri Jesu Christi Testamentum Syriacum, cum versione Latina; curâ & studio J. Leusden et C. Schaaf editum...4°, Lugduni Batavorum, 1709.
H; M/B6 – 8; Tr/NQ.8.6.

218 — *Concordance.* A concordance to the Holy Scriptures...By S. N. [i.e. Samuel Newman]. 2nd ed. corrected and enlarged. [The Cambridge Concordance.] F°, Cambridge, 1672.
    H; not M; ?

219 — — 4th ed. very much enlarged. F°, Cambridge, 1698.
    H; M/D8 – 16; ?

220 BIBLIOTHECA chemica contracta ex delectu & emendatione N. Albinei in gratiam & commodum artis chemicæ studiosorum. 8°, Genevæ, 1673.
    H; M/E3 – 3; Tr/NQ.16.112.

BIBLIOTHÈQUE ancienne et moderne. Pour l'année MDCCXIV–MDCCXXII. Vols. 1–18. 1714–22. *See* 928.

— Bibliothèque choisie, pour servir de suite à la Bibliothèque universelle. Année MDCCVIII–MDCCXIII. Par J. Le Clerc. Vols. 16–27. 1708–13. *See* 929.

221 — Bibliothèque des philosophes (chymiques), ou Recueil des oeuvres des auteurs les plus approuvez qui ont écrit de la pierre philosophale ...Par Le Sieur S. D. E. M. [i.e. W. Salmon]. 2 vols. 12°, Paris, 1672–8.
    H; M/G2 – 35 & 36; Tr/NQ.16.94 [vol. 1 only].

BIDDLE, John. The Faith of One God... [By J. Biddle.] 1691. *See* 604.

222 BIDLOO, Govard. De oculis et visu variorum animalium observations physico-anatomicæ. 4°, Lugduni Batavorum, 1715.
    Not H; M (Tracts) J6 – 15; Tr/NQ.8.52².

223 — Exercitationum anatomico-chirurgicarum decas. 8°, Lugduni Batavorum, 1704.
    Not H; M/A4 – 9; ?

224 BINIUS, Severinus. Concilia generalia, et provincialia, Græca et Latina quæcunque reperiri potuerunt...4 vols. in 5. F°, Coloniæ Agrippinæ, 1618.
    H; M/E7 – 8 to 12; Thame/997; ?

BIRD, Benjamin. Sermons. *See* 1493.

BIRRIUS, Martinus. Tres tractatus de metallorum transmutatione... Nunc primum in lucem edi curavit M. Birrius. 1668. *See* 1641.

225 BIZOT, Pierre. Histoire metallique de la République de Hollande. Nouv. éd. augmentée. 2 vols. & suppl. (3 vols.) 8°, Amsterdam, 1688–90.
    H; M/G10 – 29 to 31; Tr/NQ.9.149 to 151 [a few signs of dog-earing in the set].

226 BLACKALL, Offspring. Fourteen sermons preach'd upon several occasions. 2nd ed. 8°, London, 1706.
    Not H; M/H3 – 22; ?

227 BLACKERBY, Samuel. The Justice of Peace his companion, or, Summary of the Acts of Parliament, to June 12, 1711. 12°, London, 1711.
Not H; M/G9 – 35; ?

228 BLACKMORE, *Sir* Richard. Essays upon several subjects. 2 vols. 8°, London, 1716–17.
H; M/J3 – 43 & 44; Tr/NQ.9.121 & 122.

229 BLAIR, Patrick. Botanick essays. In 2 parts. . .8°, London, 1720.
H; M/J7 – 26; Memorial Library, University of Wisconsin – Madison.

BLANCHINUS, Franciscus. *See* BIANCHINI, Francesco.

BLOME, Richard. The art of heraldry. [By R. Blome.] In 2 pts. 2nd ed. 1693. *See* 88.

230 BLONDEL, David. Apologia pro sententia Hieronymi de episcopis et presbyteris. 4°, Amsteledami, 1646.
H; M/B1 – 26; Tr/NQ.16.178 ['pret. 8ˢ.' in Newton's hand on flyleaf; a few signs of dog-earing].

231 BOCHART, Samuel. Geographia sacra, cujus pars prior Phaleg de dispersione gentium & terrarum divisione facta in ædificatione turris Babel; pars posterior Chanaan de coloniis & sermone Phœnicum agit. . .4°, Francofurti ad Moenum, 1681.
H; M/B1 – 9; Tr/NQ.8.27 [extensively dog-eared with 58 pages still turned and many other similar signs].

232 — Hierozoicon, sive Bipertitum opus de animalibus Sacræ Scripturæ . . .2 vols. in 1. F°, Londini, 1663.
H; M/A8 – 6; ?

233 BOECLERUS, Johannes Henricus. De scriptoribus Græcis & Latinis, ab Homero ab initium sæculi post Christum natum decimi sexti, commentatio postuma. Ed. 2ª. 8°, Ultrajecti, 1700.
H; M/A2 – 10; Tr/NQ.16.68.

234 BOERHAAVE, Herman. Aphorismi de cognoscendis et curandis morbis, in usum doctrinæ domesticæ digesti. Ed. 2ª auctior. 8°, Lugduni Batavorum, 1715.
H; M/A3 – 23; Tr/NQ.16.151.

235 — Institutiones medicae in usus annuae exercitationis domesticos. Ed. 2ª primâ longê auctior. 8°, Lugduni Batavorum, 1713.
H; M/A3 – 22; Tr/NQ.16.154.

236 — Sermo academicus, de comparando certo in physicis; quem habuit in Academia Lugduno-Batava. . .MDCCXV. . .4°, Lugduni Batavorum, 1715.
Not H; M (Orationes) C6 – 19; Tr/NQ.10.35⁴.

237 BOIZARD, Jean. Traité des monoyes, de leurs circonstances & dépendances. 12°, Paris, 1692.
H; not M; Ekins 5 – 0; ?

238 — — Nouvelle éd.. . .2 vols. 12°, Paris, 1714.
H; M/G7 – 15 & 16; Tr/NQ.9.70 & 71.

239 BONJOUR, Guilelmus. In monumenta Coptica seu Ægyptiaca Bibliothecæ Vaticanæ brevis exercitatio. 4°, (Romæ, 1699).
Not H; M (Tracts) E6 – 35; Tr/NQ.10.34².

240 BOOK OF COMMON PRAYER. The Book of Common Prayer: and administration of the Sacraments. . .8°, London, 1639.
Not H; not M; Tr/Adv.d.1.10¹ [bound with 188 & 1560; see note on the provenance of the volume at 188].

241 — The Book of Common Prayer. . .8°, London, 1679.
Not H; M/G9 – 57; ?

242 — *Greek.* Βίβλος τῆς δημοσίας εὐχῆς. . .8°, Cambridge, [1665].
Not H; M/A1 – 3; ?

243 — *Latin.* Liturgia, seu Liber precum communium, et administrationis Sacramentorum, aliorumque rituum atque ceremoniarum Ecclesiæ juxta usum Ecclesiæ Anglicanæ. . .12°, Londini, 1681.
Not H; M/A1 – 14; Tr/NQ.16.27.

244 — *Spanish.* La liturgia ynglesa, o El Libro de la oracion comun. . . Hispanizado por F. de Alvarado. . .Ed. 2ª corregida. . .8°, Londres, 1715.
Not H; M/H2 – 8 (described as 'calf, gilt leaves'); ?; presented to Newton by the translator (*Correspondence*, VII, 377–8).

245 BOOT, Anselmus Boetius de. Gemmarum et lapidum historia. Quam olim edidit A. B. de Boot. Nunc vero recensuit, à mendis repurgavit, commentariis. . .A. Toll. 8°, Lugduni Batavorum, 1636.
Not H; M/E4 – 36; Tr/NQ.16.85 [a few signs of dog-earing].

BORCH, Oluf. *See* BORRICHIUS, Olaus.

246 BOREL, Pierre. Bibliotheca chimica. Seu Catalogus librorum philosophicorum hermeticorum. . .12°, Parisiis, 1654.
Not H; M/E1 – 22; ?

247 BORELLI, Giovanni Antonio. De motionibus naturalibus a gravitate pendentibus, liber. 4°, Regio Iulio, 1670.
H; M/E6 – 42; Tr/NQ.10.65 [p. 424 turned down, pp. 449, 528 up, and a few other signs of dog-earing]; sent to Newton by Collins, 5 July 1671 (*Correspondence*, I, 66, 68).

248 — De motu animalium. Opus posthumum. Pars 1. 4°, Romæ, 1680.
H; M/C1 – 1; ?

249 — Theoricae mediceorum planetarum ex causis physicis deductae. . . 4°, Florentiae, 1666.
H; M (Tracts) D5 – 24; Tr/NQ.16.79¹ [a few signs of dog-earing].

250 BORRICHIUS, Olaus. De ortu, et progressu chemiæ, dissertatio. 4°, Hafniæ, 1668.
Not H; M/B1 – 24; Tr/NQ.16.92 [several signs of dog-earing].

251 — Lingua pharmacopoeorum, sive De accuratâ vocabulorum in pharmacopoliis usitatorum pronunciatione. 4°, Hafniæ, 1670.
Not H; M/J6 – 25; Tr/NQ.10.63.

252 BOSMAN, Willem. Voyage de Guinée, contenant une description nouvelle & très-exacte de cette côte...12°, Utrecht, 1705.
H; M/G7 – 13; Tr/NQ.10.133.

253 BOYER, Abel. The Royal dictionary abridged. In 2 pts. I. French and English. II. English and French...2nd ed....4°, London, 1708.
Not H; M/G10 – 35; ?

254 BOYLE, Robert. The aerial noctiluca: or Some new phœnomena, and a process of a factitious self-shining substance...8°, London, 1680.
H; M/E2 – 18; Tr/NQ.16.146¹ [bound with 257 & 470].

255 — Certain physiological essays and other tracts...2nd ed....4°, London, 1669.
H; M/E4 – 3; Tr/NQ.8.48 [pp. 92, 134 turned up and several other signs of dog-earing].

256 — The Christian virtuoso...8°, London, 1690.
H; M/E2 – 3; ?

— Curiosities in chymistry...Written by a person of honour, and published by his operator, H. G[regg]. [Sometimes attributed to Boyle.] 1691. *See* 470.

257 — An essay about the origine & virtues of gems...8°, London, 1672.
H; M/E2 – 18; Tr/NQ.16.146² [bound with 254; a few signs of dog-earing].

258 — An essay of the great effects of even languid and unheeded motion ...8°, London, 1690.
H; M/E2 – 5; ?

259 — Essays of the strange subtilty, determinate nature, great efficacy of effluviums...(4 pts.) 8°, London, 1673.
H; M/E2 – 13; Memorial Library, University of Wisconsin – Madison [inscribed at foot of title-page 'For Mr Isaac Newton from the Authour' in Oldenburg's hand; many signs of dog-earing]; Newton acknowledged to Oldenburg receipt of Boyle's gift 14 Sept. 1673 (*Correspondence*, I, 305); Sotheby, April 1926/579; Tregaskis **952** (1926) 162; Duveen, p. 94 & pl. XIV.

260 — Experimenta & observationes physicæ: wherein are briefly treated of several subjects relating to natural philosophy in an experimental way...8°, London, 1691.
H; M/E2 – 12; ?

261 — Experimentorum novorum physico-mechanicorum continuatio secunda...8°, Londini, 1680.

Not H; M/E2 – 7; Babson 402 ['Isaac Newton Donum nobilissimi Auctoris' in Newton's hand on fly-leaf].

262 — Experiments and considerations about the porosity of bodies, in two essays. 8°, London, 1684.

H; M/E2 – 14; ?

263 — Experiments, notes, &c. about the mechanical origine or production of divers particular qualities...(12 pts.) 8°, London, 1675.

H; M/E2 – 8; Tr/NQ.16.199 [inscribed at foot of title-page 'For Mr Isaac Newton from the Author' in Oldenburg's hand; a few signs of dog-earing].

264 — The general history of the air. 4°, London, 1692.

H; M/E4 – 6; Tr/NQ.8.78 [several signs of dog-earing].

Note by Newton in margin of p. 100 'Dr Wyberd found it [i.e. the weight of a cubic inch of water] 253⅛gr·. Put it 254gr· & a sphære of water of an inch in diameter will weigh 134gr·.'

265 — Medicina hydrostatica: or, Hydrostaticks applyed to the materia medica...8°, London, 1690.

H; M/E2 – 4; Turner Collection, University of Keele Library.

'A Table of refractions given to me by Mr. Halley' in Newton's hand on fly-leaves, notes and corrections by him on unnumbered pages at back of book, relating to a 'Table of specific gravities'; 24 pages turned down or up.

266 — Memoirs for the natural history of humane blood, especially the spirit of that liquor. 8°, London, 168¾.

H; M/E2 – 17; John Crerar Library, Chicago ['Is. Newton. Donum Honorb Authoris' in Newton's hand on fly-leaf]; Sotheby, April 1926/583.

267 — New experiments, and observations, made upon the icy noctiluca... 8°, London, 168½.

H; M/E2 – 15; ?

268 — New experiments and observations touching cold, or An experimental history of cold...(2 pts.) 8°, London, 1665.

H; M/E2 – 10; Tr/NQ.8.126 [several signs of dog-earing].

269 — New experiments physico-mechanicall, touching the spring of the air, and its effects...8°, Oxford, 1660.

H; M/E2 – 9; Tr/NQ.10.69 [a few signs of dog-earing].

270 — The sceptical chymist: or Chymico-physical doubts & paradoxes... (2 pts.) 8°, Oxford, 1680.

H; M/E2 – 2; Tr/NQ.16.84 [pt 1 has pp. 40, 134 turned down and up, pp. 213, 399, 405 up, pp. 260, 279, 359, 367, 369, 400, 402, 409 down; pt 2 has pp. 18, 118, 152, 235 up, pp. 19, 259 down, and many other signs of dog-earing in the volume].

271 — Some considerations about the reconcileableness of reason and religion. By T. E. A lay-man. To which is annex'd...A discourse of Mr. Boyle, About the possibility of the Resurrection. 8°, London, 1675.

Not H; M/E2 – 16; Fulton Collection, Yale Medical Library, New Haven, Conn. ['Isaac Newton. Donum Authoris' in Newton's hand on fly-leaf]; Sotheran **789** (1924) 5728.

272 — Some considerations touching the usefulnesse of experimental naturall philosophy...2 vols. 4°, Oxford, 1664–71.

H; M/E4 – 4 & 5; Tr/NQ.8.47 [vol. 2 only; pp. 21, 24 turned up and many other signs of dog-earing].

273 — Tracts consisting of observations about the saltness of the sea... 8°, London, 1673.

H; M/E2 – 6; Whipple Science Museum Library, Cambridge University ['Ex dono Authoris Is°. Newton' in Newton's hand on fly-leaf].

274 — Tracts: containing I. Suspicions about some hidden qualities of the air...8°, London, 1674.

H; M/E2 – 11; ?

275 — Tracts, containing new experiments, touching the relation betwixt flame and air. And about explosions. An hydrostatical discourse occasion'd by some objections of Dr. Henry More...8°, London, 1672.

H; M/E2 – 1; Tr/NQ.10.100 [a few signs of dog-earing].

276 — Works...epitomiz'd. By R. Boulton. Vols. 1–3. 8°, London, 1699–1700.

H; M/E4 – 7 to 9; University of Chicago Library; Sotheby, 12 July 1965/103.

277 BRACESCO, Giovanni. De alchemia, dialogi duo nunquam ante hac conjunctim sic editi, correcti, & emaculati...8°, Hamburgi, 1673.

H; M/A3 – 18; Tr/NQ.16.125.

278 BRADLEY, Richard. The history of succulent plants...Engraved, from the originals, on copper-plates. With their descriptions, and manner of culture. Decade 1. 4°, London, 1716.

Not H; M (Tracts) J6 – 15; Tr/NQ.8.52³.

279 — New improvements of planting and gardening...5th ed....8°, London, 1726.

Not H; M/J7 – 16; ?

280 — A philosophical account of the works of nature...4°, London, 1721.

H; M/C2 – 18; ?

281 — The plague at Marseilles consider'd: with remarks on the plague in general...8°, London, 1721.

H; M/J7 – 25; ?

282 BRAGGE, Francis. Of undissembled and persevering religion: in several sermons...8°, London, 1713.
Not H; M/H4 – 15; ?

283 — Practical discourses upon the Parables of our Blessed Saviour...
2 vols. (Vol. 1, 2nd ed.) 8°, London, 1702–4.
Not H; M/H4 – 10 & 11; ?

284 — Practical observations upon the Miracles of our Blessed Saviour.
2 vols. 8°, London, 1702–6.
Not H; M/H4 – 12 & 13; ?

285 — A practical treatise of the regulation of the passions. 8°, London, 1708.
Not H; M/H4 – 14; ?

286 — Prayers and meditations. 8°, London, 1712.
Not H; M/H4 – 16; ?

287 BRAND, Adam. Relation du voyage de Mr Evert Isbrand envoyé de sa Majesté Czarienne à l'Empereur de la Chine, en 1692, 93, & 94. Avec une lettre de Monsieur ***, sur l'état présent de la Moscovie. 8°, Amsterdam, 1699.
H; M/G7 – 12; Tr/NQ.7.65.

288 BRANDT, Geeraert. The history of the Reformation and other ecclesiastical transactions in, and about, the Low-Countries...
[Transl.] Vol. 1. 8°, London, 1719.
H; not M; ?

289 BRAY, Thomas. A course of lectures upon the Church Catechism.
Vol. 1. [*No more published.*] F°, Oxford, 1696.
Not H; M/G11 – 8; ?

290 BRENDEL, Zacharias. Chymia in artis formam redacta et publicis prælectionibus philiatris in Academia Jenensi communicata...24°, Jenæ, 1630.
Not H; not M; Sotheran **800** (1926) 10375, **804** (1927) 3796; ?

291 — Chimia in artis formam redacta...Consilio W. Rolfinck...12°, Lugduni Batavorum, 1671.
Not H; M/E1 – 32; Sotheran **800** (1926) 10376.

292 BRENT, Charles. An essay concerning the nature and guilt of lying.
8°, London, 1702.
Not H; M/H4 – 31; ?

293 BREREWOOD, Edward. Elementa logicæ. In gratiam studiosæ juventutis in Academiâ Oxoniensi. 12°, Londini, 1649.
Not H; M/E1 – 46; Tr/NQ.9.166² [bound with 1531].

294 — Enquiries touching the diversity of languages and religions, through the chief parts of the world. 8°, London, 1674.
H; M/H1 – 7; Tr/NQ.9.55 [p. 156 turned up and a few other signs of dog-earing].

295 BREWSTER, *Sir* Francis. Essays on trade and navigation. In 5 parts.
8°, London, 1695.
H; M/J8 – 40; Tr/NQ.16.31 [p. 16 was formerly turned down].

296 BREYDENBACH, Bernhard von. Peregrinatio in Terram Sanctam.
F°, Moguntiae, 1486.
H; M/C2 – 5; Tr/NQ.11.32² [bound with 1416; see p. 73 above].

BRIGGS, Henry. Chiliades logarithmorum viginti...1633. *See* 1694.

297 BRIGGS, William. Ophthalmo-graphia, sive Oculi ejusq; partium
descriptio anatomica...Ed. 2ª. 8°, Londini, 1685.
H; M/E3 – 36; ?

298 BRISSON, Barnabé. De Regio Persarum principatu libri III...8°,
[Heidelberg], 1595.
H; not M; ?

299 BRITANNIA languens, or A discourse of trade...[By W. Petyt.] 8°,
London, 1680.
H; M/J1 – 44; Tr/NQ.8.125¹ [bound with 1201; several signs of
dog-earing].

300 BROUGHTON, John. The great apostasy from Christianity, with its
evil influence on the civil state...8°, London, 1718.
H; M/H3 – 31; ?

301 BROWN, George. Arithmetica infinita, or The accurate accomptants
best companion...8°, [Edinburgh?], 17$\frac{17}{18}$.
H; M/G1 – 46; ?

302 BROWN, William. Astraeæ abdicatæ restauratio. Or Advice to Justices
of the Peace...12°, [London], 1695.
Not H; M/J8 – 48; Tr/NQ.8.92 [p. 300 turned up].

303 BROWNE, John. Myographia nova: or, A graphical description of
all the muscles in the humane body...F°, London, 1698.
H; M/B2 – 24; ?

304 BUCANUS, Gulielmus. Institutiones theologicæ, seu Locorum communium Christianæ religionis, ex Dei Verbo, et præstantissimorum
theologorum orthodoxo consensu expositorum...Ed. 2ª, auctior &
emendatior...8°, [Bern], 1604.
Not H; M/F2 – 28; Tr/NQ.7.58.

305 BUCHANAN, George. Opera omnia, historica, chronologica, juridica,
politica, satyrica & poetica...curante T. Ruddimanno...2 vols.
4°, Lugduni Batavorum, 1725.
H; M/F6 – 4 & 5; Tr/NQ.16.175 & 176.

306 BUCKINGHAM, George Villiers, *2nd Duke of*. The Rehearsal: as it is
now acted at the Theatre-Royal. 9th ed. 12°, [London], 1718.
Not H; M (Plays) H8 – 10; Tr/NQ.10.136¹.

307 BULL, George, *Bp of St David's*. Defensio fidei Nicænæ, ex scriptis, quæ extant, Catholicorum doctorum, qui intra tria prima Ecclesiæ Christianæ secula floruerunt. 4°, Oxonii, 1685.
H; M/C6 – 17; Tr/NQ.10.38.

308 BURGH, Thomas. A method to determine the areas of right-lined figures universally...4°, Dublin, 1724.
Not H; M (Tracts) J6 – 1; Tr/NQ.16.76⁵.

309 BURNET, Gilbert, *Bp of Salisbury*. A discourse of the pastoral care. 8°, London, 1692.
Not H; M/H2 – 21; ?

310 — An exposition of the Church Catechism, for the use of the Diocese of Sarum. 8°, London, 1710.
Not H; M/H3 – 37; ?

311 — An exposition of the Thirty-nine Articles of the Church of England. 2nd ed. corrected. F°, London, 1700.
H; M/D8 – 14; ?

312 BURNET, Thomas. Archæologiæ philosophicæ: sive Doctrina antiqua de rerum originibus. Libri II. 4°, Londini, 1692.
H; M/C1 – 30; Tr/NQ.8.56 [p. 155 turned down, pp. 158, 356 up and many other signs of dog-earing].

313 — De fide et officiis Christianorum. 4°, Londini, 1722.
H; not M; ?

314 — De statu mortuorum et resurgentium liber. Accessit Epistola circa Libellum de archæologiis philosophicis. 4°, Londini, 1720.
H; M/B6 – 26; Tr/NQ.16.150c ['Is. Newton Ex dono Dⁿⁱˢ Ri. Meade' in Newon's hand on fly-leaf; a specially bound presentation copy from a privately printed limited edition].

315 — Telluris theoria sacra: orbis nostri originem & mutationes generales ...2 vols. 4°, Londini, 1681–9.
H; M/C1 – 28 & 29; Tr/NQ.16.150a & b ['Isaac Newton Ex dono Authoris' in Newton's hand on fly-leaf of vol. 1; a few signs of dog-earing in both vols.].

316 — The theory of the earth...3rd ed. review'd by the author. (2 pts.) F°, London, 1697.
Not H; M/C2 – 9; ?

BUSBY, Richard. Græcæ grammatices rudimenta, in usum Scholæ Westmonasteriensis. [By R. Busby.] 1663. *See* 680.

317 — Grammatica Busbeiana auctior & emendatior, i.e. Rudimentum grammaticæ Græco-Latinæ metricum. In usum Scholæ Regiæ Westmonasteriensis...8°, Londini, 1722.
H; M/A4 – 6; Tr/NQ.9.27 [a few signs of dog-earing].

318 BUSSY-RABUTIN, Roger, *comte de.* Histoire en abrégé de Louis le Grand, quatorzième du nom, roy de France. 12°, Amsterdam, 1700.
H; M/G2 – 33; ?
BUTLER, Lilly. Sermons. *See* 1493.

319 BUTLER, Samuel. Hudibras. Adorn'd with cuts. 12°, London, 1716.
H; M/G9 – 42; ?

320 — Posthumous works in prose and verse...12°, London, 1715.
Not H; M/G9 – 41; ?

321 BUXTORF, Johann. Epitome grammaticæ Hebrææ...Ed. 5ᵃ. 8°, Basileæ, 1629.
Not H; M/F1 – 3; ?

322 — Lexicon Hebraicum et Chaldaicum complectens...Ed. 3ᵃ ab auctore recognita. 8°, Basileæ, 1621.
H; M/F2 – 27; Keynes Collection, King's College, Cambridge ['Isaac Newton' on fly-leaf and on end-paper; pp. 45, 237, 636 turned up, pp. 164, 593 down]; bought by J. M. Keynes from Thorp of Guildford, March 1921, for £3. 3s. 0d.

323 — Synagoga Iudaica...Addita est...Iudæi cum Christiano disputatio de Messia nostro...Ed. 3ᵃ....12°, Hanoviæ, 1622.
Not H; M/G2 – 8; ?

324 BYTHNER, Victorinus. Lingua eruditorum sive Methodica institutio linguæ sanctæ...Ed. novissima...12°, Londini, 1650.
H; M/A1 – 19; ?

325 CAESAR, Gaius Julius. C. Iulii Cæsaris rerum ab se gestarum commentarii, hac postrema editione...repurgati...16°, Lugduni, 1614.
Not H; M/G1 – 34; ?

326 — C. Iulii Cæsaris quæ exstant, cum selectis variorum commentariis, quorum plerique novi, opera et studio A. Montani...8°, Amstelodami, 1670.
H; M/F4 – 5; ?

327 — C. Julii Cæsaris quæ extant...Accesserunt annotationes S. Clarke ...F°, Londini, 1712.
H; M/B5 – 9; Tr/NQ.18.31.

328 — C. Julii Cæsaris quæ extant...Accesserunt annotationes S. Clarke ...8°, Londini, 1720.
H; M/F4 – 4; ?

329 — C. Julius Cæsar's Commentaries of his wars in Gaul, and civil war with Pompey...With the author's life. Adorn'd with sculptures from the designs of the famous Palladio. Made English...by M. Bladen. 2nd ed. improv'd...8°, London, 1706.
H; M/J6 – 17; Tr/NQ.8.136.

330 — The eyght bookes of Caius Iulius Cæsar...translated oute of Latin into English by A. Goldinge. 8°, London, 1565.
Not H; M/G9 – 18; ?

331 CAESIUS, Bernardus. Mineralogia, sive Naturalis philosophiæ thesauri...F°, Lugduni, 1636.
H; M/A8 – 10; Tr/NQ.18.4 [pp. 66, 201 turned down, p. 189 up].

332 CALAMY, Benjamin. Sermons preached upon several occasions... 2nd ed., corrected. 8°, London, 1690.
Not H; M/H7 – 35; ?

333 CALEPINUS, Ambrosius. Dictionarium (octolingue), quanta maxima fide ac diligentia accurate emendatum...Ed. novissima. F°, Lugduni, 1647.
H; M/B3 – 3; ?

334 CALLIMACHUS. Hymni, epigrammata, et fragmenta ex recensione T. J. G. F. Graevii...Accedunt...commentarius et annotationes E. Spanhemii...2 vols. 8°, Ultrajecti, 1697.
H; M/F4 – 1 & 2; ?

335 CALVIN, Jean. Institutio Christianæ religionis, in libros quatuor nunc primùm digesta...8°, [Geneva] 1561.
Not H; M/D1 – 14; Tr/NQ.16.188 ['Isaac Newton Trin. Coll. Cant 1661' on fly-leaf, see plate 6].

336 CAMBRIDGE. *University*. Mœstissimæ ac lætissimæ Academiæ Cantabrigiensis carmina funebria & triumphalia. Illis...Reginam Annam repentina morte abreptam deflet...F°, Cantabrigiæ, 1714.
H; M/B3 – 9; ?
— *Trinity College*. A true and impartial account of the present differences between the Master (Richard Bentley) and Fellows of Trinity College in Cambridge, consider'd...1711. *See* 1643.

337 CAMDEN, William. Britannia, sive Florentissimorum regnorum, Angliæ, Scotiæ, Hiberniæ, et insularum adiacentium ex intima antiquitate chorographica descriptio. 8°, Londini, 1586.
H; M/A2 – 6; Tr/NQ.16.145 ['Is. Newton. pret. 1ˢ. 6ᵈ' on fly-leaf].
— Græcæ grammatices rudimenta in usum Scholæ Westmonasteriensis. [By W. Camden.] 1682. *See* 681.

338 — The history of...Princess Elizabeth, late Queen of England... 4th ed., revised...F°, London, 1688.
H; M/J9 – 7; ?

339 CAMUS, Jean Pierre. Homelies sur la Passion de Nostre Seigneur... 12°, Paris, 1616.
Not H; M/G2 – 27; Tr/NQ.9.167.

340 CANEPARIUS, Petrus Maria. De atramentis cujuscunque generis. . . In sex descriptiones digestum. 4°, Londini, 1660.
H; M/C1 – 21; Tr/NQ.7.56 [pp. 193, 517 turned down and many other signs of dog-earing].

341 CAPPELLUS, Ludovicus. Historia Apostolica illustrata, ex Actis Apostolorum et Epistolis Paulinis. . .4°, Genevæ, 1634.
Not H; M/F6 – 33; ?

342 —— [Another ed.] 4°, Salmurii, 1682.
H; M/E6 – 23; ?

343 CARELLIS, Vincentius de. De auri essentia, et eius facultate in medendis, ac sanandis morbis compendium. . .8°, Venetiis, 1646.
Not H; M/E3 – 12; Tr/NQ.16.127.

344 CARION, Johannes. Chronicon Carionis, expositum et auctum, multis & veteribus, & recentibus historiis. . .A P. Melancthone, & C. Pevcero. Postrema ed.. . .8°, Genevæ, (1625).
Not H; M/E2 – 20; Tr/NQ.8.124 [pp. 205, 549, 578, 590, 714, Index Tttii, v turned up, pp. 323, 388, 396, 410, 552, 553, 711, 822, 851 down, and a few other signs of dog-earing].
CAROLUS, a Sancto Paulo. See VIALART, Charles, Bp of Avranches.

345 CARR, William. The travellour's guide, and historian's faithful companion. 12°, London, 1697.
Not H; M/G9 – 47; ?
CARTER, Samuel. Legal provisions for the poor. . .By S. C[arter]. 4th ed.. . .1718. See 933.

346 CARYOPHILUS, Blasius. Dissertationum miscellanearum pars prima . . .4°, Romæ, 1718.
H; M/F6 – 15; Tr/NQ.10.59 [a few signs of dog-earing].

347 CASALI, Giovanni Baptista. De urbis ac Romani olim Imperii splendore opus eruditionibus, historiis, ac animadversionibus. . . illustratum. . .F°, Romæ, 1650.
H; M/D8 – 11; Tr/NQ.11.15 ['Is. Newton pret 11ˢ.' on fly-leaf; p. 98 turned down and a few other signs of dog-earing].

348 CASAUBON, Isaac. Animadversionum in Athenæi Dipnosophistas libri xv. . .F°, Lugduni, 1600.
H; M/C3 – 8; ?
CASIMIR, poet. See SARBIEWSKI, Maciej Kazimierz.

349 CASSANDER, Georgius. De articulis religionis inter Catholicos et Protestantes controversis consultatio. . .8°, Lugduni, 1608.
Not H; M/A2 – 11; Tr/NQ.9.161¹ [bound with 144 & 730].
CASSIANUS, Johannes. Ioannis Cassiani Eremitæ non prorsus dissimilis argumenti libri aliquot. . .(Greek & Latin.) (1575.) See 854.

350 CASSINI, Giovanni Domenico. Abregé des observations & des réflexions sur la comète qui a paru au mois de décembre 1680, & aux mois de janvier, février & mars de cette année 1681. 4°, Paris, 1681.

H; M/E6 – 34; Butler Library, Columbia University, New York [bound with 1368].

350a — Observations sur la comète qui a paru au mois de décembre 1680. et en janvier 1681. 4°, Paris, 1681.

H; M/E6 – 34; Butler Library, Columbia University, New York [with marginal notes in Halley's (?) hand; bound with 1368].

CASSINI, Jacques. De la grandeur et de la figure de la terre. [By J. Cassini.] 1720. *See* 1252.

351 CASTELL, Edmund. Lexicon heptaglotton, Hebraicum, Chaldaicum, Syriacum [etc.]. . .2 vols. F°, Londini, 1686.

H; M/C5 – 8 & 9; Thame/970; ?

352 CATALOGI librorum manuscriptorum Angliæ et Hiberniæ in unum collecti. . .[By E. Bernard.] F°, Oxoniæ, 1697.

H; M/C4 – 14; ?

CATROU, François. Histoire générale de l'Empire du Mogol depuis sa fondation. Sur les mémoires portugais de M. Manouchi, Vénitien. Par F. Catrou. 2 vols. 1705 & 1 vol. ed. 1708. *See* 1025 & 1026.

353 CATULLUS, Gaius Valerius. Catullus, Tibullus, Propertius cum C. Galli (vel potius Maximiani) fragmentis. Serio castigati. 8°, Amstæledami, 1686.

Not H; M/G1 – 29; ?

354 CAUSA Dei asserta per justitiam ejus, cum cæteris ejus perfectionibus, cunctisque actionibus conciliatam. [By G. W. von Leibniz.] 8°, Amstælodami, 1710.

Not H; M/G10 – 36; Tr/NQ.8.82² [bound with 571; a few signs of dog-earing].

355 CAUSSIN, Nicolas. De symbolica Ægyptiorum sapientia. . .8°, Coloniæ Agrippinæ, 1631.

Not H; M/A2 – 17; ?

356 — Thesaurus Græcæ poeseos, ex omnibus Græcis poetis collectus. Libri II. 8°, Moguntiæ, 1614.

Not H; M/A1 – 13; ?

CAVE, William. Antiquitates Christianæ: or, The history of the life and death of the Holy Jesus. . .2 pts. (Pt 1 by J. Taylor, pt 2 by W. Cave.) 8th ed. 1684. *See* 1598.

357 — Primitive Christianity: or, The religion of the ancient Christians in the first ages of the Gospel. 6th ed. 3 pts. 8°, London, 1702.

Not H; M/H4 – 22; ?

358 CAWDRY, Daniel. The inconsistencie of the independent way with Scripture and it self; manifested in a threefold discourse...4°, London, 1651.
Not H; M/H2 – 37; ?

359 CAWLEY, William. The laws of Q. Elizabeth, K. James, and K. Charles the First. Concerning Jesuites, Seminary Priests, Recusants, &c....F°, London, 1680.
H; M/J9 – 5; ?

360 CAWOOD, Francis. Navigation compleated; being a new method never before attain'd to by any...4°, London, 1710.
H; M/F5 – 18; Tr/NQ.8.133.

361 CELSUS, Aulus Cornelius. De medicina libri VIII, ex recognitione J. A. van der Linden. Ed. 2ª. 16°, Lugduni Batavorum, 1665.
H; M/G1 – 5; Thame/979; ?

362 CENSORINUS. De die natali. H. Lindenbrogius recensuit, et notis illustravit...8°, Cantabrigiæ, 1695.
H; M/E3 – 10; ?

363 CERTAIN sermons or homilies appointed to be read in churches, in the time of Queen Elizabeth...and now...reprinted...F°, London, 1683.
Not H; M/G11 – 12; ?

364 CEVA, Giovanni. Replica in difesa delle sue dimostrazioni, e ragioni, per le quali non debbasi introdurre Reno in Pò, contro la risposta datasi dal Sig. Dottor E. Manfredi. 4°, Mantova, 1717.
Not H; M (Tracts) D6 – 5; Tr/NQ.10.28⁴.

365 CHAMBERLAYNE, John. Magnæ Britanniæ notitia: or, The present state of Great-Britain; with divers remarks upon the ancient state thereof. 25th ed. 2 pts. 8°, London, 1718.
H; M/J4 – 3; Tr/NQ.8.98.

— Oratio Dominica in diversas omnium fere gentium linguas versa... Editore J. Chamberlaynio. (2 pts.) 1715. *See* 984.

CHAPPUZEAU, Samuel. The history of jewels and of the principal riches of the East and West...[Transl. from the French of S. Chappuzeau.] 1671. *See* 777.

366 CHARDIN, *Sir* John. Travels into Persia and the East Indies...F°, London, 1686.
H; M/D8 – 17; Thame/965; ?

367 — Voyages en Perse, et autres lieux de l'Orient. 3 vols. 4°, Amsterdam, 1711.
H; M/B6 – 1 to 3; Thame/968; ?

368 CHEYNE, George. An essay of health and long life. 3rd ed. 8°, London, 1725.
H; M/J7 – 28; ?

369 — Fluxionum methodus inversa; sive Quantitatum fluentium leges
generaliores...4°, Londini, 1703.
H; M/E6 – 37; ?

370 — Philosophical principles of natural religion: containing the elements
of natural philosophy, and the proofs for natural religion, arising
from them. (3 pts.) 8°, London, 1705.
H; M/G10 – 14; Tr/NQ.8.65.

371 — Rudimentorum methodi fluxionum inversæ specimina: quæ re-
sponsionem continent ad animadversiones A. de Moivre in librum
G. de Cheynæi. [By G. Cheyne.] 4°, Londini, 1705.
Not H; M (Tracts) J6 – 1; Tr/NQ.16.76¹.

372 CHILD, *Sir* Josiah. A new discourse of trade...2nd ed. 8°, London,
1694.
H; M/J8 – 14; Tr/NQ.7.76 [a few signs of dog-earing in the
preface].

373 CHOMEL, Noel. Dictionnaire œconomique, contenant divers moyens
d'augmenter son bien et de conserver sa santé...F°, Lyon, 1709.
H; M/H9 – 1; ?

374 CHRISTIAN V, *King of Denmark*. Regis Christiani Quinti Jus Danicum, Latine redditum ab H. Weghorst. 4°, Hafniæ, 1698.
H; M/C1 – 10; Tr/NQ.16.56.

375 CHRISTIANITY no enthusiasm: or, The several kinds of inspirations
and revelations pretended to by the Quakers, tried, and found
destructive...[By T. Comber.] 8°, London, 1678.
Not H; M/H2 – 20; ?

376 CHRONICON Saxonicum. Ex MSS codicibus nunc primum
integrum edidit, ac Latinum fecit E. Gibson. 4°, Oxonii, 1692.
H; M/F6 – 26; Tr/NQ.10.37 [pp. 76, 129 turned down and several
other signs of dog-earing].

377 CHRYSOSTOM, *St* John, *Abp of Constantinople*. Expositio perpetua in
Novum Iesu Christi Testamentum Græcè ac Latinè...F°, [Heidel-
berg], 1603.
H; M/D8 – 13; Tr/NQ.17.16 ['Is. Newton pret 5ˢ. 6ᵈ.' on front
paste-down].
*6-line Latin note by Newton on Chrysostom (probably written 1677–8) on slip
of paper attached at end of book.*

378 CHYMICAL, medicinal, and chyrurgical addresses; made to S.
Hartlib Esquire...12°, London, 1655.
Not H; M/G9 – 34; ?

379 CICERO, Marcus Tullius. De officiis libri tres, & in illos S. Rachelii
commentarius philosophico-juridicus...8°, Amstelædami, 1686.
H; M/D1 – 28; Tr/NQ.16.138.

380 — Mar. Tullij Ciceronis Epistolæ familiares cum commento Hubertini Crescentinatis: & Martini Philetici super epistolis electis: & Georgij Merulæ Alexandrini. Fº, (Lugduni, 1505).
H; M/J9 – 27; Tr/NQ.8.4 [wants title-page].

381 — Opera omnia quæ exstant...emendata studio atcȝ industria J. Gulielmi & J. Gruteri...5 vols. in 2. Fº, Hamburgi, 1618–19.
H; M/C4 – 20 & 21; ?

382 — Opera omnia. 25 vols. [Probably a made-up set of 19 vols. from the edition of J. G. Graevius and 6 vols. from that of J. Davies.] 8º, [Amsterdam, etc., c. 1677–1720].
Not H; M/A5 – 1 to 25; ?

383 — Orationum volumen 1. Ex emendatione D. Lambini. (Opera omnia, 2.) 8º, Lutetiæ, 1572.
H; not M; ?

384 CLAGETT, William. A discourse concerning the operations of the Holy Spirit...2nd ed. 3 pts. 8º, London, 1680.
Not H; M/H7 – 32; ?

385 CLAPHAM, Jonathan. A full discovery and confutation of the wicked and damnable doctrines of the Quakers...4º, London, 1656.
Not H; M/H2 – 27; Tr/NQ.9.63³ [bound with 1004].

CLARKE, James. An essay, wherein a method is humbly propos'd for measuring equal time with the utmost exactness...[By J. Clarke.] 1714. See 579.

—The mercurial chronometer improv'd...[By J. Clarke.] 1715. See 1074.

386 CLARKE, John. An enquiry into the cause and origin of evil... Being the substance of eight sermons preached...in the year 1719, at the lecture founded by R. Boyle. 8º, London, 1720.
H; M/H3 – 11; ?

387 — An enquiry into the cause and origin of moral evil...Being the substance of eight sermons preached...in the year 1720, at the lecture founded by R. Boyle. 8º, London, 1721.
H; not M; ?

CLARKE, Samuel. A collection of papers, which passed between the late learned Mr. Leibnitz, and Dr. Clarke, in the years 1715 and 1716...(2 pts.) 1717. See 935.

388 — A demonstration of the Being and attributes of God...Being the substance of eight sermons...2nd ed., corrected. 8º, London, 1706.
H; M/H3 – 13; ?

— Observations on Dr. Waterland's second defense of his queries... [By S. Clarke.] 1724. See 1198.

— Recueil de diverses pièces...Par Mrs. Leibniz, Clarke, Newton, & autres autheurs célèbres. [Ed. by P. Des Maizeaux.] 2 vols. 1720. See 1379 & 1380.

389 — A reply to the objections of Robert Nelson...against Dr Clarke's
Scripture-doctrine of the Trinity...8°, London, 1714.
H; M/H3 – 17; ?

390 — The Scripture doctrine of the Trinity. In 3 parts...8°, London,
1712.
H; M/H3 – 16; ?

391 — — 2nd ed. 8°, London, 1719.
H; M/H3 – 16; ?

392 — XVII sermons on several occasions...8°, London, 1724.
H; M/H3 – 14; ?

393 — Three practical essays, on baptism, confirmation and repentance...
3rd ed. 12°, London, 1710.
H; M/H1 – 40; ?
— Tracts in divinity by Dr Clarke and others. 4 vols. 1706. *See* 1625.

394 CLAUBERG, Johann. De cognitione Dei et nostri, quatenus naturali
rationis lumine, secundum veram philosophiam, potest comparari,
exercitationes centum. Ed. novissima...8°, Harlingæ, 1685.
Not H; M/A3 – 30; Tr/NQ.9.162.

395 CLAUDER, Gabriel. Dissertatio de tinctura universali, vulgò Lapis
philosophorum dictâ...4°, Altenburgi, 1678.
Not H; M/B1 – 23; ?

396 CLAUDIANUS, Claudius. Cl. Claudiani quæ exstant: ex emendatione
N. Heinsii. 24°, Amstelodami, 1677.
H; M/G1 – 33; ?

397 CLAVEUS, Gasto, *called Dulco.* Apologia Chrysopoeiæ et Argyropoeiæ,
adversus T. Erastum...De triplici auri et argenti præparatione.
8°, Ursellis, 1602.
Not H; M/A2 – 15; Tr/NQ.16.152 [p. 207 turned down and a
few other signs of dog-earing].
— Le filet d'Ariadne...[By G. Claveus?] 1695. *See* 619.

398 CLEMENS, *Alexandrinus.* Opera Græce et Latine quæ exstant...
Accedunt diversæ lectiones & emendationes...a F. Sylburgio
collectæ...F°, Lutetiæ Parisiorum, 1641.
H; M/D7 – 10; ?

399 CLEMENS, *Romanus.* Ad Corinthios epistola prior. Ex laceris reliquiis
...Latinè vertit, & notis brevioribus illustravit P. Junius. 4°,
Oxonii, 1633.
H; M/F6 – 30 [bound with 118]; ?

400 — De constitutionibus Apostolicis libri viii...I. C. Bovio interprete.
Eiusdem scholia...8°, Parisiis, 1564.
H; not M; ?

401 CLERGYMAN'S (The) vade-mecum: containing the canonical codes of the Primitive and Universal Church...[By J. Johnson.] 3rd ed. 2 vols. 12°, London, 1723.
Not H; M/J8 – 30 & 31; ?

402 CLIPSHAM, Robert. The grand expedient for suppressing Popery examined. Or The project of exclusion proved to be contrary to reason and religion. 8°, London, 1685.
Not H; M/H2 – 33; Tr/NQ.10.104.

CLODOVEUS. Libelli seu decreta a Clodoveo et Childeberto, & Clothario...[1550?] See 955.

403 CLUVERIUS, Philippus. Introductionis in universam geographiam... libri vi...4°, Amstelædami, 1682.
Not H; M/C6 – 15; Sotheran **795** (1925) 7520; ?

404 COCKBURN, William. An account of the nature, causes, symptoms and cure of loosnesses. 2nd ed. 8°, London, [1706].
H; M/G10 – 26; ?

405 — Doctor Cockburn's Solution of his problem for determining the proper doses of purging, and vomiting medicins in every age of a man...4°, [London, 1705?].
Not H; M (Tracts) J6 – 15; Tr/NQ.8.52⁷.

406 — The practice of purging and vomiting medicines, according to Dr. Cockburn's Solution of his problem...4°, [London, 1705?].
Not H; M (Tracts) J6 – 15; Tr/NQ.8.52⁶.

407 — Sea diseases: or, A treatise of their nature, causes, and cure...2nd ed., corrected...8°, London, 1706.
H; M/J7 – 30; ?

COCKMAN, Thomas. Free-thinking rightly stated...[By T. Cockman.] 1713. See 636.

408 CODEX canonum Ecclesiæ Africanæ. C. Iustellus ex MSS. codicibus edidit, Græcam versionem adiunxit, & notis illustravit. 8°, Lutetiae Parisiorum, 1614.
H; M/A2 – 9; Tr/NQ.16.11.

COENDERS VAN HELPEN, Barent. L'escalier des sages, ou La philosophie des anciens...[By B. Coenders van Helpen.] 1689. See 570.

COLBATCH, John. An account of the Court of Portugal, under the reign of the present King Dom Pedro II....[By J. Colbatch.] 1700. See 5.

409 COLES, Elisha. A dictionary, English–Latin, and Latin–English... 4th ed., enlarged. 8°, London, 1699.
H; M/J7 – 18; Tr/NQ.9.47.

COLET, John. A short introduction of grammar...[By W. Lily and J. Colet.] (2 pts.) 1692. See 1510.

410 COLLECTANEA chymica: a collection of ten several treatises in chymistry, concerning the liquor Alkahest, the mercury of philosophers, and other curiosities...8°, London, 1684.
H; M/G9 – 2; ?

411 COLLECTION (A) of articles, injunctions, canons, orders, ordinances, & constitutions ecclesiastical, with other publick records of the Church cf England... [By A. Sparrow.] 4th impr. with additions ...4°, London, 1684.
Not H; M/C6 – 25; Tr/NQ.10.49.

412 — A collection of prophetical warnings of the Eternal Spirit, pronounc'd by the following persons, viz. Mary Aspinal, Mary Beer, aged 13 [etc.]. 8°, London, 1708.
Not H; M/H1 – 27; ?

413 — A collection of the several statutes, and parts of statutes, now in force, relating to high treason, and misprision of high treason. 12°, London, 1709.
H; M/J8 – 54; Tr/NQ.9.65^1 [bound with 625].

414 COLLIER, Arthur. Clavis universalis: or, A new inquiry after truth. Being a demonstration of the non-existence, or impossibility of an external world. 8°, London, 1713.
Not H; M (Tracts) J3 – 24; Tr/NQ.9.125^2.

415 COLLIER, Jeremy. Essays upon several moral subjects. In 2 parts. 5th ed. 8°, London, 1703.
Not H; M/J7 – 31; Tr/NQ.9.118.

416 — Essays upon several moral subjects. Parts 3 & 4. 8°, London, 1705–9.
Not H; M/J7 – 32 & 33; Tr/NQ.9.119 & 120.

417 — The great historical, geographical, genealogical and poetical dictionary...Collected...especially out of L. Morery...2nd ed. Vol. 1. F°, London, 1701.
Not H; M/C4 – 16; ?

COLLINS, Anthony. A discourse of the grounds and reasons of the Christian religion. [By A. Collins.] 1724. See 527.

COLLINS, John. Commercium epistolicum D. Johannis Collins, et aliorum de analysi promota: jussu Societatis Regiæ in lucem editum. 1712. See 422.

— Commercium epistolicum D. Johannis Collins, et aliorum, de analysi promota, jussu Societatis Regiæ in lucem editum: et jam...iterum impressum. 1722. See 423.

— Extrait du livre intitulé Commercium epistolicum Collinii & aliorum, de analysi promota...[c. 1715.] See 424.

418 — A plea for the bringing in of Irish cattel, and keeping out of fish caught by foreigners...4°, London, 1680.
Not H; M (Tracts) J3 – 41; Tr/NQ.8.111^{10}.

419 COLSON, Lancelot. Philosophia maturata: an exact piece of philosophy, containing the practick and operative part thereof in gaining
the philosophers stone...24º, London, 1668.
Not H; M/G9 – 53; Memorial Library, University of Wisconsin –
Madison [a few signs of dog-earing]; Thame/973; Duveen, p. 141.

COMBACHIUS, Ludovicus. Tractatus aliquot chemici singulares
summum philosophorum arcanum continentes...[Ed. by L.
Combachius.] 1647. *See* 1623.

COMBER, Thomas. Christianity no enthusiasm...[By T. Comber.]
1678. *See* 375.

420 — Friendly and seasonable advice to the Roman Catholics of England.
3rd ed....12º, London, 1677.
Not H; M/G9 – 13; ?

COMES, Natalis. *See* CONTI, Natale.

COMITIBUS, Ludovicus de. *See* CONTI, Luigi de'.

COMITIBUS, Natalis de. *See* CONTI, Natale.

421 COMMENTARIUS in Epistolam ad Hebræos. Cum indice rerum
locorumɋ Scripturæ. [By J. Schlichting.] 8º, Racoviæ, 1634.
Not H; M/F1 – 6; Tr/NQ.9.114² [bound with 458; pp. 54, 314
turned up, p. 318 down].

422 COMMERCIUM epistolicum D. Johannis Collins, et aliorum de
analysi promota: jussu Societatis Regiæ in lucem editum. 4º,
Londini, 1712.
Not H; M (Tracts) J6 – 1; Tr/NQ.16.76².

423 — Commercium epistolicum D. Johannis Collins, et aliorum, de
analysi promota, jussu Societatis Regiæ in lucem editum: et jam
unà cum ejusdem recensione præmissa...iterum impressum. 8º,
Londini, 1722.
H; M/E5 – 5; Tr/NQ.8.114.

424 — Extrait du livre intitulé Commercium epistolicum Collinii &
aliorum, de analysi promota; publié par ordre de la Société Royale,
à l'occasion de la dispute élevée entre Mr. Leibnitz & le Dr. Keil
...4º, [c. 1715].
Not H; M (Tracts) J3 – 24; Tr/NQ.9.125⁷.

425 COMPLETE (A) history of England: with the lives of all the Kings
and Queens thereof...3 vols. (Vols. 1 & 2 ed. by J. Hughes; vol. 3
ed. by W. Kennett.) Fº, London, 1706.
Not H; M/A8 – 1 to 3; ?

426 CONCIONES et orationes ex historicis Latinis excerptæ...[Ed. by
J. Veratius.] 12º, Amstelodami, 1686.
Not H; M/G1 – 2; ?

427 CONFESSION (The) of faith: together with the larger and lesser catechisms. Composed by the Reverend Assembly of Divines, then sitting at Westminster: presented to both Houses of Parliament. 3rd ed. 12°, London, 1688.
Not H; M/H1 – 43; Tr/NQ.9.66.

428 CONFUCIUS. Confucius Sinarum philosophus, sive Scientia Sinensis Latine exposita. Studia & opera P. Intorcetta [etc.]. (5 pts.) F°, Parisiis, (1686–)1687.
H; M/H9 – 16; Tr/NQ.11.16 [a few signs of dog-earing].

429 CONGREVE, William. The double-dealer, a comedy. Acted at the Theatre Royal, by Their Majesties servants. 4°, London, 1694.
Not H; M/J6 – 6; Tr/NQ.10.66¹ [bound with 430–32].

430 — Love for love: a comedy. Acted at the Theatre in Little Lincolns-Inn Fields, by His Majesty's servants. 3rd ed. 4°, London, 1697.
Not H; M/J6 – 6; Tr/NQ.10.66³ [bound with 429].

431 — The mourning bride, a tragedy. As it is acted at the Theatre in Lincoln's-Inn-Fields, by His Majesty's servants. 4°, London, 1697.
Not H; M/J6 – 6; Tr/NQ.10.66⁴ [bound with 429].

432 — The old batchelour, a comedy. As it is acted at the Theatre-Royal, by His Majesty's servants. 6th ed. corrected. 4°, London, 1697.
Not H; M/J6 – 6; Tr/NQ.10.66² [bound with 429].

433 — Poems. 8°, London, 1720. *Edition not identified.*
Not H; M/J8 – 26; ?

434 — The way of the world. A comedy. 8°, [London?], 1711.
Not H; M (Plays) H8 – 19; Tr/NQ.10.92³.

435 CONNOISSANCE des temps pour l'année MDCCXIV, MDCCXVI–MDCCXXII. 8 vols. 12°, Paris, 1713–21.
H; not M; ?

436 CONTI, Luigi de'. Clara fidelisqu. admonitoria disceptatio practicæ manualis experimento veraciter comprobata. De duobus artis, & naturæ miraculis: hoc est de liquore alchaest; nec non lapide philosophico...12°, Francofurti, 1664.
Not H; M/E1 – 23; Memorial Library, University of Wisconsin – Madison [many signs of dog-earing]: Duveen, p. 141.

437 — Discours philosophiques sur les deux merveilles de l'art et de la nature. Ou Traité de la liqueur de l'Alchaest...2ᵉ éd., reveüe & corrigée. 12°, Paris, 1678.
H; M/G2 – 44; Tr/NQ.16.110 [a few signs of dog-earing].

438 — Metallorum ac metallicorum naturæ operum ex orthophysicis fundamentis recens elucidatio...8°, Coloniæ Agrippinæ, 1665.
H; M/A3 – 6; Tr/NQ.16.87.

439 CONTI, Natale. Mythologiæ, sive explicationis fabularum libri x. . . Nuper ab ipso autore recogniti & locupletati. 8°, Coloniæ Allobrogum, 1612.

Not H; M/F3 – 22; ?

440 COOKE, Edward. A voyage to the South Sea, and round the world, perform'd in the years 1708, 1709, 1710, and 1711. . . 8°, London, 1712.

H; M/J4 – 37; Tr/NQ. 9. 6 [a few signs of dog-earing].

441 COOPER, Thomas. Thesaurus linguæ Romanæ & Britannicæ. . . F°, Londini, 1578.

H; M/A9 – 11; ?

442 COOPER, William. A catalogue of chymicall books. In 3 parts. . . 8°, London, 1675.

Not H; M/E2 – 41; Babson 403.

CORDONNIER, Hyacinthe. L'Europe savante. [By H. Cordonnier, etc.] 12 vols. in 10. 1718–20. *See* 587.

443 CORNARO, Luigi. Sure and certain methods of attaining a long and healthful life. . . made English by W. Jones. . . 2nd ed. 12°, London, 1704.

H; M/E3 – 35; ?

444 COSIN, John, *Bp of Durham*. Historia transubstantiationis papalis. Cui præmittitur, atque opponitur, tùm S. Scripturæ, tùm Veterum Patrum, & Reformatarum Ecclesiarum doctrina Catholica. . . 8°, Londini, 1675.

H; M/D1 – 19; Dr Bertram A. Lowy, New York [long note in Barrow's hand in margin of p. 13; p. 80 turned down]; see p. 11 n. 2 above.

445 COSMOPOLITE, ou Nouvelle lumière chimique, pour servir d'éclaircissement aux trois principes de la nature. . . [By M. Sendivogius.] Dernière éd., revue et augmentée. . . 2 vols. 12°, Paris, 1691.

H; M/G6 – 9 & 10; ? ; *see also* 1192 & 1485.

COTELIER, Jean Baptiste. SS. Patrum qui temporibus Apostolicis floruerunt. . . opera edita et inedita. . . J. B. Cotelerius ex MSS. codicibus eruit, ac correxit, versionibusque & notis illustravit. . . Nova ed. . . . Recensuit. . . J. Clericus. 2 vols. 1700. *See* 68.

446 COTES, Roger. Æstimatio errorum in mixta mathesi, per variationes partium trianguli plani et sphærici. 4°, [Cambridge, 1722].

Not H; M (Tracts) D6 – 4; Tr/NQ. 16. 197³ [a possible pre-publication copy of pt 2, pp. 1–71, of no. 447].

447 — Harmonia mensurarum, sive Analysis & synthesis per rationum & angulorum mensuras promotæ: accedunt alia opuscula mathematica. Edidit et auxit R. Smith. (2 pts.) 4°, Cantabrigiæ, 1722.

H; M/E6 – 20; Tr/NQ. 10. 6.

COURCELLES, Étienne de. *See* CURCELLAEUS, Stephanus.

448 COVEL, John. Some account of the present Greek Church...F°, Cambridge, 1722.
H; M/C4 – 2; ?

449 COWELL, John. The interpreter; or Book containing the signification of words. Wherein is set forth the true meaning of...such words and terms as are mentioned in the law-writers...F°, London, 1658.
Not H; M/J9 – 15; Thame/971; ?

450 COWLEY, Abraham. Works...(With an account of the life and writings of A. Cowley, by T. Sprat.) 6th ed. 4°, London, 1680.
Not H; M/C2 – 8; ?

451 COWPER, William, Myotomia reformata: or An anatomical treatise on the muscles of the human body. Illustrated with figures after the life. F°, London, 1724.
H; M/B5 – 8; Tr/NQ.17.3.

452 COZZANDUS, Leonardus. De magisterio antiquorum philosophorum libri vi. 8°, Coloniæ, 1684.
H; M/E1 – 11; ?

453 CRADOCK, Samuel. The Apostolical history: containing the Acts, labours, travels...of the Holy Apostles...F°, London, 1672.
Not H; M/J9 – 14; ?

454 CRAIGE, John. De calculo fluentium libri ii. Quibus subjunguntur libri ii de optica analytica. 4°, Londini, 1718.
H; M/E6 – 1; Tr/NQ.10.60.

455 — Methodus figurarum lineis rectis & curvis comprehensarum quadraturas determinandi. 4°, Londini, 1685.
Not H; M (Tracts) D5 – 24; Tr/NQ.16.79[5].

456 — Theologiæ Christianæ principia mathematica. 4°, Londini, 1699.
Not H; M (Tracts) D6 – 4; Tr/NQ.16.197[5].

457 — Tractatus mathematicus de figurarum curvilinearum quadraturis et locis geometricis. 4°, Londini, 1693.
Not H; M (Tracts) D5 – 25; Tr/NQ.8.51[1].

CRASSELLAME, Marc' Antonio. La lumière sortant par soy même des tenebres... [Variously attributed to M.-A. Crassellame and O. Tachenius.] 1687. See 1003.

CRAVEN, Joseph. Two letters to the Reverend Dr. Bentley...concerning his intended edition of the Greek Testament...[By J. Craven?] 1717. See 1650.

458 CRELLIUS, Johannes. Commentarius in Epistolas Pauli Apostoli ad Thessalonicenses. Ex prælectionibus I. Crellii Franci conscriptus à P. Morskovio. 8°, Racoviæ, 1636.
H; M/F1 – 6; Tr/NQ.9.114[1] [bound with 421; pp. 32, 38 turned down].

459 CRELLIUS, Samuel. Initium Evangelii S. Joannis Apostoli ex antiquitate ecclesiastica restitutum, indidemque nova ratione illustratum...per L. M. Artemonium [i.e. S. Crellius]. 8°, [Amsterdam], 1726.
H; M/A4 – 1; ?

460 CRESSENER, Drue. The judgments of God upon the Roman-Catholick Church...4°, London, 1689.
Not H; M/H2 – 4; ?

461 CROESIUS, Gerardus. Ὅμηρος ἑβραῖος, sive Historia Hebræorum ab Homero Hebraicis nominibus ac sententiis conscripta in Odyssea & Iliade...Vol. 1. 8°, Dordraci, 1704.
Not H; M/A2 – 1; Tr/NQ.9.18 [a few signs of dog-earing].

462 CROLLIUS, Oswaldus. Basilica chymica cum notis Hartmanni, et Praxis chymiatrica...4°, Francofurti, 1647.
Not H; M/A7 – 15; Tr/NQ.16.81[1] [bound with 743; a few signs of dog-earing].
CROMARTY, George MacKenzie, *1st Earl of.* Several proposals conducing to a farther Union of Britain...[By George MacKenzie, 1st Earl of Cromarty.] 1711. *See* 1500.

463 CROUSAZ, Jean Pierre de. Commentaire sur l'Analyse des infini-ment petits (de M. le Marquis de L'Hôpital). 4°, Paris, 1721.
H; M/E6 – 9; Tr/NQ.10.25.

464 — Discours sur le principe, la nature et la communication du mouve-ment. (Pièces qui ont remporté les deux prix de l'Académie Royale des Sciences, 1720.) 4°, Paris, 1721.
Not H; M (Tracts) D6 – 3; Tr/NQ.8.1[4].

465 — Reflexions sur l'utilité des mathématiques et sur la manière de les étudier. Avec un nouvel essai d'arithmétique demontrée. 8°, Amsterdam, 1715.
H; M/G2 – 34; Tr/NQ.9.52.

466 CUDWORTH, Ralph. A discourse concerning the true notion of the Lord's Supper...2nd ed. 8°, London, 1670.
Not H; M/H1 – 6; Tr/NQ.9.49 [a few signs of dog-earing].

467 CUMBERLAND, Richard, *Bp of Peterborough.* An essay towards the recovery of the Jewish measures & weights, comprehending their monies...8°, London, 1686.
H; M/J7 – 41; ? .

468 — Origines gentium antiquissimæ; or, Attempts for discovering the times of the first planting of nations. In several tracts. Publish'd... by S. Payne. 8°, London, 1724.
H; M/J4 – 6; Tr/NQ.9.75.

— Sanchoniatho's Phœnician history, transl. from the first book of
Eusebius De præparatione Evangelica...By R. Cumberland. 1720.
See 1441.

469 CURCELLAEUS, Stephanus. Opera theologica, quorum pars præcipua Institutio religionis Christianæ. F°, Amstelodami, 1675.
H; M/D 8 – 15; Tr/NQ.11.46 [pp. 41, 885 turned up, 79 down, and
a few other signs of dog-earing].

470 CURIOSITIES in chymistry: being new experiments and observations
concerning the principles of natural bodies. Written by a person
of honour, and published by his operator, H. G[regg]. [Sometimes
attributed to Robert Boyle.] 8°, London, 1691.
H; M/E2 – 18; Tr/NQ.16.143³ [bound with 254].

471 CURSE (The) of popery, and popish princes to the civil government,
and Protestant Church of England...8°, London, 1716.
Not H; M/H4 – 29; ?

472 CUSPINIANUS, Johannes. De Turcarum origine, religione, ac
immanissima eorum in Christianos tyrannide...12°, Lugduni
Batavorum, 1654.
H; M/G1 – 11; ?

473 CYDER. A poem. In 2 books. [By J. Philips.] 12°, London, 1720.
Not H; M/J8 – 25; Tr/NQ.8.109³ [bound with 1299 & 1501].

474 CYPRIAN, *St, Bp of Carthage.* Opera. N. Rigaltii observationibus ad
veterum exemplarium fidem recognita et illustrata. F°, Lutetiæ
Parisiorum, 1648.
H; not M; ?

475 — Opera recognita & illustrata per Joannem [Fell] Oxoniensem
Episcopum. Accedunt Annales Cyprianici...brevis historia chrono-
logice delineata per Joannem [Pearson] Cestriensem. F°, Oxonii,
1682.
H; M/E7 – 14; Tr/NQ.18.1.
— De idolorum vanitate...1678. *See* 1083.

476 CYRIL, *St, Abp of Jerusalem.* Opera, quæ reperiuntur. Ex variis
bibliothecis, præcipue Vaticana, Græcè nunc primùm in lucem
edita, cum Latina interpret. I. Grodecii. Plerisque in locis aucta &
emendata, studio & opera I. Prevotii. (*With* Synesii Episcopi
Cyrenes opera quæ extant omnia. Nunc denuò Græcè et Latinè
coniunctim edita. Interprete D. Petavio...) (2 pts.) F°, Lutetiæ
Parisiorum, 1631.
                                                    lb s d
H; M/E7 – 15; Tr/NQ.11.48 ['pret 1. 0. 0.' in Newton's hand on fly-
leaf; pt 2, p. 220 turned down and a few other signs of dog-
earing].

477 DADINUS ALTESERRA, Antonius. Notæ et observationes in Anastasium De vitis Romanorum Pontificum. 4°, Parisiis, 1680.
H; M/C6 – 14; ?

478 DAILLÉ, Jean. Adversus Latinorum de cultus religiosi objecto traditionem, disputatio...4°, Genevæ, 1664.
H; M/A7 – 2; ?

479 — De cultibus religiosis Latinorum libri ix...4°, Genevæ, 1671.
H; M/A7 – 1; ?

480 — De imaginibus libri iv. 8°, Lugd. Batavor., 1642.
Not H; M/F1 – 15; Tr/NQ.10.73.¦

481 — De jejuniis et quadragesima liber. 8°, Daventriæ, 1654.
H; M/A2 – 20; ?

482 — A treatise concerning the right use of the Fathers, in the decision of the controversies that are at this day in religion...[Transl.] (2 pts.) 4°, London, 1675.
Not H; M/H2 – 11; Tr/NQ.9.23 [several signs of dog-earing].

483 DALE, Antonius van. Dissertationes de origine ac progressu idolo-latriæ et superstitionum: de vera ac falsa prophetia; uti et de divinationibus idololatricis Judæorum. 4°, Amstelodami, 1696.
H; M/J6 – 12; Tr/NQ.8.45 [pp. 81, 213 turned down, p. 210 up and a few other signs of dog-earing].

484 DALE, Thomas. Dissertatio medico-botanica inauguralis de pareira brava, et serapia off....4°, Lugduni Batavorum, 1723.
Not H; M (Orationes) C6 – 19; Tr/NQ.10.35[7].
DALLAEUS, Joannes. See DAILLÉ, Jean.

485 DAMASUS I, Pope. Opera quæ extant et vita ex codicibus MSS. cum notis M. M. Sarazanii. 8°, Parisiis, 1672.
H; M/D1 – 20; Sjögren Library, Royal Swedish Academy of Engineering Sciences, Stockholm [inscribed 'Is. Newton Donum R[ndi] amici D. Moor S.T.D.' (presumably Henry More) in Newton's hand on fly-leaf].
DAMETO, Juan Bautista. The ancient and modern history of the Balearick Islands...Transl. from the original Spanish [of J. B. Dameto]. 1716. See 43.

486 DAMPIER, William. A voyage to New Holland, &c. In the year, 1699 ...Vol. 3. 8°, London, 1703.
H; not M; ?

487 DARY, Michael. Dary's miscellanies: being, for the most part, a brief collection of mathematical theorems, from divers authors...8°, London, 1669.
Not H; M/E1 – 16; ?

488 — — [Another copy.]

Not H; M (Tracts) H8 – 20; Tr/NQ.10.143[1]; Newton acknowledged receipt of copy from the author, 18 February 1669/70 (*Correspondence*, I, 27).

Mathematical calculations and notes by Newton in margin of pp. 4, 6, 19, 36.

489 — Interest epitomized, both compound and simple...4°, London, 1677.

Not H; M/E5 – 41; Tr/NQ.16.62 [a note added at end of Preface states that the printed errata corrections have been written in the text 'by the Author'].

490 DAVENANT, Charles. Discourses on the publick revenues and on the trade of England...2 vols. 8°, London, 1698.

H; M/J4 – 28 & 29; ?

DAVENANT, *Sir* William. The Tempest...A comedy. First written by Mr. William Shakespear, & since altered by Sr. William Davenant, and Mr. John Dryden. 1710. *See* 1505.

491 DAVIDSON, William. Philosophia pyrotechnica, seu Cursus chymiatricus nobilissima illa & exoptatissima medicinæ parte pyrotechnica instructus...2 vols. in 1. 8°, Parisiis, 1640.

Not H; M/F3 – 30; ?

492 DAVIES, Myles. Athenæ Britannicæ: or, A critical history of the Oxford and Cambrige [*sic*] writers and writings...3 vols. 8°, London, 1716.

H; M/J1 – 53 to 55; Tr/NQ.8.120 to 122.

493 DE alchimia opuscula complura veterum philosophorum, quorum catalogum sequens pagella indicabit. (Pt 1.) 4°, (Francoforti, 1550).

Not H; M/C1 – 11; Sir Geoffrey Keynes, Brinkley [many signs of dog-earing].

494 — De finibus virtutis Christianæ. The ends of Christian religion...By R. S[harrock]. LL.D. 4°, Oxford, 1673.

Not H; M/F5 – 4; Tr/NQ.9.74 [a few signs of dog-earing].

495 — De Iesu Christi Filii Dei natura sive essentia...disputatio, adversus A. Volanum...Secundò edita. [*Preface signed* F. S., i.e. F. Socinus.] 8°, Racoviæ, 1627.

H; M/A3 – 14; ?

496 — De Unigeniti Filii Dei existentia, inter Erasmum Iohannis, & Faustum Socinum disputatio...[By E. Johannis.] 8°, 1595.

Not H; M/A3 – 12; NQ.9.54[2] [bound with 1385].

497 DEE, Arthur. Fasciculus chemicus, abstrusæ Hermeticæ scientiæ, ingressum, progressum, coronidem, verbis apertissimis explicans...12°, Parisiis, 1631.

Not H; M/E1 – 20; Sotheran **800** (1926) 10608; ?

DEFOE, Daniel. An essay on the South-Sea trade...[By D. Defoe.] 1712. *See* 576.

— Religious courtship: being historical discourses, on the necessity of marrying religious husbands and wives only...[By D. Defoe.] 1722. *See* 1390.

498 DELLON, Charles. Nouvelle relation d'un voyage fait aux Indes Orientales...12°, Amsterdam, 1699.
H; M/G7 – 37; Tr/NQ.9.169.

499 DEMOSTHENES. Demosthenis orationes selectæ Gr. Lat. 8°, 1660.
Edition not identified.
Not H; M/E1 – 29; ?

500 DERHAM, William. Astro-theology: or, A demonstration of the being and attributes of God, from a survey of the heavens. 8°, London, 1715.
H (2 copies); M (1 only) J7 – 3; ?

501 — Physico-theology: or, A demonstration of the being and attributes of God, from his works of Creation...8°, London, 1713.
H; M/J7 – 4; Lathrop C. Harper, New York, **214** (1974) 239.

502 — Dimostrazione della essenza, ed attributi d'Iddio dall'opere della sua creazione...tradotta dall'idiome inglese. 4°, Firenze, 1720.
H; M/F6 – 24; ?

503 DESAGULIERS, Henri. La science des nombres par rapport au commerce en général, & particulièrement à celui qui se pratique en Hollande...2 vols. 8°, Amsterdam, 1701.
H; M/G2 – 30 & 31; Tr/NQ.10.85 & 86 [vol. 1 has p. 252 turned up, vol. 2, p. 219 up, and a few other signs of dog-earing in both vols.; see p. 25 above and plate 5].

504 — Traité général de la réduction des changes et monnoyes des principales places de l'Europe...8°, Amsterdam, 1701.
H; M/F6 – 17; ?

505 DESAGULIERS, Jean Théophile. Physico-mechanical lectures. Or, An account of what is explain'd and demonstrated in the course of mechanical and experimental philosophy...8°, London, 1717.
H; M/E5 – 25; Tr/NQ.16.91[1] [bound with 1701].

506 DESCARTES, René. Geometria, à Renato Des Cartes anno 1637 Gallicè edita; nunc autem cum notis F. de Beaune, in linguam Latinam versa, & commentariis illustrata, operâ atque studio F. à Schooten. 4°, Lugduni Batavorum, 1649.
H (one of the 6 'Books that has Notes of Sir Is. Newtons'); Portsmouth (1888); ULC/Adv.d.39.1 [inscribed 'Sum e libris Gulielmi Sherlock Petrensis' (Sherlock was at Peterhouse, 1657–61) from whom Newton presumably acquired the book; its brief MS annotations were mistakenly attributed to Newton by compilers of Huggins List and Portsmouth catalogue].

507 — Geometria, à Renato Des Cartes anno 1637 Gallicè edita; postea autem unà cum notis F. de Beaune...in Latinam linguam versa, & commentariis illustrata, operâ atque studio F. à Schooten... (Ed. 2ᵃ.) 2 pts. 4°, Amstelædami, 1659(–1661).
Not H; not M; Tr/NQ.16.203 [with bookplate of Robert Smith, Master of Trinity, who bequeathed it in 1768].
Marginal annotations by Newton on pp. 7, 11, 12, 17, 24, 25, 38, 65, 67, 80, 81, 84, 85, 97 relating to his MS (of c. 1680) on 'Errores Cartesij Geometriæ' (ULC Add.3961 (4), 23r–24r, see 'Math. papers', IV, 336–44, VII, 194 n. 46, see also pp. 14–15 above; a few signs of dog-earing.

508 — Meditationes de prima philosophia, in quibus Dei existentia, & animæ humanæ à corpore distinctio, demonstrantur. His adjungitur Tractatus de initiis primæ philosophiæ juxta fundamenta clarissimi Cartesii, tradita in ipsius meditationibus...Authore L. Velthusio. 8°, Londini, 1664.
Not H; M/E2 – 25; Tr/NQ.9.115 [p. 74 turned down, p. 209 formerly turned].

509 — Principia philosophiæ. Nunc demum hac editione diligenter recognita, & mendis expurgata. 4°, Amstelodami, 1656.
Not H; M/A6 – 19; Tr/NQ.9.116 [a few signs of dog-earing].

DES COMTES, Louis. *See* CONTI (Luigi de').

510 DESCRIPTION historique du Royaume de Macaçar. Divisée en 3 livres, augmentée de diverses pièces curieuses. [By N. Gervaise.] 8°, Ratisbonne, 1700.
H; M/G7 – 38; Tr/NQ.10.79.

DES MAIZEAUX, Philippe. An historical and critical account of the life and writings of...John Hales, Fellow of Eton College...[By P. Des Maizeaux.] 1719. *See* 732.

— Recueil de diverses pièces...Par Mrs. Leibniz, Clarke, Newton, & autres autheurs célèbres. [Ed. by P. Des Maizeaux]. 2 vols. 1720. *See* 1379 & 1380.

511 DEUX traitez nouveaux sur la philosophie naturelle. Contenant Le tombeau de Semiramis nouvellement ouvert aux sages, et La refutation de l'anonyme Pantaleon, soy disant Disciple d'Hermés. 8°, Paris, 1689.
Not H; M/G6 – 22; Tr/NQ.16.30; Ekins 1–6.

512 DIALOGUE (A) betwixt two Protestants, in answer to a Popish Catechism, called, A short Catechism against all Sectaries...[By J. Rawlet.] 2 pts. 8°, London, 1685.
Not H; M/H7 – 29; Tr/NQ.7.46.

513 DICKINSON, Edmund. Epistola ad Theodorum Mundanum philosophum adeptum. De quintessentia philosophorum et de vera physiologia...8°, Oxoniæ, 1686.
H; M/E5 – 20; Tr/NQ.16.89.
Notes by Newton consisting of references to other alchemical works in margins of pp. 70, 180, 187; many signs of dog-earing.

514 DICTIONARY, *Dutch–Latin*. Biglotton sive Dictionarium Teuto-Latinum novum. 8°, [Amsterdam, *c*. 1680].
Not H; M/F2 – 11; Tr/NQ.10.98² [wants title-page; bound with 1038].

515 DICTYS, *Cretensis*. De Bello Trojano...Cum notis J. Merceri [etc.].
8°, Argentorati, 1691.
H; M/A2 – 5; ?,

516 DIGBY, *Sir* Kenelm. Two treatises: in the one of which, The nature of bodies; in the other, The nature of mans soul, is looked into...
(2 pts.) 4°, London, 1658.
Not H; M/F5 – 29; ?

517 DIODORUS, *Siculus*. Bibliothecæ historicæ libri xv, de xl...Studio & labore L. Rhodomani. (*Greek & Latin*.) F°, Hanoviæ, 1604.
H; M/C3 – 1; ?

518 — The historical library of Diodorus the Sicilian...Made English, by G. Booth. F°, London, 1700.
H; M/B2 – 7; ?

519 DIOGENES, *Laertius*. De vitis dogmatis et apophthegmatis...libri x. T. Aldobrandino interprete, cum annotationibus eiusdem. Quibus accesserunt annotationes H. Stephani, & utriusque Casauboni; cum uberrimis Ægidii Menagii observationibus. (*Greek & Latin*.) F°, Londini, 1664.
H; M/C3 – 17; Tr/NQ.17.21 [pt 2, p. 146 turned down and up, and a few other signs of dog-earing].

520 DION CASSIUS. Ex Dione Nicæo excerpta & in epitomes formã redactæ vitæ Pompeii Magni & Cæsarum, usque ad Alexandrum Mãmææ filium, per I. Xiphilinum. (2 pts.) (*Greek & Latin*.) (Varii historiæ Romanæ scriptores, 2.) 8°, [Geneva], 1568.
H; M/F1 – 4; Tr/NQ.16.1a.

521 — Historiæ Romanæ libri xlvi...I. Leunclavii studio tam aucti quam expoliti...F°, Hanoviæ, 1606.
H; M/B3 – 5; ?

522 DIONYSIUS, *Halicarnassensis*. Dionysii Halicarnassei scripta, quæ extant, omnia, et historica, et rhetorica...cum Latina versione... Opera & studio F. Sylburgii. 2 vols. in 1. F°, Lipsiæ, 1691.
H; M/B3 – 4; ?

523 DIONYSIUS, *Periegetes*. Orbis descriptio; cum commentariis Eustathii,
Archiepiscopi Thessalonicensis. (*Greek & Latin.*) 8°, Oxoniæ, 1710.
H; M/F4 – 28; Tr/NQ.9.36.

524 DIOPHANTUS, *Alexandrinus*. Arithmeticorum libri VI. Et de numeris
multangulis liber I. Nunc primùm Græcè & Latinè editi, atque
absolutissimis commentariis illustrati. Auctore C. G. Bacheto. F°,
Lutetiæ Parisiorum, 1621.
H; M/C3 – 16; Tr/NQ.11.19 [a few signs of dog-earing].

525 DIRECTORY (A) for the publique worship of God, throughout the
three kingdoms of England, Scotland, and Ireland...4°, London,
1644.
Not H; M/H7 – 25; ?

526 DISCOURSE (A) of local motion. Undertaking to demonstrate the
laws of motion, and withall to prove, that of the seven rules delivered
by M. Des-Cartes on this subject, he hath mistaken six. By A. M.
[i.e. I. G. Pardies.] Englished out of French. 8°, London, 1670.
H; M/E2 – 49; Hale Observatories, Pasadena; Sotheran **804**
(1927) 3807; possibly the book for which Newton thanked Oldenburg, 16 March 1671/2 (*Correspondence*, I, 120).

527 — A discourse of the grounds and reasons of the Christian religion.
[By A. Collins.] 8°, London, 1724.
H; M/H3 – 29; ?

528 — Discourse on Divine Providence. 8°, 1700. *Not identified.*
Not H; M/H1 – 30; ?

529 — A discourse on the judicial authority belonging to the office of
Master of the Rolls in the High Court of Chancery...[By Philip
Yorke, Earl of Hardwicke.] 8°, London, 1727.
H; not M; ?

530 DITTON, Humphry. The new law of fluids: or, A discourse con-
cerning the ascent of liquors, in exact geometrical figures, between
two nearly contiguous surfaces...8°, London, 1714.
H; M/E4 – 11; Tr/NQ.9.158.
— A new method for discovering the longitude both at sea and land...
By W. Whiston and H. Ditton. 2nd ed....1715. *See* 1732.

531 DIVERS traitez de la philosophie naturelle. Sçavoir, La turbe des
philosophes...La parole delaissée de Bernard Trevisan. Les deux
traitez de Corneille Drebel Flaman. Avec Le très-ancien duel des
Chevaliers...12°, Paris, 1672.
Not H; M/G2 – 42; Tr/NQ.16.118 [p. 17 turned down and a
few other signs of dog-earing]; Ekins 2 – 6.

532 DIVINE (The) right of Church Government asserted, by sundry
ministers in London. 4°, London, 1646.
Not H; M/H2 – 30; ?

533 DODWELL, Henry. Dissertationes Cyprianicæ. 8°, Oxoniæ, 1684.
H; M/A4 – 4; ?

534 — The works of the learned Mr H. Dodwell abridg'd: with an account of his life by Mr Brokesby. 2nd ed. 8°, London, 1723.
H; M/H4 – 36; ?

535 DOMINIS, Marc'Antonio de, *Abp of Spalatro*. De radiis visus et lucis in vitris perspectivis et iride tractatus. Per I. Bartolum in lucem editus. 4°, Venetiis, 1611.
Not H; M (Tracts) D5 – 25; Tr/NQ.8.51⁵ [wants title-page].

536 DORN, Gerard. Lapis metaphysicus, aut philosophicus, qui universalis medicina vera fuit patrum antiquorum...8°, [Frankfurt?], 1570.
Not H; M/E3 – 21; Tr/NQ.16.137³ [bound with 85].

537 DOUGLAS, James. Bibliographiæ anatomicæ specimen: sive Catalogus omnium penè auctorum, qui ab Hippocrate ad Harveum rem anatomicam ex professo, vel obiter, scriptis illustrârunt... 8°, Londini, 1715.
H; M/A4 – 7; Tr/NQ.8.117.

538 DRURY, William. Dramatica poemata. Ed. 2ᵃ...12°, Duaci, 1628.
Not H; M/G1 – 14; ?

DRYDEN, John. The Tempest...A comedy. First written by Mr. William Shakespear, & since altered by Sr. William Davenant, and Mr. John Dryden. 1710. *See* 1505.

DU CHESNE, Joseph. Quercetanus redivivus, hoc est, Ars medica dogmatico-hermetica, ex scriptis J. Quercetani digesta opera J. Schröderi...Ed. 2ᵃ. 3 vols. in 1. 1679. *See* 1473.

539 — Recueil des plus curieux et rares secrets touchant la medecine metallique & minerale...8°, Paris, 1648.
H; M/G10 – 38; Tr/NQ.16.100; Ekins 5 – 0.

540 — Traicté de la matière, préparation et excellente vertu de la medecine balsamique des anciens philosophes...8°, Paris, 1626.
H; M/G8 – 8; Tr/NQ.16.132; Ekins 3 – 6.

541 DUCTOR historicus: or, A short system of universal history, and an introduction to the study of that science...Partly transl. from the French of M. de Vallemont, but chiefly composed anew by W. J. M. A. [i.e. T. Hearne]. 8°, London, 1698.
Not H; M/J4 – 31; Tr/NQ.9.62.

542 DUEZ, Nathanael. Dictionnaire françois–allemand–latin, et allemand–françois–latin. Reveu, corrigé, & augmenté...en cette 3ᵉ éd. 2 vols. (*Vol. 2 with title* Dictionarium Germanico–Gallico–Latinum. Teutsch, Frantzosisch, und Lateinisch Dictionarium. 3. Druck.) 8°, Genève, 1663.
H; M/G8 – 3 & 4; Tr/NQ.8.80 & 81.

543 DU HAMEL, Jean Baptiste. Regiæ Scientiarum Academiæ historia. Ed. 2ª...4º, Parisiis, 1701.
H; M/D6 – 6; ?
DULCO, Gasto. *See* CLAVEUS, Gasto, *called Dulco*.
DUMONT, Jean. Les soupirs de l'Europe &c. or, The groans of Europe at the prospect of the present posture of affairs...Made English from the original French [of J. Dumont]. 1713. *See* 1539.
DU MOULIN, Pierre. The history of the English & Scotch Presbytery ...Written in French, by an eminent divine [P. Du Moulin]... and now Englished. 1659. *See* 781.

544 DU PIN, Louis Ellies. Bibliothèque universelle des historiens...4º, Amsterdam, 1708.
H; M/D6 – 2; ?

545 — A compendious history of the Church...2nd ed. 4 vols. 12º, London, 1715–16.
H; M/H1 – 21 to 24; ?
DUPUY, Jacques. Catalogus Bibliothecæ Thuanæ a P. & I. Puteanis [i.e. P. & J. Dupuy]...2 vols. 1679. *See* 1613.
DUPUY, Pierre. Catalogus Bibliothecæ Thuanæ a P. & I. Puteanis [i.e. P. & J. Dupuy]...2 vols. 1679. *See* 1613.

546 ECCLESIASTICA historia, integram Ecclesiæ Christi ideam, quantum ad locum, propagationem, persecutionem, tranquillitatem... secundum singulas centurias, perspicuo ordine complectens...Per aliquot studiosos & pios viros in Urbe Magdeburgica...Centuriæ 1–13. 7 vols. Fº, Basileæ, (1560–74). $\overset{\text{lb}}{}\ \overset{\text{s}}{}$
H; M/G11 – 13 to 19; Tr/NQ.11.38 to 44 ['pret. 2. 15' in Newton's hand on fly-leaf of vol. 1; considerable dog-earing in the set with 28 columns still turned up or down, and several other signs].

547 ECHARD, Laurence. An exact description of Ireland...12º, London, 1691.
H; M/G9 – 43; ?

548 — A general ecclesiastical history from the Nativity of Our Blessed Saviour to the first establishment of Christianity by human laws...2nd ed. 2 vols. 8º, London, 1710.
H; M/H5 – 24 & 25; Tr/NQ.8.62 & 63 [vol. 2 has a few signs of dog-earing].

549 — A most compleat compendium of geography...12º, London, 1691.
Not H; M/G9 – 44; ?

550 EDWARDS, Thomas. Diocesan episcopacy proved from Holy Scripture: with a letter to E. Calamy...8º, London, 1705.
Not H; M/H2 – 17; ?

551 ELLIS, John. Articulorum xxxix Ecclesiæ Anglicanæ defensio...
Ed. 4ᵃ. 12º, Amstelodami, 1700.
Not H; M/J7 – 13; ?

552 ELMACINUS, Georgius. Historia Saracenica...a Muhammede...
usque ad initium imperij Atabacæi...Latinè reddita operâ ac
studio T. Erpenii. Accedit & R. Ximenez Historia Arabum...
(2 pts.) Fº, Lugduni Batavorum, 1625.
H; M/A9 – 10; Tr/NQ.17.35 [on fly-leaf in Newton's hand
'E libris Is. Newton 1680, Pret 9ˢ, valet 25ˢ'; on title-page, scored
through 'Trin: Coll. Cant. Aº Dñi 1668. Ex dono Mg̃ri Thomæ
Gale huius Collegij Socij', see p. 4 above; a few signs of dog-
earing].

553 EMENDATIONES in Menandri et Philemonis reliquias, ex nupera
editione J. Clerici...auctore Phileleuthero Lipsiensi [R. Bentley].
8º, Trajecti ad Rhenum, 1710.
H; M/A4 – 3; Tr/NQ.9.38.

554 ENARRATIO methodica Trium Gebri medicinarum, in quibus
continetur Lapidis philosophici vera confectio. Autore Anonymo
sub nomine Æyrenæi Philalethes [i.e. G. Starkey?]. 8º, [London],
1678.
H; M/E3 – 11; Tr/NQ.16.86.
Note in Newton's hand on title-page: 'Vide Borelli Bibliothecam Chemicam
p. 20' (*where there is a reference to this work*); p. 212 turned up and
several other signs of dog-earing.

555 ENCHIRIDION legum: a discourse concerning the beginnings,
nature, difference, progress and use, of laws in general...8º,
London, 1673.
Not H; M/G9 – 23; ?

556 ENGLAND'S black tribunall. Set forth in the triall of K. Charles I.
at the pretended Court of Justice at Westminster Hall, Jan. 20...
1648. 8º, London, 1660.
H; M/J8 – 16; ?

557 ENIEDINUS, Georgius. Explicationes locorum Veteris & Novi
Testamenti, ex quibus Trinitatis dogma stabiliri solet. 4º, [Gro-
ningen, 1670].
H; M/A6 – 14; Tr/NQ.8.23 ['Is. Newton pret. 6ˢ.' on fly-leaf].

558 ENQUIRY (An) after happiness. [By R. Lucas.] Vol. 1. [*No more
published.*] 8º, London, 1685.
Not H; M/H1 – 12; ?

559 — An enquiry into the constitution, discipline, unity & worship of the
Primitive Church...By an impartial hand [i.e. Peter King].
2 pts. 8º, London, 1692.
H; not M; ?

560 — — [Another ed.] 8°, London, 1712–13.
    H; M/H1 – 4; ?

561 EPICTETUS. Enchiridion Latinis versibus adumbratum per E. Ivie.
    Ed. 2ª...12°, Oxoniæ, 1723.
    Not H; M/F1 – 21; ?

562 — Epictetus et Cebes. Gr. Lat. 8°, 1620. *Edition not identified.*
    Not H; M/G1 – 47; ?

563 — Epicteti Enchiridion, Cebetis Thebani Tabula, Theophrasti
    Characteres ethici [etc.]...Cum versione Latina...12°, Oxonii,
    1680.
    Not H; M/E1 – 3; ?

564 — The life and philosophy of Epictetus with the Embleme of human
    life by Cebes. Rendred into English; by J. Davies. 8°, London,
    1670.
    H; M/G9 – 22; ?

565 EPIPHANIUS, *Bp of Constantia.* Opera omnia...D. Petavius...
    recensuit, Latine vertit et animadversionibus illustravit...Ed.
    nova...cui accessit Vita D. Petavii ab H. Valesio. (*Greek &*
    Latin.) 2 vols. F°, Coloniæ, 1682.
    H; M/E8 – 16 & 17; ?

566 EPISTOLA Archimedis ad Regem Gelonem, Albæ Græcæ reperta...
    1688. [By A. Pitcairne.] 8°, [London, *c.* 1710].
    Not H; M (Tracts) H8 – 13; Tr/NQ.16.99[1].

567 ERASMUS, Desiderius. Enarratio in Psalmum I. Beatus vir, &c. 12°,
    Lugduni Batavorum, 1644.
    Not H; M/G1 – 28; ?

568 — Erasmus on the New Testament. 2 vols. F°, London, 1579. *Edition*
    not identified.
    Not H; M/G11 – 20 & 21; ?

569 — Paraphrasis in Servatoris et Domini Nostri Jesu Christi Novum
    Testamentum...Studio & curâ H. Deichmanni. 2 vols. 4°,
    Hannoveræ, 1668.
    Not H; M/A6 – 15 & 16; Tr/NQ.7.51 & 52.

    ERCKER, Lazarus. Assays in v. books...transl. into English...
    1683. *See* 1291.

570 ESCALIER (L') des sages, ou La philosophie des anciens. Avec des
    belles figures...[By B. Coenders van Helpen.] F°, Groningue,
    1689.
    Not H; M/J9 – 8; ?

    ESPAGNET, Jean d'. La philosophie naturelle restablie en sa pureté
    ...[By J. d'Espagnet.] 1651. *See* 1311.

571 ESSAIS de Théodicée sur la bonté de Dieu, la liberté de l'homme et l'origine du mal. [By G. W. von Leibniz.] 2 vols. in 1. 8°, Amsterdam, 1710.

H; M/G10 – 36; Tr/NQ.8.82¹ [bound with 354 & 834; a few signs of dog-earing].

572 ESSAY d'analyse sur les jeux de hazard. [By P. Rémond de Monmort.] 4°, Paris, 1708.

H; not M; ?; sent to Newton as gift from author 16 Feb. 1708/9 (*Correspondence*, IV, 533–4).

573 —— 2ᵉ éd. revûe & augmentée de plusieurs lettres. 4°, Paris, 1713.

H; M/E6 – 6; Tr/NQ.10.26 [a few signs of dog-earing including marking of Newton's name in text of p. 396, see p. 27 above]; presentation inscription on fly-leaf 'Pour Monsieur Newton Par son tres humble serviteur Remond de Monmort'.

574 — An essay for discharging the debts of the nation, by equivalents: in a letter to The Right Honᵇˡᵉ Charles, Earl of Sunderland. And the South-Sea scheme consider'd; in a letter to the Right Honᵇˡᵉ Robert Walpole, Esq. 8°, London, 1720.

Not H; M (Tracts) J2 – 38; Tr/NQ.16.194³.

575 — An essay on the ancient and modern use...of physical necklaces for distempers in children...8th ed. 8°, London, 1719.

Not H; M/J7 – 37; Tr/NQ.10.102.

576 — An essay on the South-Sea trade...[By D. Defoe.] 8°, London, 1712.

Not H; M (Tracts) J2 – 38; Tr/NQ.16.194⁶.

577 — An essay on the usefulness of mathematical learning, in a letter from a gentleman in the City to his friend in Oxford. [By M. Strong.] 8°, Oxford, 1701.

Not H; M (Tracts) J3 – 24; Tr/NQ.9.125¹.

578 — An essay towards a new method to shew the longitude at sea; especially near the dangerous shores. [By E. Place.] 8°, London, 1714.

Not H; M (Tracts) J8 – 11; Tr/NQ.10.119².

579 — An essay, wherein a method is humbly propos'd for measuring equal time with the utmost exactness...[By J. Clarke.] 8°, London, 1714.

Not H; M (Tracts) H8 – 14; Tr/NQ.16.95¹.

580 EUCLID. Elementorum libri xv. Græcè & Latinè...8°, Parisiis, 1573.

H; M/E3 – 6; Tr/NQ.16.60.

581 — Elementorum libri xv. breviter demonstrati, operâ I. Barrow. 12°, Cantabrigiæ, 1655.

H; M/E3 – 23; Tr/NQ.16.201¹ [bound with 585].

Copious annotations by Newton, esp. in Books II, V, VII, X; a few signs of dog-earing; Sotheby, Apr. 1926/597; Sotheran **804** (1927) 3794, **828** (1931) 3603, **843** (1935) 1496; *Math. papers*, I, 12 n. 28; see p. 51 above.

582 — Elementorum libri xv. breviter demonstrati, operâ I. Barrow. Et prioribus mendis typographicis nunc demum purgati. (2 pts.) 8°, Londini, 1678.
H; M/E1 – 26; ?

583 — Elementorum libri xv. breviter demonstrati. Opera I. Barrow. 12°, Londini, 1711.
H; M/E3 – 22; Tr/NQ.10.78.

584 — Elements of geometry: the first VI. books...By T. Rudd. 4°, London, 1651.
H; M/E4 – 34; Tr/NQ.7.82 [imperfect, wanting all before sign. O.1; 'Isaacus Newton' on end-paper]; Sotheran **828** (1931) 3602.
Extensive notes and corrections in Newton's hand on p. 22.

585 — Euclidis data succinctè demonstrata: unà cum emendationibus quibusdam & additionibus ad Elementa Euclidis nuper edita. Operâ I. Barrow. 12°, Cantabrigiæ, 1657.
Not H; M/E3 – 23; Tr/NQ.16.201² [bound with 581].
Corrections and insertions by Newton at pp. 1, 38, 46–7.

586 EURIPIDES. Tragœdiæ quæ extant. Cum Latina G. Canteri interpretatione...2 vols. 4°, [Geneva], 1602.
H; M/A7 – 8 & 9; ?

587 EUROPE (L') savante. [By H. Cordonnier, etc.] 12 vols. in 10. 8°, La Haye, 1718–20.
H; M/G6 – 29 to 38; ?

588 EUSEBIUS PAMPHILI, *Bp of Caesarea.* De demonstratione Evangelica libri x...omnia studio R. M[ontacutii]. Latinè facta...Ed. nova. (*Greek & Latin.*) F°, Coloniæ, 1688.
H; M/B4 – 8; Tr/NQ.18.22.

589 — Hystoria ecclesiastica Eusebij Cesariensis per Magistrum G. Boussardam...8°, Parisiis, (1525).
Not H; M/E3 – 7; Tr/NQ.16.167 ['Is. Newton pret 6ᵈ' on fly-leaf].

590 — Historia ecclesiastica Eusebii Pamphili, Socratis Scholastici, Hermiæ Sozomeni, Theodoriti Episcopi Cyri et Evagrii Scholastici, ab H. Valesio in linguam Latinam conversa...3 vols. F°, Parisiis, 1678.
H; M/E8 – 4 to 6; ?

591 — Præparatio Evangelica. F. Vigerus...recensuit, Latinè vertit, notis illustravit. Ed. nova. (*Greek & Latin.*) F°, Coloniæ, 1688.
H; M/B4 – 7; Tr/NQ.18.23 [a few signs of dog-earing].

592 — Thesaurus temporum, chronicorum canonum omnimodæ historiæ libri II, interprete Hieronymo...Opera ac studio J. J. Scaligeri ...Ed. 2ᵃ...F°, Amstelodami, 1658.
H; M/B4 – 9; Tr/NQ.17.25 [pt 2, p. 7, pt 4, p. 66, pt 5, p. 160 turned down and a few other signs of dog-earing].

EVANGELIUM Infantiæ. Vel Liber apocryphus de infantia Servatoris. Ex manuscripto edidit...H. Sike. (2 pts.) (*Arabic & Latin.*) 1697. *See* 59.

EVANS, Abel. The apparition. A poem [by A. Evans]. 1710. *See* 69.

593 EVELYN, John. Numismata. A discourse of medals, antient and modern. Together with some account of heads and effigies... Fº, London, 1697.
H; M/B2 – 27; Tr/NQ.18.27 [p. 37 turned down and several other signs of dog-earing].

594 EVERARD, Thomas. Stereometry made easie: or, The description and use of a new gauging rod or sliding rule...12º, London, 1684.
H; M/G9 – 8; R. B. Honeyman, Jr, San Juan Capistrano, Calif. *Calculations in Newton's hand on two slips of paper loosely inserted.*

595 EXACT (An) collection of all remonstrances, declarations, votes [etc.]...betweene the Kings Most Excellent Majesty, and his High Court of Parliament...December 1641...untill March the 21, 1643...[Collected by E. Husband.] 4º, London, 1643.
H; M/F5 – 34; Tr/NQ.9.76.

596 EXAMINATION (An) of Dr. Woodward's Account of the deluge, &c. By J. A[rbuthnot]., M.D. With a letter to the author concerning an abstract of A. Scilla's book on the same subject... By W. W[otton]., F.R.S. 8º, London, 1697.
Not H; M (Tracts) J3 – 33; Tr/NQ.16.162⁵.

597 EXAMINERS (The) for the year 1711...12º, London, 1712.
Not H; M/G9 – 46; ?

EXQUEMELIN, Alexander Olivier. The history of the bucaniers of America...written in several languages...The whole newly translated into English [by A. O. Exquemelin]. 1699. *See* 780.

EYQUEM DU MARTINEAU, Mathurin. Le pilote de l'onde vive, ou Le secret du flux et reflux de la mer...2ᵉ éd. reveuë & augmentée...[By M. Eyquem du Martineau.] 1689. *See* 1316.

598 FABRE, Pierre Jean. Operum voluminibus duobus exhibitorum volumen prius [& ii]. (*Vol. 2 with title* Opera reliqua volumine hoc posteriore comprehensa.) 4º, Francofurti ad Moenum, 1652.
Not H; M/B1 – 6 & 7; Tr/NQ.9.174 & 175.
On the fly-leaf of vol. 1 Newton listed 11 of Fabre's published works under the heading 'Author opera sua hoc ordine impressit' and added at the end 'Desumpta sunt hæc ex Libris ipsis impressis in Bibliotheca Bodleiana Oxonij', i.e. 824 in this catalogue; vol. 1 has pp. 37, 558, 575, 657, 686, 692, turned down, p. 690 turned down and up, pp. 44, 555, 685, 691 up; vol. 2 pp. 51, 759, 881 down, pp. 62, 422, 854, 882 up, with many other signs of dog-earing.

599 FABRI, Honoré. Dialogi physici quorum primus est de lumine, secundus & tertius, di vi percussionis & motu, quartus, de humoris elevatione per canaliculum, quintus et sextus, de variis selectis...
8°, Lugduni, 1669.
H; M/F2 – 20; Tr/NQ.9.51; Newton acknowledged receipt of the volume from Collins, 25 May 1672 (*Correspondence*, I, 161).

600 — Synopsis geometrica, cui accessere tria opuscula, nimirum, de linea sinuum & cycloide, de maximis & minimis centuria, et synopsis trigonometriæ planæ. 8°, Lugduni, 1669.
Not H; M/E3 – 39; Turner Collection, University of Keele Library.

601 FABRICIUS, Johann Albert. Bibliographia antiquaria, sive Introductio in notitiam scriptorum...4°, Hamburgi, 1713.
Not H; M/D5 – 11; Tr/NQ.10.64 [title-page defective; p. 291 turned down and a few other signs of dog-earing].

602 — Bibliotheca Græca, sive Notitia scriptorum veterum Græcorum... Græce & Latine, cum brevibus notis. 9 vols. in 10. 4°, Hamburgi, 1705–19.
H; M/D5 – 1 to 10; Tr/NQ.8.8 to 17 [vol. 1 has pp. 154, 157, 475 turned down, p. 371 up, and several other signs of dog-earing; vol. 4 has p. 255 turned down].

603 — Bibliotheca Latina, sive Notitia auctorum veterum Latinorum, quorumcunque scripta ad nos pervenerunt, distributa in libros IV...8°, Hamburgi, 1708.
H; M/F3 – 21; Tr/NQ.8.85.
— Codex Apocryphus Novi Testamenti, collectus, castigatus testimoniisque, censuris & animadversionibus illustratus à J. A. Fabricio. 1703. *See* 57.
— Codex Pseudepigraphus Veteris Testamenti, collectus, castigatus testimoniisque, censuris et animadversionibus, illustratus a J. A. Fabricio. 2 vols. 1713–23. *See* 58.

604 FAITH (The) of One God, who is only the Father...asserted and defended, in several tracts...[By J. Biddle.] 4°, London, 1691.
H; M/H2 – 13; Tr/NQ.9.32.

605 FAME (The) and confession of the Fraternity of R:C: commonly, of the Rosie Cross...By Eugenius Philalethes [i.e. T. Vaughan]. 8°, London, 1652.
Not H; M/G9 – 15; Beinecke Library, Yale ['Is. Newton. Donum M^{ri} Doyley' in Newton's hand on fly-leaf].
Notes by Newton on the Rosicrucians on fly-leaf and in margin of p. 4, see I. Macphail, *Alchemy and the occult*, II, 347–9, where the notes are reproduced in full; Thame/973.
FANTET DE LAGNY, Thomas. *See* LAGNY, Thomas Fantet de.

FATIO DE DUILLIER, Nicolas. Fruit-walls improved, by inclining them to the horizon...[By N. Fatio de Duillier.] 1699.
See 642.

606 — Lettre à M. Cassini...touchant une lumière extraordinaire qui
paroît dans le ciel depuis quelques années. 12°, Amsterdam, 1686.
Not H; M/G2 – 24; ?

607 — Lineæ brevissimi descensus investigatio geometrica duplex...4°,
Londini, 1699.
Not H; M/C6 – 5; Tr/NQ.10.33² [bound with 642].

FAZELLO, Tommaso. Rerum Sicularum scriptores ex recentioribus
præcipui, in unum corpus nunc primum congesti, diligentiq;
recognitione plurimis in locis emendati [by T. Fazello]...1579.
See 1394.

608 FEATLEY, Daniel. Καταβάπτισται κατάπτυστοι. The Dippers dipt:
or, The Anabaptists duck'd and plung'd over head and ears...
7th ed. augmented...4°, London, 1660.
Not H; M/H2 – 31; ?

609 FEGUERNEKINUS, Isaacus L. Enchiridii locorum communium
theologicorum, rerum, exemplorum, atq; phrasium sacrarum...
Ed. 5ª. 8°, Basileæ, 1604.
Not H; M/E2 – 36; Tr/NQ.10.151 [wants title-page; 'Isaac
Newton Trin Coll: Cant: 1661' on blank opposite p. 1].

FEILD, Richard. *See* FIELD, Richard.

610 FELTON, Henry. A dissertation on reading the classics, and forming
a just style...12°, London, 1713.
H; M/G9 – 4; ?

611 FENICE, Giovanni Antonio. Dictionnaire françois & italien, profitable
et necessaire...(2 pts.) 8°, Morges, 1585.
Not H; M/J8 – 33; Tr/NQ.10.138.

612 FENNER, William. The works of W. Fenner in four treatises...To
which is annexed his Catechism...4°, London, 1656.
Not H; M/H2 – 26; ?

613 FENTON, Elijah. Mariamne. A tragedy. Acted at the Theatre Royal
in Lincoln's-Inn-Fields. 2nd ed. 12°, London, 1726.
Not H; M (Plays) H8 – 9; Tr/NQ.10.83².

614 FERRARIUS, Philippus. Lexicon geographicum...Ed. nova, multo
quam prior accuratior...F°, Londini, 1657.
Not H; M/C2 – 2; Tr/NQ.11.36.

615 FEUILLÉE, Louis. Journal des observations physiques, mathématiques et botaniques, faites...sur les côtes orientales de l'Amérique
Meridionale, & dans les Indes Occidentales, depuis l'année 1707,
jusques en 1712. 2 vols. 4°, Paris, 1714.
H; M/E6 – 4 & 5; Tr/NQ.10.23 & 24.

616 FIDDES, Richard. A letter in answer to one from a free thinker: occasion'd by the late Duke of Buckinghamshire's epitaph...8°, London, 1721.
Not H; M (Tracts) J3 – 7; Tr/NQ.7.1⁷.

617 — The life of Cardinal Wolsey. 2nd ed., corrected by the author. F°, London, 1726.
Not H; M/H9 – 12; ?

618 FIELD, Richard. Of the Church, five bookes. 4°, London, 1606.
Not H; M/C6 – 16; Tr/NQ.10.61.

619 FILET (Le) d'Ariadne, pour entrer avec seureté dans le labirinthe de la philosophie hermetique. [By Gasto Claveus?] 12°, Paris, 1695.
H; M/G7 – 17; Tr/NQ.8.60; Ekins 3 – 0.

620 FISHER, John, *Bp of Rochester.* The funeral sermon of Margaret Countess of Richmond and Derby, mother to King Henry VII... (Reprinted.) With a preface containing some further account of her charities and foundations...(2 pts.) 8°, London, 1708.
Not H; M/H2 – 25; ?

621 FIVE treatises of the philosopher's stone. Two of Alphonso King of Portugall...One of John Sawtre a Monke...Another written by Florianus Raudorff...Also a treatise of the names of the philosopher's stone, by W. Gratacolle...To which is added the Smaragdine Table...By the paines and care of H. P. 4°, London, 1652.
H; M/F5 – 13; ?

FLAMEL, Nicolas. Philosophie naturelle de trois anciens philosophes renommez: Artephius, Flamel, & Synesius...Dernière éd.... 1682. *See* 1309 & 1310.

622 FLAMSTEED, John. Historiæ cœlestis libri II quorum prior exhibet catalogum stellarum fixarum Britannicum novum & locupletissimum una cum earundem planetarumque omnium observationibus sextante, micrometro, &c. habitis. Posterior transitus syderum per planum arcus meridionalis et distantias eorum a vertice complectitur. F°, Londini, 1712.
H; M/C4 – 4; Tr/NQ.17.29.

FLEETWOOD, William, *Bp of Ely.* The life and miracles of St. Wenefrede, together with her litanies...[By W. Fleetwood.] 1713. *See* 1746.

623 FLOYER, *Sir* John. L'oriuolo da polso de medici...tradotta da un cavaliere inglese...(3 pts.) 4°, Venezia, 1715.
H; M/F6 – 28; Tr/NQ.8.34.

624 FONTENELLE, Bernard Le Bovier de. Histoire du renouvellement de l'Académie Royale des Sciences en M.DC.XCIX. et les éloges historiques de tous les Académiciens morts depuis ce renouvellement. 2 vols. 12°, Paris, 1717.

H; M/G7 – 20 & 21; Tr/NQ.10.131 & 132 [a few signs of dog-earing in both vols including marking of Newton's name in text of vol. 2, p. 43, see p. 27 above]; presentation inscription on fly-leaf 'Pour Monsieur Newton'.

625 FORM (A) and method of trial of commoners, in cases of high treason, and misprision of high treason...12°, London, 1709.
Not H; M/J8 – 54; Tr/NQ.9.65² [bound with 413].

626 FORTESCUE, Sir John. The difference between an absolute and limited monarchy; as it more particularly regards the English constitution...8°, London, 1714.
H; M/G10 – 17; ?

627 FOSTER, Samuel. Miscellanea: sive Lucubrationes mathematicæ... Omnia in lucem edita...opera & studio J. Twysden. F°, Londini, 1659.
H; M/J9 – 19; ?

FOTHERGILL, George. Sermons. See 1496.

628 FOWLER, Edward, Bp of Gloucester. The design of Christianity...8°, London, 1671.
Not H; M/H7 – 34; ?

629 FOX MORCILLO, Sebastian. De naturæ philosophia, seu de Platonis, & Aristotelis consensione, libri v...8°, Parisiis, 1560.
Not H; not M; Tr/Adv.d.1.13 ['Isaac Newton Trin: Coll: Cant: 1661' on verso of title-page]; presented to Trinity by John Laughton, Librarian 1679–86.

630 FOY-VAILLANT, Jean. Numismata Imperatorum Romanorum præstantiora a Julio Cæsare ad Postumum et tyrannos. Ed. 3ᵃ emendatior & plurimis rarissimis nummis auctior...2 vols. in 1. 4°, Lutetiæ Parisiorum, 1694.
H; M/C1 – 9; Tr/NQ.7.55.

631 — Nummi antiqui familiarum Romanarum perpetuis interpretationibus illustrati. 2 vols. F°, Amstelædami, 1703.
H; M/A9 – 12 & 13; Tr/NQ.11.35 [vol. 1 only, lacks plates].

FRANCE. Académie Royale des Inscriptions et Belles Lettres. See PARIS. Académie Royale des Inscriptions et Belles Lettres.
— Académie Royale des Médailles et des Inscriptions. See PARIS. Académie Royale des Médailles et des Inscriptions.
— Académie Royale des Sciences. See PARIS. Académie Royale des Sciences.

632 — Cour des Monnoyes. [A collection of 180 official documents relating to the French currency issued by the Cour des Monnoyes, 1689–1701, with titles beginning 'Declaration du Roy...'] 4°, Paris, 1689–1701.
H; M/F6 – 9; Tr/NQ.10.43 [several signs of dog-earing throughout].

633 FRANCKE, August Hermann. Manuductio ad lectionem Scripturæ
Sacræ. Omnibus theologiæ sacræ cultoribus commendata a P.
Allix...12°, Londini, 1706.
Not H; M/A2 – 16; ?

634 FRANCKENBERGER, Reinhold. Chronologiæ Scaligero-Petavianæ
breve compendium, in scientiæ forma per cognoscendi principia
& breves canones...4°, Wittebergæ, 1661.
Not H; M/J3 – 40; Tr/NQ.9.8.

635 FRANÇOIS MARIE, Père. Nouvelle découverte sur la lumière,
pour la mesurer & en compter les degrés. 12°, Paris, 1700.
Not H; M (Tracts) H8 – 17; Tr/NQ.16.161^2; Ekins 1 – 6.

636 FREE-THINKING rightly stated; wherein a discourse, falsly so
called, is fully consider'd. [By T. Cockman.] 8°, London, 1713.
Not H; M (Tracts) J3 – 7; Tr/NQ.7.1^6.

637 FREIND, John. Prælectiones chymicæ: in quibus omnes fere operationes
chymicæ ad vera principia & ipsius naturæ leges rediguntur:
anno 1704 Oxonii, in Museo Ashmoleano habitæ. Ed. 2^a, priore
emendatior...8°, Londini, 1726. [Dedicated to Newton.]
H (2 copies); M (1 only) E5 – 22; Tr/NQ.16.105.

638 FRÉMONT D'ABLANCOURT, Nicolas. Mémoires...contenant
l'histoire de Portugal...12°, La Haye, 1701.
H; M/G6 – 17; ?

639 FRENICLE DE BESSY, Bernard. Traité des triangles rectangles en
nombres...12°, Paris, 1676.
Not H; M/G2 – 40; ?; Newton thanked Collins for sending the book,
5 Sept. 1676 (*Correspondence*, II, 95).

640 FRESH (A) suit against independency: or The national church-way
vindicated, the independent church-way condemned...[By T.
Lamb.] 8°, London, 1677.
Not H; M/H1 – 9; ?

641 FROMONDUS, Libertus. Meteorologicorum libri VI. 8°, [Oxford],
1639.
Not H; M/E3 – 43; ?

642 FRUIT-WALLS improved, by inclining them to the horizon: or,
A way to build walls for fruit-trees...By a Member of the Royal
Society [N. Fatio de Duillier]. 4°, London, 1699.
H; M/C6 – 5; Tr/NQ.10.33^1 [bound with 607].

643 FULL (A) account of the proceedings in relation to Capt. Kidd. In
2 letters. Written by a person of quality to a kinsman of the Earl
of Bellomont in Ireland. 4°, London, 1701.
Not H; M (Tracts) J3 – 41; Tr/NQ.8.111^5.

644 — A full and impartial account of all the late proceedings in the University of Cambridge against Dr. Bentley. By a member of that University [i.e. C. Middleton]. 8°, London, 1719.
Not H; M (Tracts) J3 – 7; Tr/NQ.7.1³.

645 — A full answer to the depositions, and to all other the pretences and arguments whatsoever, concerning the birth of the pretended Prince of Wales. . .8°, [London], 1711.
Not H; M (Tracts) J3 – 13; Tr/NQ.9.153³.

646 FULLER, Thomas. Pharmacopœia extemporanea, sive Præscriptorum chilias. . .Ed. 5ª. . .8°, Londini, 1714.
H; M/J7 – 27; ?

FULLER, William. Original letters of the late King's [James II], and others. . .Published by command, by W. Fuller. 1702. *See* 846.

647 GAFFAREL, Jacques. Unheard-of curiosities: concerning the talismanical sculpture of the Persians. . .Englished by E. Chilmead. 8°, London, 1650.
Not H; M/E4 – 42; Thame/987; ?

GALE, Thomas. Historiæ poeticæ scriptores antiqui. . .Græcè & Latinè. . .[Ed. by T. Gale.] (3 pts.) 1675. *See* 772.

— Opuscula mythologica, ethica et physica. Græce & Latine. . . [Ed. by T. Gale.] 1671. *See* 1207.

GALILEO, Galilei. Sidereus nuncius. . .1682. *See* 651.

648 — Systema cosmicum. . .Accessit alterâ hâc ed. præter conciliationem locorum S. Scripturæ cum terræ mobilitate, ejusdem Tractatus de motu, nunc primum ex Italico sermone in Latinum versus. (2 pts.) 4°, Lugduni Batavorum, 1699.
H; M/J6 – 13; Tr/NQ.8.42.

GALTRUCHIUS, Petrus. *See* GAUTRUCHE, Pierre.

649 GARLAND, John. Compendium alchimiæ: cum dictionario eiusdem artis. . .Omnia nunc primum in lucem edita. 8°, Basileæ, 1560.
H; M/E3 – 21; Tr/NQ.16.137² [bound with 85; a few signs of dog-earing].

GAROFALO, Biagio. *See* CARYOPHILUS, Blasius.

650 GARTH, Samuel. The Dispensary: a poem. In six cantos. 2nd ed., corrected by the author. 8°, London, 1699.
Not H; M/J3 – 3; ?

651 GASSENDI, Pierre. Institutio astronomica, juxta hypotheses tam veterum quam recentiorum. Cui accesserunt Galilei Galilei Nuntius Sidereus, et J. Kepleri Dioptrice. 3ª ed. priori correctior. 8°, Amsterdami, 1682.
Not H: M/E5 – 33; Tr/NQ.7.41.

652 GASTRELL, Francis, *Bp of Chester*. The Bishop of Chester's case with relation to the wardenship of Manchester...F°, Oxford, 1721.
Not H; M/A8 – 11; ?

653 — The Christian Institutes, or, The sincere Word of God...8°, London, 1707.
Not H; M/H1 – 45; ?

654 GAUGER, Nicolas. Fires improv'd: being a new method of building chimneys so as to prevent their smoking...Made English and improved, by J. T. Desaguliers. 12°, London, 1715.
H; M/G9 – 1; ?
— La méchanique du feu...contenant Le traité de nouvelles cheminées...Par Mr. G*** [i.e. N. Gauger]. 1714. *See* 1058.

655 GAUTRUCHE, Pierre. Mathematicæ totius...clara, brevis & accurata institutio. 8°, Cantabrigiæ, 1668.
Not H; M/E1 – 4; ?

656 GAY, John. Poems by Gay and others. 8°, 1720. *Edition not identified*.
Not H; M/J3 – 12; ?

657 GEBER. Chimia, sive Traditio summæ perfectionis et investigatio magisterii, innumeris locis emendata à C. Hornio...Accessit ejusdem Medulla alchimiæ Gebricæ, omnia edita à G. Hornio. 24°, Lugduni Batavorum, 1668.
Not H; M/E1 – 28; S. I. Barchas, Sonoita, Arizona; Sotheran **800** (1926) 10868.
2 closely written pages of Latin notes by Newton on the fly-leaves, giving his rendering of Geber's terms into contemporary versions.

658 — Works of Geber, the most famous Arabian Prince and philosopher ...Englished by R. Russel. 8°, London, 1678.
H; not M; ?

659 — The works of Geber, the most famous Arabian Prince and philosopher of the investigation and perfection of the philosophers-stone. 8°, London, 1686.
Not H; M/E2 – 32; Memorial Library, University of Wisconsin – Madison; Thame/976.

660 GELASIUS, *Cyzicenus*. Commentarius actorum Nicæni Concilii, cum Corollario Theodori Presbyteri, de Incarnatione Domini. Nunc primùm Gręcè prodeunt, ex opt. Bibliothecis, interprete R. Balforeo, cum eiusdem notis. 8°, Lutetiæ, 1599.
Not H; M/F3 – 33; Tr/NQ.10.90.

661 GELLIBRAND, Henry. An epitome of navigation...8°, London, 1698.
H; M/G9 – 10; ?

662 GELLIUS, Aulus. Noctium Atticarum. Libri xix...8°, (Argentinæ, 1521).
H; M/F2 – 29; Tr/NQ.8.72 [a few signs of dog-earing].

663 GENERAL (The) delusion of Christians, touching the ways of God's revealing Himself, to, and by the Prophets. . . [By J. Lacy.] 4 pts. 8°, London, 1713.

H; M/H4 – 35; Tr/NQ.16.189 [at top of title-page 'Ex dono Amici F: M.', not in Newton's hand, 'F. M.' not identified; a few signs of dog-earing].

664 — General maxims in trade, particularly applied to the commerce between Great Britain and France. [By Sir T. Janssen.] 8°, London, 1713.

Not H; M (Tracts) J2 – 38; Tr/NQ.16.194⁵.

665 GEOFFREY, *of Monmouth, Bp of St Asaph*. The British history, translated into English. . .By A. Thompson. 8°, London, 1718.

H (2 copies); M (1 only) G10 – 18; ?

666 GEOGRAPHIA classica: the geography of the ancients so far describ'd as it is contain'd in the Greek and Latin classicks, in 29 maps of the Old World. . . [By H. Moll.] 2nd ed. 4°, London, 1717.

H; M/J7 – 38; Tr/NQ.16.70.

667 GERHARD, Johann. Decas quæstionum physico-chymicarum selectiorum & graviorum. . .Cui adjuncta est Medulla Gebrica. De lapide philosophorum. 8°, Tubingæ, 1643.

Not H; M (Tracts) H8 – 13; Tr/NQ.16.99².

GERVAISE, Nicolas. Description historique du Royaume de Macaçar . . . [By N. Gervaise.] 1700. *See* 510.

668 GIBBON, John. Introductio ad Latinam Blasoniam. An essay to a more correct blason in Latine than formerly hath been used. . .8°, London, 1682.

H; M/J7 – 36; Tr/NQ.9.83.

GIBSON, Edmund, *Bp of London*. Chronicon Saxonicum. Ex MSS codicibus nunc primum integrum edidit, ac Latinum fecit E. Gibson. 1692. *See* 376.

669 GILBERT, Samuel. The florist's vade mecum. . .3rd ed., enlarged. 12°, London, 1702.

Not H; M/J8 – 53; ?

GILLMORE, Joseph. An account of the fish-pool. . .by Sir Richard Steele, and J. Gillmore. 1718. *See* 1555.

GLANVILL, Joseph. Lux Orientalis, or An enquiry into the opinion of the Eastern sages, concerning the præexistence of souls. . . [By J. Glanvill.] 1662. *See* 1006.

670 GLASER, Christophe. Traité de la chymie. . .Nouvelle éd., reveuë & augmentée. . .18°, Lyon, 1676.

H; M/G2 – 43; Sotheran **800** (1926) 10892, **804** (1927) 3799; ?

671 GODFREY, Ambrose. An account of the new method of extinguishing fires by explosion and suffocation. Introduced by A. Godfrey... 8°, [London], 1724.

Not H; M (Tracts) J3 – 33; Tr/NQ.16.162².

672 GODWIN, Francis, *Bp of Hereford*. A catalogue of the Bishops of England, since the first planting of Christian religion in this Island...8°, London, 1615.

Not H; M/F5 – 25; ?

673 GODWIN, Thomas. Moses and Aaron. Civil and ecclesiastical rites, used by the ancient Hebrewes; observed, and at large opened... 4th ed. 4°, London, 1631.

Not H; M/F5 – 26; Babson 405 [bound with 674 & 1424].

674 — Romanæ historiæ anthologia recognita et aucta. An English exposition of the Roman antiquities...Newly revised and inlarged by the author. 4°, London, 1658.

Not H; M/F5 – 26; Babson 405 [bound with 673].

675 GONZÁLEZ DE MENDOZA, Juan. Nova et succincta, vera tamen historia de...regno China...Ex Hispanica...in Latinam linguam conversa, operâ M. Henningi...12°, Francofurdi ad Mœnum [1589?].

H; M/F1 – 16; Tr/NQ.16.9³ [bound with 1712].

GOODDAY, Bartholomew. Mercator improved, or, The description and use of a new instrument...By B. G[oodday]. Philomath. 1725 *See* 1070.

GOODMAN, John. Winter-evening conference between neighbours. [By J. Goodman.] 3rd ed. corrected. 3 pts. 1686. *See* 1748.

676 GORDON, George. An introduction to geography, astronomy, and dialling...8°, London, 1726.

H; M/E5 – 19; ?

677 — Remarks upon the Newtonian philosophy, as propos'd by Sir Isaac Newton, in his Principia philosophiæ naturalis; and by Dr. Gregory, in his Principia astronomiæ physicæ...12°, London, 1719.

H; M/E2 – 40; Tr/NQ.10.84.

GORDON, Thomas. The Independent Whig. [By J. Trenchard and T. Gordon.] January 20, 1720 – January 4, 1721. *See* 832.

GOTHER, John. Nubes testium: or A collection of the Primitive Fathers...[By J. Gother.] 1686. *See* 1193.

GOTHOFREDUS, Dionysius. Antiquæ historiæ ex xxvii. authoribus contextæ libri vi....D. Gothofredi operâ...(2 pts.) 1590–91. *See* 52.

678 GRAAF, Abraham de. De beginselen van de algebra of stelkonst, volgens de manier van R. Des Cartes, verklaart met uytgelezene voorbeelden...4°, Amsterdam, 1672.
Not H; M/Bl – 3; Tr/NQ.7.50 ['Is. Newton Ex dono Viri amicissimi Johannis Collinsij' in Newton's hand on fly-leaf].

GRABE, Johann Ernst. Septuaginta interpretum tomus I, IV...summa cura edidit J. E. Grabe. 1707–20. See 197.

679 — Spicilegium SS. Patrum, ut et hæreticorum...Ed. 2ª...2 vols. 8°, Oxoniæ, 1714.
H; M/A4 – 11 & 12; ?

680 GRÆCÆ grammatices rudimenta, in usum Scholæ Westmonasteriensis. [By R. Busby.] 8°, Londini, 1663.
H; M/E2 – 21; Tr/NQ.9.88.

681 — Græcæ grammatices rudimenta in usum Scholæ Westmonasteriensis. [By W. Camden.] 8°, Londini, 1682.
Not H; M/Al – 18; ?

GRAEF, Abraham de. See GRAAF, Abraham de.

682 GRANDAMI, Jacques. Chronologia Christiana de Christo nato et rebus gestis ante et post eius Nativitatem. Ed. 2ª. 3 vols. in 1. 4°, Parisiis, 1668.
H; M/C6 – 9; Tr/NQ.10.8 [vol. 3, p. 180 turned down and several other signs of dog-earing esp. in vol. 1].

683 GRANDI, Guido. De infinitis infinitorum, et infinite parvorum ordinibus disquisitio geometrica...4°, Pisis, 1710.
Not H; M (Tracts) D6 – 3; Tr/NQ.8.1¹.

684 — Geometrica demonstratio theorematum Hugenianorum circa logisticam, seu logarithmicam lineam...4°, Florentiæ, 1701.
H; M/E6 – 22; Tr/NQ.8.5¹ [bound with 685]; Newton thanked Grandi in June 1704 for presenting him with this and with 685 (*Correspondence*, VII, 434–5).

685 — Geometrica demonstratio Vivianeorum problematum quæ in exercitatione geometrica...anno 1692. edita...4°, Florentiæ, 1699.
Not H; M/E6 – 22; Tr/NQ.8.5² [bound with 684].

686 — Prostasis ad exceptiones Cl. Varignonii libro de infinitis infinitorum ordinibus oppositas...4°, Pisis, 1713.
Not H; M (Tracts) D6 – 4; Tr/NQ.16.197⁴ [a few signs of dog-earing].

GRASSHOFF, Johann. Harmoniæ inperscrutabilis chymico-philosophicæ...decas I(–II). [By J. Grasshoff and J. Rhenanus.] (2 pts.) 1625. See 740.

687 GRAVESANDE, Willem Jacob 's. Essai de perspective. (*With* Usage de la chambre obscure pour le dessein.) (2 pts.) 12°, La Haye, 1711.
H; M/G7 – 41; Tr/NQ.9.143.

688 — Essai d'une nouvelle théorie du choc des corps, fondée sur l'expérience. 8°, La Haye, 1722.
Not H; M (Tracts) H8 – 17; Tr/NQ.16.161³ [misbound; 'Is. Newton' in 's Gravesande's (?) hand at top of title-page].

689 — Mathematical elements of natural philosophy, confirmed by experiments, or An introduction to Sir Isaac Newton's philosophy ...Transl. into English by J. T. Desaguliers. 2 vols. 8°, London, 1720–21. [Dedicated to Newton.]
H; M/E6 – 11 & 12; ?

690 — Matheseos universalis elementa. Quibus accedunt specimen commentarii in Arithmeticam universalem Newtoni...8°, Lugduni Batavorum, 1727.
H; not M; ?

691 — Oratio inauguralis, de matheseos, in omnibus scientiis, præcipue in physicis, usu, nec non de astronomiæ perfectione ex physica haurienda...4°, Lugduni Batavorum, 1717.
Not H; M (Orationes) C6 – 19; Tr/NQ.10.35⁶.

692 — Philosophiæ Newtonianæ institutiones, in usus academicos. 12°, Lugduni Batavorum, 1723.
H; M/F3 – 25; Tr/NQ.9.146.

693 — Physices elementa mathematica, experimentis confirmata. Sive Introductio ad philosophiam Newtonianam. 2 vols. 4°, Lugduni Batavorum, 1720–21.
H (2 copies); M (1 only) E6 – 11 & 12; Tr/NQ.10.44 & 45.

694 — Supplementum physicum, sive Addenda & corrigenda in prima editione, tomi primi, libri editi Lugd. Bat. anno MDCCXXI. cui titulus Physices elementa mathematica...4°, Lugduni Batavorum, 1725.
H; M/E6 – 13; Tr/NQ.16.177.

695 — Remarques sur la possibilité du mouvement perpetuel. 12°, [*c.* 1720].
Not H; M (Tracts) H8 – 17; Tr/NQ.16.161⁴.

696 GREAT (The) propitiation: or, Christ's satisfaction. And man's justification by it, upon his faith...[By J. Truman.] 8°, London, 1669.
Not H; M/G9 – 28; ?

697 GREAVES, John. A discourse of the Romane foot, and denarius... 8°, London, 1647.
H; M/E4 – 40 [bound with 698]; ?

698 — Pyramidographia: or A description of the Pyramids in Ægypt. 8°, London, 1646.
H; M/E4 – 40 [bound with 697]; ?

699 GREENE, Robert. The principles of natural philosophy... (*With* Geo-
metria solidorum, sive materiæ...) (2 pts.) 8°, Cambridge, 1712.
H; M/E5 – 14; Turner Collection, University of Keele Library
[inscribed 'To the Honoured & Much Esteemed Sir Isaac Newton
These Principles of Natural Philosophy are with the Greatest
Deference & Submission Presented'].

700 GREENWOOD, James. An essay towards a practical English grammar. Describing the genius and nature of the English tongue...
12°, London, 1711.
H; M/J8 – 28; Tr/NQ.9.111.

GREGG, Hugh. Curiosities in chymistry...Written by a person of
honour, and published by his operator, H. G[regg]. [Sometimes
attributed to Robert Boyle.] 1691. *See* 470.

701 GREGORY I, *Pope, St, the Great.* Opera omnia...Studio & labore
Monachorum Ordinis Sancti Benedicti, è Congregatione Sancti
Mauri. 4 vols. F°, Parisiis, 1705.
H; M/C5 – 11 to 14; Tr/NQ.18.32 to 35 [vol. 2, col. 501 turned
down, col. 1244 up, and a few other signs of dog-earing].

702 GREGORY, *St, Bp of Nyssa.* Opera. Nunc denuo correctius et accura-
tius edita, aucta, & notis...ornata...(*Greek & Latin.*) 3 vols. F°,
Parisiis, 1638.
H; M/F7 – 11 to 13; Tr/NQ.17.22 to 24 [vol. 3 has p. 281 turned
down and a few other signs of dog-earing].

703 GREGORY, *St, Bp of Tours.* Historiae Francorum libri x...(2 pts.)
8°, Basileae, 1568.
H; not M; ?

704 GREGORY, *St, Patriarch of Constantinople, Nazianzen.* Opera. Jac.
Billius Prunæus....cum MSS. Regijs contulit, emendavit, inter-
pretatus est...(*Greek & Latin.*) 2 vols. F°, Parisiis, 1630.
         lb s d
H; M/D7 – 7 & 8; Tr/NQ.11.20 & 21 ['Pret. 2. 10. 0' in Newton's
hand on the 1st fly-leaf of vol. 1; the signature 'Jacobus Duport',
Regius Professor of Greek at Cambridge (d. 1679) on 2nd fly-
leaf; vol. 1, p. 609 turned up and down and a few other signs
of dog-earing in both vols.].

705 — Tractatus, sermones, & libri aliquot, Ruffino presbytero, & Petro
Mosellano interpretibus...F°, Parisiis, 1532.
Not H; M/D8 – 2; Tr/NQ.17.20² [bound with 1212].

706 GREGORY, David. Astronomiæ physicæ & geometricæ elementa. F°,
Oxoniæ, 1702.
H (2 copies); M (1 only) F8 – 19; Tr/NQ.18.21.

707 — Catoptricæ et dioptricæ sphæricæ elementa. 8°, Oxonii, 1695.
Not H; M/G10 – 15; Tr/NQ.8.33.

708 — De curva catenaria demonstrationes geometricæ. 4°, Oxoniæ, 1697.
Not H; M (Tracts) D5 – 25; Tr/NQ.8.51³.

709 — Exercitatio geometrica de dimensione figurarum, sive Specimen methodi generalis. Dimetiendi quasvis figuras. 4°, Edinburgi, 1684.
Not H; M (Tracts) D5 – 24; Tr/NQ.16.79⁴; Gregory sent Newton a copy 9 June 1684 (*Correspondence*, II, 396).

710 GREGORY, Francis. Ἐτυμολογικὸν μικρὸν, sive, Etymologicum parvum, ex magno illo Sylburgii, Eustathio, Martinio excerptum, digestum, explicatum...8°, Londoni, 1654.
H; M/F2 – 19; ?

711 GREGORY, James. Exercitationes geometricæ. 4°, Londini, 1668.
H; M/E5 – 4; Tr/NQ.9.48³ [bound with 714].

712 — Geometriæ pars universalis, inserviens quantitatum curvarum transmutationi & mensuræ. 4°, Patavii, 1668.
H; M/E5 – 4; Tr/NQ.9.48² [bound with 714; p. 20 turned up and a few other signs of dog-earing].

713 — Optica promota, seu Abdita radiorum reflexorum & refractorum mysteria, geometrice enucleata...4°, Londini, 1663.
H; M/J3 – 34; Turner Collection, University of Keele Library [pp. 19, 41, 132 turned up, pp. 22, 26, 60, 78, 93 down]; Newton told Oldenburg that he had received a copy, 4 May 1672 (*Correspondence*, I, 153).

714 — Vera circuli et hyperbolæ quadratura, in propria sua proportionis specie, inventa, & demonstrata. 4°, Patavii, 1667.
H; M/E5 – 4; Tr/NQ.9.48¹ [bound with 711 & 712].

715 GRETTON, Phillips. A vindication of the doctrines of the Church of England, in opposition to those of Rome...8°, London, 1725.
H; M/H7 – 2; Tr/NQ.8.94 [p. 218 turned up, pp. 239, 253 turned down].

716 GREW, Nehemiah. Musæum Regalis Societatis. Or A catalogue & description of the natural and artificial rarities belonging to the Royal Society and preserved at Gresham Colledge. F°, London, 1681.
H (2 copies); M (1 only) B2 – 13; Babson 406.

717 GREWER, Jodocus. Secretum; et Alani philosophi dicta de lapide philosophico. Item alia nonnulla eiusdem materiæ, pleraque iam primùm edita a Iusto a Balbian. 12°, [Leyden], 1599.
Not H; M/E3 – 32; Babson 407; Sotheran **804** (1927) 3800.

GRILLET, Jean. Voyages and discoveries in South-America...1698.
See 1700.

718 GRONINGIUS, Johannes. Bibliotheca universalis. Seu Codex operum variorum...(6 pts.) 8°, Hamburgi, 1701.
H; M/A3 – 10; Tr/NQ.9.89.
Newton added 'omnia' in the margin of pt 6, p. 115, line 4, to follow the printed word 'corpora'; pt 6, p. 73 turned up and a few other signs of dog-earing.

719 GROTIUS, Hugo. De iure belli ac pacis libri III...Ed. nova cum annotatis auctoris...8°, Amstelædami, 1667.
Not H; M/F4 – 21; ?

720 — De veritate religionis Christianæ. Ed. nova, additis annotationibus in quibus testimonia. 12°, Parisiis, 1640.
Not H; M/G1 – 7; ?

721 — De veritate religionis Christianæ. Ed. novissima...12°, Amstelodami, 1669.
H; M/G2 – 6; ?

722 — Historia Gotthorum, Vandalorum, & Langobardorum...8°, Amstelodami, 1655.
H; M/D1 – 18; ?
GRUMMET, Christoph. Sanguis naturæ, or, A manifest declaration of the sanguine and solar congealed liquor of nature. By Anonimus [i.e. C. Grummet]. 1696. *See* 1445 & 1446.

723 GUEVARA, Antonio de. Spanish letters: historical, satyrical and moral...Recommended by Sir R. L'S[trange], and made English...by Mʳ Savage. 8°, London, 1697.
Not H; M/J4 – 43; ?

724 GUICCIARDINI, Francesco. Historie: containing the warres of Italie and other partes...Reduced into English by G. Fenton. F°, London, 1599.
Not H; M/J9 – 30; Tr/NQ.10.32 [imperfect: wants title-page and sign. A1–4].

725 GUISCARD, Antoine, *marquis de.* Mémoires du Marquis de Guiscard. Dans lesquels est contenu le récit des entreprises qu'il a faites dans le Roiaume & hors du Roiaume de France...Pt 1. 12°, Delft, 1705.
H; M/G6 – 14; Tr/NQ.10.87.

726 GUMBLE, Thomas. The life of General Monck, Duke of Albermarle, &c. With remarks upon his actions. 8°, London, 1671.
Not H; M/J7 – 48; Tr/NQ.9.112 [a few signs of dog-earing].

727 GUNNING, Peter, *Bp of Ely.* The Paschal or Lent-fast apostolical & perpetual...4°, London, 1662.
Not H; M/H2 – 12; ?

728 GUNTER, Edmund. The description and use of the sector, cross-staffe, and other instruments: with a canon of artificiall sines and tangents...2nd ed. much augmented (2 pts.) 4°, London, 1636.
H; M/F5-31; Tr/NQ.9.160 [p. 12 of *Canon triangulorum* turned down and a few other signs of dog-earing]; bought by Newton in 1667 for 5s., see Fitzwilliam Notebook.

HAAK, Theodorus. Le teinturier parfait...[By T. Haak.] 1708. *See* 1600.

729 HACKE, William. A collection of original voyages...3 pts. 8°, London, 1699.
H; M/J4 – 39; ?

730 HAGEN, Johann Ludwig ab. Disputatio theologica, de fide hæreticis servanda...8°, Moguntiæ, 1607.
Not H; M/A2 – 11; Tr/NQ.9.161² [bound with 349].

731 HALE, *Sir* Matthew. The original institution, power and jurisdiction of Parliaments. 2 pts. 8°, London, 1707.
Not H; M/J4 – 32; Tr/NQ.16.51.

— Pleas of the Crown...[By Sir Matthew Hale.] 1678. *See* 1326.

732 HALES, John. An historical and critical account of the life and writings of...John Hales, Fellow of Eton College...[By P. Des Maizeaux.] 8°, London, 1719.
Not H; M/J3 – 8 [bound with 1746]; ?

733 HALIFAX, Charles Montagu, *Earl of*. The works and life of the Right Honourable Charles, late Earl of Halifax. Including the history of his Lordship's times. 8°, London, 1715.
H; M/J4 – 30; Tr/NQ.9.159.
A note by Newton on striking of medals at end of Dedication; see p. 15 above where it is reproduced in full.

734 HALLEY, Edmond. Astronomiæ cometicæ synopsis. [3 leaves.] F°, [1705].
Not H; M/F8 – 4; Babson 408 [bound with 1009]; Sotheran **804** (1927) 3804.

735 — Catalogus stellarum Australium sive Supplementum Catalogi Tychonici exhibens longitudines & latitudines stellarum fixarum ...4°, Londini, 1679.
Not H; M (Tracts) J6 – 2; Tr/NQ.16.77³ [a series of numbers written by Halley in the margins and in the text relating to an earlier catalogue of stars].

— Miscellanea curiosa. Being a collection of some of the principal phænomena in nature...[Ed. and in part written by Halley.] Vols. 1, 2. 1705–6. *See* 1086.

736 HAMMOND, Henry. A paraphrase and annotations upon all the books of the New Testament...4th ed. corrected. F°, London, 1675.
H; M/B4 – 1; ?

737 — A practical catechism. 8th ed....8°, London, 1668.
H; M/H2 – 39; ?

738 HANBURY, Nathaniel. Supplementum analyticum ad æquationes Cartesianas. 4°, Cantabrigiæ, 1691.
H; M/E6 – 39; Stanford University Library, Stanford, Calif.; Sotheran **783** (1923) 1652, **804** (1927) 3801.

739 HARDOUIN, Jean. Chronologiæ ex nummis antiquis restitutæ prolusio de nummis Herodiadum. 4°, Parisiis, 1693.
H; M/A6 – 5; Tr/NQ.8.132 [a few signs of dog-earing].

HARDWICKE, Philip Yorke, *1st Earl of*. A discourse on the judicial authority belonging to the office of Master of the Rolls in the High Court of Chancery...[By Philip Yorke, Earl of Hardwicke.] 1727. *See* 529.

740 HARMONIÆ inperscrutabilis chymico-philosophicæ, sive Philosophorum antiquorum consentientium...decas ɪ(–ɪɪ). [By J. Grasshoff and J. Rhenanus.] (2 pts.) 8°, Francofurti, 1625.
Not H; M/F3 – 36; Tr/NQ.16.122.
Notes and references by Newton all relating to 'Theatrum chemicum', vols. 4 & 5, in margin of pt 1, pp. 205, 228, pt 2, pp. 324, 338; pt 2, p. 326 turned down and several other signs of dog-earing.

741 HARRIS, John. The description and use of the celestial and terrestrial globes...8°, London, 1703.
H; M/E5 – 17; ?

742 — Remarks on some late papers, relating to the universal deluge: and to the natural history of the earth. 8°, London, 1697.
H; M/E5 – 30; Tr/NQ.9.72 [a few signs of dog-earing].

HARTLIB, Samuel. Chymical, medicinal, and chyrurgical addresses: made to S. Hartlib Esquire...1655. *See* 378.

743 HARTMANN, Johann. Praxis chymiatrica. Edita à I. Michaelis et G. E. Hartmanno. 4°, Moguntiæ, 1647.
Not H; M/A7 – 15; Tr/NQ.16.81² [bound with 462].

744 HATTON, Edward. An intire system of arithmetic: or Arithmetic in all its parts...(2 pts.) 4°, London, 1721.
H; M/J6 – 14; Tr/NQ.8.61.

745 HAUKSBEE, Francis. A course of mechanical, optical, hydrostatical, and pneumatical experiments. To be perform'd by F. Hauksbee; and the explanatory lectures read by W. Whiston. 4°, [London?, 1713].
Not H; M (Tracts) J6 – 1; Tr/NQ.16.76⁶.

746 — Physico-mechanical experiments on various subjects...4°, London, 1709.
    H; M/E5 – 1; Tr/NQ.9.68.

747 — Esperienze fisico-meccaniche sopra varj soggetti...Opera tradotta dall'idioma inglese. 4°, Firenze, 1716.
    H; M/E6 – 21; Tr/NQ.10.40.

748 HAWKINS, Isaac. An essay for the discovery of the longitude at sea, by several new methods fully and particularly laid before the publick. 8°, London, 1714.
    Not H; M (Tracts) J8 – 11; Tr/NQ.10.119⁴.
    HEARNE, Thomas. Ductor historicus: or, A short system of universal history...Partly transl. from the French of M. de Vallemont, but chiefly composed anew by W. J. M.A. [i.e. T. Hearne.] 1698. *See* 541.

749 HEGESIPPUS. De bello Iudaico, et Urbis Hierosolymitanæ excidio, libri v...8°, Coloniæ, 1559.
    Not H; M/A3 – 21; ?

750 HELIODORUS, *Emesenus, Bp of Tricca.* Æthiopicorum libri x...cura et labore D. Parei. (*Greek & Latin.*) 8°, Francofurti, 1631.
    Not H; M/F3 – 28; Tr/NQ.16.14 [a few signs of dog-earing].
    HELMONT, Francis Mercurius, *baron van.* Seder Olam: or, The order of ages...Transl. out of Latin, by J. Clark, upon the leave and recommendation of F. M. Baron of Helmont. 1694. *See* 1479.

751 HELMONT, Jan Baptista van. Ortus medicinæ, id est initia physicæ inaudita...edente F. M. van Helmont...Ed. 4ᵃ. F°, Lugduni, 1667.
    H; M/F7 – 15; ?

752 HELVETIUS, Johann Friedrich. The golden calf which the world adores, and desires: in which is handled the most rare and incomparable wonder of nature, in transmuting metals...12°, London, 1670.
    Not H; M/E1 – 9; Memorial Library, University of Wisconsin – Madison [several signs of dog-earing]; Thame/974; Duveen, p. 288.

753 HELVICUS, Christophorus. Theatrum historicum et chronologicum ...Nunc continuatum et revisum a I. B. Schuppio. Ed. 6ᵃ...F°, Oxoniæ, 1662.
    H; M/J9 – 16; Tr/NQ.10.57.

754 HENRY VIII, *King of England.* Love-letters from King Henry VIII. to Anne Boleyn: some in French, and some in English...8°, London, 1714.
    Not H; M (Tracts) J3 – 13; Tr/NQ.9.153¹.

755 HERBELOT, Barthélemy d'. Bibliothèque orientale, ou Dictionaire universel contenant généralement tout ce qui regarde la connoissance des peuples de l'Orient...F°, Paris, 1697.

H; M/C4 – 15; Tr/NQ.18.28 [pp. 215, 234, 273, 455, 931 turned up, p. 232 down, and several other signs of dog-earing].

756 HERMANN, Jakob. Phoronomia, sive De viribus et motibus corporum solidorum et fluidorum libri ii. 4°, Amstelædami, 1716.

H; M/E6 – 36; Millikan Memorial Library, California Institute of Technology, Pasadena [pp. 113 & 394 turned down, the latter marking a reference to Proposition xxxvii, Book ii of the *Principia*].

757 HERODIANUS. Historiarum libri 8. recogniti & notis illustrati. (*Greek & Latin.*) 4°, Oxoniæ, 1704.

H; M/F4 – 22; Tr/NQ.16.66.

758 HERODOTUS. Historiarum libri ix...Cum Vallæ interpret. Latina ...ab H. Stephano recognita & spicilegio F. Sylburgii. F°, [Geneva], 1618.

H; M/B3 – 11; ?

759 HESYCHIUS, *Alexandrinus*. Dictionarium. (*Greek.*) F°, (Hagenoæ, 1521).

H; M/B2 – 20; Tr/NQ.11.5 [col. 587 turned up].

760 HEYDON, John. The wise-mans crown: or, The glory of the rosie-cross...(3 pts.) 8°, London, 1664–5.

H; M/E2 – 30; ?

761 HICKERINGILL, Edmund. Miscellaneous tracts, essays, satyrs, &c. in prose and verse. 4°, London, 1707.

H; M/F5 – 27; ?

762 HIEROCLES, *Alexandrinus*. Commentarius in aurea Pythagoreorum carmina. I. Curterio interprete...(*Greek & Latin.*) 12°, Parisiis, 1583.

H; M/E1 – 12; Tr/NQ.9.78 [p. 312 turned up].

763 HIG (The) Dutch Minerva à-la-mode, or A perfect grammar never extant before whereby the English may...learne the neatest dialect of the German mother-language...[By C. Miller?] 12°, [London, 1670?].

Not H; M/G2 – 47; ?

764 HILARY, *St, Bp of Poitiers*. Quotquot extant opera, nostro fere seculo literatorum quorundam non mediocri labore conquisita...F°, Parisiis, 1652.

H; M/D7 – 9; Tr/NQ.18.20 ['1680 pret. 14ˢ. 6ᵈ' in Newton's hand on fly-leaf; col. 147 turned down and a few other signs of dog-earing].

765 HIND, Thomas. The history of Greece. Vol. 1. 8°, London, 1707.

H (2 copies); M (1 only) J4 – 30; ?

HIND, Thomas. Sermons. *See* 1492.

HISTOIRE de l'Académie Royale des Inscriptions et Belles Lettres, depuis son établissement jusqu'à présent...Vols. 1–4. 1717–23. *See* 1247.

— Histoire de l'Académie Royale des Sciences. Année M.DCCVII, M.DCCVIII, M.DCCX...3 vols. 1708–13 & 13 vols. 1712–24. *See* 1249 & 1250.

766 — Histoire du règne de Louis XIV. Roi de France et de Navarre... par H. P. D. L. D. E. D. [i.e. H. P. de Limiers.] 7 vols. 12°, Amsterdam, 1717.

H; M/G8 – 15 to 21; Tr/NQ.7.68 to 74.

767 — Histoire fabuleuse et généalogique des Dieux et des héros de l'antiquité payenne. Ou Nouvelle histoire poëtique...12°, Paris, 1670.

Not H; M/G2 – 28; Tr/NQ.9.163.

768 HISTORIA ab orbe condito ad an. 1655. 8°, 1661. *Not identified.*

Not H; M/G2 – 15; ?

769 — Historia et concordia Evangelica...Opera & studio Theologi Parisiensis [i.e. A. Arnauld]. Ed. 2ª...8°, Parisiis, 1660.

Not H; M/E1 – 13; ?

770 — Historia rerum in Oriente gestarum ab exordio mundi et orbe condito ad nostra haec usque tempora...F°, Francof. ad Mœnum, 1587.

H; M/E7 – 16; Butler Library, Columbia University, New York ['Is. Newton, pret 4ˢ. 6ᵈ.' on title-page]; Sotheran **789** (1924) 5729.

771 HISTORIÆ Augustæ scriptores VI...Cum integris notis I. Casauboni, C. Salmasii & J. Gruteri...2 vols. 8°, Lugduni Batav., 1671.

H; M/F4 – 14 & 15; ?

772 — Historiæ poeticæ scriptores antiqui. Apollodorus Atheniensis. Conon Grammaticus. Ptolemæus Hephæst [etc.]. Græcè & Latinè... [Ed. by T. Gale.] (3 pts.) 8°, Parisiis, 1675.

H; M/F2 – 26; ?

773 — Historiæ rei nummariæ veteris scriptores aliquot insigniores...[By M. Hostus, J. Selden, P. Labbe, etc.] 2 vols. 4°, Lugd. Batavor., 1695.

H; M/A5 – 6 & 7; ?

774 — Historiæ Romanæ epitomæ L. I. Flori, C. V. Paterculi, S. Aur. Victoris [etc.]...16°, Amsterodami, 1647.

H; M/G1 – 25; ?

775 HISTORIARUM et chron. totius mundi epitome. 8°, 1538. *Not identified.*

Not H; M/E3 – 26; ?

776 HISTORY and antiquities of the Counties in England. 2 vols. 4°, 1721.
Not identified.
Not H; M/D5 – 15 & 16; ?

777 — The history of jewels and of the principal riches of the East and West. . .[Transl. from the French of S. Chappuzeau.] 8°, London, 1671.
Not H; M/G9 – 32; Thame/973; ?

778 — The history of religion. Written by a person of quality [Sir Robert Howard]. 8°, London, 1694.
Not H; M/H2 – 38; ?

779 — The history of the Apostles Creed: with critical observations on its several articles. [By Peter, 1st Lord King.] 2nd ed. 8°, London, 1703.
H; M/H4 – 20; Tr/NQ.8.97 [a few signs of dog-earing].

780 — The history of the bucaniers of America. . .written in several languages. . .The whole newly translated into English [by A. O. Exquemelin]. 8°, London, 1699.
H; M/J4 – 9; Tr/NQ.9.46.

781 — The history of the English & Scotch Presbytery. . .Written in French, by an eminent divine [P. Du Moulin]. . .and now Englished. 8°, Villa Franca pr. [London], 1659.
Not H; M/J8 – 12; ?

782 — The history of the Royal Family: or, A succinct account of the marriages and issue of all the Kings and Queens of England, from the Conquest. . .8°, London, 1713.
H; M/J4 – 27; Tr/NQ.9.82.

783 — The history of the Treaty of Utrecht. . .8°, London, 1712.
H; M/J4 – 24; Tr/NQ.16.52.

784 HOADLY, Benjamin, *Bp of Winchester*. An answer to the representation drawn up by the Committee of the Lower-House of Convocation concerning several dangerous positions and doctrines. . .8°, London, 1718.
H; M/H4 – 26; Tr/NQ.9.44.
— Tracts on the Bangorian controversy. 4 vols. 1718. *See* 1635.

785 HOBBES, Thomas. De mirabilibus Pecci: being the wonders of the Peak in Darby-shire. . .In English and Latine. . .8°, London, 1678.
Not H; M/J1 – 52; ?
HOBBS, William. A new discovery for finding the longitude. 1714?
See 1636.

786 HOFMANN, Kaspar. De locis affectis libri III. . .12°, Noribergæ, 1642.
Not H; M/G1 – 16; ?

HOLDER, William. Introductio ad chronologiam...Ed. 2ª, priori emendatior. [By W. Holder.] 1704. *See* 836.

787 HOLLANDUS, Johann Isaac. Opera mineralia, sive De lapide philosophico, omnia, duobus libris comprehensa...Nunc primùm ...in Latinum sermonem translata, à P. M. G. 8°, Middelburgi, 1600.
H; M/E3 – 44; Sotheran **804** (1927) 3802; ?

788 HOMERUS. Batrachomyomachia Græce ad veterum exemplarium fidem recusa...[Ed. by M. Maittaire.] 8°, Londini, 1721.
Not H; M/G10 – 19; ?

789 — Ilias, et veterum in eam scholia, quæ vulgò appellantur Didymi ...(*Greek & Latin.*) 4°, Cantabrigiæ, 1689.
Not H; M/F6 – 23; ?

790 — The Iliad. Translated by Mr Pope. 6 vols. 4°, London, 1715–20.
H (lists vols. 4–6 only); M (6 vols.) C2 – 25 to 30; ?

791 — — [Another issue.] 6 vols. 12°, London, 1720.
H; M/J1 – 3 to 8; ?

792 — Odyssea. Cum interpretatione Lat. ad verbum, post alias omnes editiones repurgata plurimis erroribus...partim ab H. Stephano, partim ab alijs...Ed. postrema diligenter recognita per I. T. P. 8°, Amstelredami, 1648.
H; M/E1 – 1; Tr/NQ.9.79 [several signs of dog-earing].

793 — Poemata duo, Ilias et Odyssea, sive Ulyssea...Cum interpretatione Lat....partim ab H. Stephano, partim ab aliis...16°, [Paris], 1589.
Not H; not M; Sir Geoffrey Keynes, Brinkley [on fly-leaf 'Isaac Newton Trin: Coll: Cam: 1661' and under this, also in Newton's hand, 'H: N. Feb: 25 1688' presumably indicating a gift to Humphrey Newton]; for provenance of the volume see G. Keynes, *Bibliotheca bibliographici* (1964) 3294.

793a — — [Another copy.]
Not H; M/G1 – 12; ?

794 HOOKE, Robert. An attempt to prove the motion of the earth from observations made by R. Hooke. 4°, London, 1674.
Not H; M (Tracts) J6 – 2; Tr/NQ.16.77².

795 — Lectures and collections made by R. Hooke...Cometa...Microscopium...4°, London, 1678.
Not H; M (Tracts) J6 – 2; Tr/NQ.16.77¹.

796 — Philosophical experiments and observations...publish'd by W. Derham. 8°, London, 1726.
H; M/E4 – 10; ?

797 — Posthumous works...Publish'd by R. Waller. F°, London, 1705.
H; M/J9 – 11; ?

798 HOOKER, Richard. The works of...R. Hooker, in eight books of ecclesiastical polity. With...an account of his life and death. F°, London, 1676.
Not H; M/A8 – 9; ?

799 HOPKINS, Ezekiel, *Bp of Derry*. An exposition on the Lord's Prayer ...To which is added some sermons...2nd ed., corrected. 8°, London, 1698.
Not H; M/H4 – 17; ?

800 — An exposition on the Ten Commandments: with other sermons. 4°, London, 1692.
Not H; M/H7 – 3; ?

801 HORATIUS FLACCUS, Quintus. Eclogæ, una cum scholiis perpetuis, tam veteribus quam novis...Adjecit etiam, ubi visum est, & sua, textumque ipsum plurimis locis, vel corruptum, vel turbatum restituit W. Baxter. 8°, Londini, 1701.
H; M/F4 – 24; Tr/NQ.9.124.

802 — Horatius sine notis. 8°, 1668. *Edition not identified.*
Not H; M/G1 – 31; ?

803 — The Odes and Epodon of Horace, in 5 books. Transl. into English by J. H. Esq. 8°, London, 1684.
Not H; M/J7 – 45; ?

804 — Opera, interpretatione et notis illustravit L. Desprez...in usum... Delphini...Ed. 6ᵃ. 8°, Londini, 1717.
Not H; M/F4 – 25; ?

805 — Poëmata, scholiis et argumentis ab H. Stephano illustrata...Ed. 3ᵃ. (3 pts.) 8°, [Paris], 1600.
Not H; M/F1 – 25; Tr/NQ.10.139 [pp. 45, 121 turned down and a few other signs of dog-earing].

806 — Poemata, scholiis sive annotationibus, quæ brevis commentarii vice esse possint, à J. Bond illustrata. 6ᵃ ed. recognita...8°, Londini, 1637.
Not H; M/G2 – 1; ?

807 — Q. Horatius Flaccus, ex recensione & cum notis atque emendationibus R. Bentleii. (2 pts.) 4°, Cantabrigiæ, 1711.
H; M/F6 – 11; Tr/NQ.8.3.

808 HORROCKS, Jeremiah. Opera posthuma; viz. Astronomia Kepleriana, defensa & promota. Excerpta ex Epistolis ad Crabtræum suum. Observationum cœlestium catalogus. Lunæ theoria nova. Accedunt G. Crabtræi, Observationes cœlestes. Quibus accesserunt, J. Flamstedii, De temporis æquatione diatriba. Numeri ad Lunæ theoriam Horroccianam. In calce adjiciuntur, nondum editæ, J. Wallisii Exercitationes tres; viz. De cometarum distantiis investigandis. De rationum & fractionum reductione. De periodo Juliana. 4°, Londini, 1678.

H; M/C1 – 2; Tr/NQ.8.19 [p. 323 turned up, p. 393 down and a few other signs of dog-earing].

HOSTUS, Matthaeus. Historiæ rei nummariæ veteris scriptores aliquot insigniores...[By M. Hostus, J. Selden, P. Labbe, etc.] 2 vols. 1695. *See* 773.

809 HOUGHTON, Thomas. Rara avis in terris: or The compleat miner, in two books...12°, London, 1681.
H; M/G9 – 40; Tr/NQ.16.108.

810 HOWARD, Edward. Copernicans of all sorts, convicted: by proving, that the earth hath no diurnal or annual motion...8°, London, 1705.
H; M/E2 – 39; Tr/NQ.10.153.
Extensive notes and observations by Newton on astronomy and magnetic variations on pp. 79, 81, 86, 121; see plate 1.

HOWARD, *Hon. Sir* Robert. The history of religion. Written by a person of quality [Sir Robert Howard]. 1694. *See* 778.

811 HOWELL, James. Instructions and directions for forren travell... 12°, London, 1650.
Not H; M/G9 – 33; ?

812 HOWELL, Laurence. A compleat history of the Holy Bible contain'd in the Old and New Testament...3 vols. 8°, London, 1718.
H; M/H5 – 26 to 28; ?

813 — Synopsis canonum Ecclesiæ Latinæ: qua canones spurii, epistolæ adulterinæ, et decreta supposititia istius Ecclesiæ Conciliorum in lucem proferuntur...F°, Londini, 1710.
H; M/A9 – 3; Tr/NQ.11.25.

814 — Synopsis canonum S.S. Apostolorum, et Conciliorum œcumenicorum & provincialium, ab Ecclesia Græca receptorum...F°, Londini, 1708.
H; M/A9 – 2; Tr/NQ.11.24.

HOWELL, William. Medulla historiæ Anglicanæ. Being a comprehensive history of the lives and reigns of the monarchs of England...[By W. Howell.] 3rd ed....1687. *See* 1059.

815 HUE (An) and cry after Doctor S—T. [Swift]; occasion'd by a true and exact copy of part of his own Diary...[By J. Smedley.] 8°, London, 1714.
Not H; M (Plays) H8 – 10; Tr/NQ.10.136⁴.

816 HUET, Pierre Daniel, *Bp of Avranches*. Histoire du commerce et de la navigation des anciens. 2ᵉ éd. revûë...8°, Paris, 1716.
H; M/G6 – 8; Tr/NQ.10.93 [several signs of dog-earing].
— Mémoires sur le commerce des Hollandois, dans tous les états et empires du monde...[By P. D. Huet.] 1717. *See* 1064.

— Memoirs of the Dutch trade...Done from the French [of P. D. Huet]. [c. 1700.] See 1067.

HUGHES, John. A complete history of England...3 vols. (Vols. 1 & 2 ed. by J. Hughes; vol. 3 ed. by W. Kennett.) 1706. See 425.

817 HUNT, William. Clavis stereometriæ: or, A key to the art of gauging ...12°, London, 1691.

Not H; M/G9 – 45; British Library/C.122.c.25; Thame/979.

HUSBAND, Edward. An exact collection of all remonstrances, declarations, votes [etc.]...December 1641...untill March the 21, 1643...[Collected by E. Husband.] 1643. See 595.

HUTCHESON, Francis. An inquiry into the original of our ideas of beauty and virtue; in two treatises...[By F. Hutcheson.] 1725. See 835.

818 HUYGENS, Christiaan. Astroscopia compendiaria, tubi optici molimine liberata. 4°, Hagæ-Comitum, 1684.

H; M/E6 – 34; Butler Library, Columbia University, New York [bound with 1368].

819 — — [Another copy.]

Not H; M (Tracts) J6 – 2; Tr/NQ.16.77⁶.

— De ratiociniis in aleæ ludo. 1657. See 1471.

820 — Horologium oscillatorium, sive De motu pendulorum ad horologia aptato demonstrationes geometricæ. F°, Parisiis, 1673.

H; M/B2 – 12; S. I. Barchas, Sonoita, Arizona [inscribed 'Pour Monsieur New[ton' – clipped by binder] in Huygens's hand on title-page]; sent to Newton via Oldenburg, 4 June 1673 (Correspondence, I, 284, 290); Thame/988; see also J. Carter & P. H. Muir, Printing and the mind of man (1967), p. 93, no. 154.

821 — Κοσμοθεωρός, sive De terris cœlestibus, earumque ornatu, conjecturæ. 4°, Hagæ-Comitum, 1698.

H; M/B1 – 27; ?

822 — Traité de la lumière. Où sont expliquées les causes de ce qui luy arrive dans la reflexion, & dans la refraction...Par C. H[uygens]. ...Avec un discours de la cause de la pesanteur. 4°, Leide, 1690.

H; M/E6 – 15; Tr/NQ.16.186 [large paper copy; on fly-leaf in Newton's hand 'Is. Newton Donum Nobilissimi Authoris']; sent to Newton via Fatio de Duillier, 24 Feb. 1689/90 (Correspondence, III, 390); a few signs of dog-earing; Sotheran 828 (1931) 3602.

823 — — [Another copy.]

H; M/E5 – 2; Tr/NQ.16.113 [standard size copy; a few signs of dog-earing].

824 HYDE, Thomas. Catalogus impressorum librorum Bibliothecæ Bodlejanæ in Academia Oxoniensi. 2 vols. in 1. F°, Oxonii, 1674.

H; M/B3 – 2; ?

825 HYGINUS, Gaius Julius. Fabularum liber, ad omnium poëtarum lectionem mirè necessarius. 12°, Lugd. Bat., 1670.
Not H; M/E1 – 39; ?

826 HYPOTHESIS physica nova...Autore G. G. L. L. [i.e. G. W. von Leibniz]. 12°, Londini, 1671.
H; not M; ?

827 IAMBLICHUS, *Chalcidensis*. De mysteriis liber. Præmittitur Epistola Porphyrii ad Anebonem Ægyptium, eodem argumento. T. Gale Græce nunc primum edidit, Latine vertit, & notas adjecit. F°, Oxonii, 1678.
H; M/B2 – 16; Tr/NQ.11.26.

828 — De vita Pythagorica liber, Græce & Latine...notisque perpetuis illustratus à L. Kustero...Accedit Malchus, sive Porphyrius, De vita Pythagoræ: cum notis L. Holstenii, & C. Rittershusii. Itemque Anonymus apud Photium De vita Pythagoræ. (3 pts.) 4°, Amstelodami, 1707.
H; M/A6 – 18; Tr/NQ.8.75 [p. 195 turned up and down, p. 213 turned down and a few other signs of dog-earing].

829 IBBOT, Benjamin. Thirty discourses on practical subjects. 2 vols. 8°, London, 1726.
H; M/H4 – 38 & 39; ?

830 IGNORAMUS. Comœdia coram Rege Jacobo...[By G. Ruggle.] Ed. 4ª...16°, Londini, 1668.
Not H; M/G1 – 13; ?

831 IMPARTIAL (An) enquiry into the management of the War in Spain, by the ministry at home, and into the conduct of those generals...abroad...8°, London, 1712.
H; M/J4 – 42; Tr/NQ.9.34.

832 INDEPENDENT (The) Whig. [By J. Trenchard and T. Gordon.] January 20, 1720 – January 4, 1721. 8°, London, 1721.
Not H; M/G10 – 21; ?

833 INDEX expurgatorius librorum qui hoc sæculo prodierunt...16°, Antverpiæ, 1584.
Not H; M/G1 – 32; ?

834 INNYS, William, *bookseller*. A catalogue of some books sold by William Innys, at the Prince's-Arms in St. Paul's Church-yard. 8°, [London, 1711?].
Not H; M/G10 – 36; Tr/NQ.8.32³ [bound with 571].

835 INQUIRY (An) into the original of our ideas of beauty and virtue; in two treatises...[By F. Hutcheson.] 8°, London, 1725.
H; M/H3 – 6; ?

836 INTRODUCTIO ad chronologiam: sive Ars chronologica in epitomen redacta. Ed. 2ª, priori emendatior. [By W. Holder.] 8°, Oxoniæ, 1704.

H; M/F1 – 14; Tr/NQ.7.2.

837 INTRODUCTION (An) of algorisme (for to lerne to recken with the pen). Newly oversene and corrected. 8°, London, 1574.

Not H; M/E1 – 14; ?

838 INTROITUS apertus ad occlusum regis palatium: autore Anonymo Philaletha [i.e. G. Starkey]...nunc primum publicatus, curante J. Langio. 8°, Amstelodami, 1667.

Not H; M/E4 – 28; ?; see also 1478.

IRELAND. A collection of all the Statutes now in use in the Kingdom of Ireland, with notes...1678. See 1261.

839 IRENAEUS, St, Bp of Lyons. Adversus Valentini, & similium Gnosticorum hæreses, libri v....Cum scholiis & annotationibus J. Billii, F. Ducæi, & F. Feu-ardentii....F°, Lutetiæ Parisiorum, 1675.

H; M/E8 – 12; Tr/NQ.18.10 ['Is. Newton. pret 14ˢ' on fly-leaf; p. 246 turned down and a few other signs of dog-earing].

840 — Contra omnes hæreses libri v...Omnia notis variorum, & suis illustravit J. E. Grabe. (Greek & Latin.) F°, Oxoniæ, 1702.

H; M/C4 – 1; ?

841 — Fragmenta anecdota, quæ ex Bibliothecâ Taurinensi eruit, Latinâ versione notisque donavit...8°, Hagæ Comitum, 1715.

H; M/A4 – 19; ?

842 ISAAC ABENDANA. Discourses of the ecclesiastical and civil polity of the Jews...8°, London, 1706.

H; M/H7 – 36; Tr/NQ.8.106.

843 ISAACSON, Henry. Saturni ephemerides, sive Tabula historico-chronologica...F°, London, 1633.

H; M/B5 – 21; ?

ISAACUS, Johannes, called Hollandus. See HOLLANDUS, Johann Isaac.

JĀBIR IBN ḤAYYĀN. See GEBER.

844 JACCHAEUS, Gilbertus. Institutiones medicæ. Ed. postrema, ab autore recognita. 12°, Lugduni Batavorum, 1631.

Not H; M/E1 – 45; ?

845 JACKSON, John. A collection of queries. Wherein the most material objections...against Dr Clarke's Scripture-doctrine of the Trinity...are proposed, and answered...8°, London, 1716.

H; M/H3 – 26; ?

— A reply to Dr. Waterland's Defense of his queries...By a clergyman in the country [J. Jackson]. 1722. See 1392.

846 JAMES II, *King of England*. Original letters of the late King's [James II], and others, to his greatest friends in England...Published by command, by W. Fuller. 8°, London, 1702.

Not H; M (Tracts) J3 – 13; Tr/NQ.9.153².

JAMES, *Prince of Wales, the Old Pretender*. A full answer to the depositions, and to all other the pretences and arguments whatsoever, concerning the birth of the pretended Prince of Wales... 1711. *See* 645.

JANSSEN, *Sir* Theodore. General maxims in trade, particularly applied to the commerce between Great Britain and France. [By Sir T. Janssen.] 1713. *See* 664.

847 JANSSONIUS, Johannes. Novus atlas, sive Theatrum orbis terrarum. Vol. 6. F°, Amstelodami, 1656.

H; M/B5 – 6; ?

848 JAY, Stephen. Τὰ Καννάκου: the tragedies of sin contemplated...2 pts. 8°, London, 1689.

Not H; M/H1 – 11; ?

849 JENKS, Benjamin. Prayers and offices of devotion for families...8°, London, 1697.

Not H; M/H1 – 10; ?

850 JESSOP, Francis. Propositiones hydrostaticæ ad illustrandum Aristarchi Samii systema destinatæ, et quædam phænomena naturæ generalia. 4°, Londini, 1687.

Not H; M (Tracts) J6 – 2; Tr/NQ.16.77⁴.

851 JEWEL, John, *Bp of Salisbury*. Apologia Ecclesiæ Anglicanæ. Priorum editionum collatione castigatior. 12°, Cantabrigiæ, 1683.

Not H; M/G2 – 22; ?

852 — A defense of the Apologie of the Churche of Englande. Conteininge an answeare to a certaine booke lately set foorthe by M. Hardinge ...F°, London, 1570.

Not H; M/J9 – 29; ?

853 JEWISH (The) Kalendar...(An almanack for the year 1693.) 12°, Oxford, [1693].

Not H; M/G9 – 49; ?

JOBERT, Louis. The knowledge of medals...Written by a Nobleman of France [L. Jobert]. Made English by an eminent hand. 2nd ed. ...1715. *See* 891.

— La science des medailles antiques et modernes...Nouv. éd., revuë, corrigée & augmentée considérablement par l'auteur [L. Jobert] ...1717. *See* 1474.

JOHANNIS, Erasmus. De Unigeniti Filii Dei existentia, inter Erasmum Iohannis, & Faustum Socinum disputatio...[By E. Johannis.] 1595. *See* 496.

854 JOHN, *St, Damascene*. Beati Ioannis Damasceni opera omnia...Item, Ioannis Cassiani Eremitæ non prorsus dissimilis argumenti libri aliquot...(*Greek & Latin*.) Fº, Basileæ, (1575).
Not H; M/G11 – 10; ?

JOHNSON, John. The clergyman's vade-mecum...[By J. Johnson.] 3rd ed....2 vols. 1723. *See* 401.

855 JOHNSON, William. Lexicon chymicum...(Ed. ultima.) (2 pts.) 8º, Francof. & Lipsiæ, 1678.
Not H; M/F3 – 35; Sotheran **804** (1927) 3803, **806** (1927) 14128, **828** (1931) 3605; ?

856 JOHNSTONE, John. Notitia regni mineralis, seu subterraneorum catalogus...12º, Lipsiæ, 1661.
Not H; M/G1 – 9; ?

857 JOLI, Guy. Mémoires: contenant l'histoire de la régence d'Anne d'Autriche, & des premières années de la majorité de Louïs XIV. jusqu'en 1666 avec les intrigues du Cardinal de Retz à la Cour. 2 vols. 8º, Amsterdam, 1718.
H; M/G6 – 23 & 24; Tr/NQ.9.140 & 141.

JONES, Henry. The Philosophical transactions, from the year 1700 to the year 1720. Abridg'd and dispos'd under general heads. By H. Jones. Vols. 4, 5. 1721. *See* 1308.

858 JONES, William. A new epitomy of the art of practical navigation... 8º, London, 1706.
H; M/G9 – 3; Tr/NQ.16.144¹ [bound with 1588]; described by Björnståhl who examined this copy in 1775 as 'rare' (*Briefe*, III (1781), 289) and marked 'scarse Book' in Musgrave Catalogue; errata list corrections made in printed text, not in Newton's hand, possibly by author.

859 — Synopsis palmariorum matheseos: or, A new introduction to the mathematics: containing the principles of arithmetic & geometry demonstrated, in a short and easie method...8º, London, 1706.
H; M/E5 – 15; Tr/NQ.9.13; described by Björnståhl as 'quite extraordinarily rare' (*Briefe*, III (1781), 289).

860 JOSEPH BEN GORION. Josephus Hebraicus...iuxta Hebraismum opera S. Munsteri uersus & annotationibus...illustratus. (2 pts.) Fº, Basileæ, 1541.
H; M/J9 – 11; ?

861 JOSEPHUS, Flavius. Opera quæ extant omnia...prolegomenis & appendice auctior redditur. (*Greek & Latin*.) Fº, Coloniæ, 1691.
H; M/D7 – 15; ?

862 — Works. With great diligence revised and amended, according to the excellent French transl. of A. D'Andilly. Fº, London, 1693.
Not H; M/C4 – 9; Tr/NQ.18.14.

863 JOURNAL (Le) des sçavans. 1665–1722. 75 vols. 12º, Amsterdam, 1679–1723.

H; M/G5 – 31 to 105?; Thame/982; ?

864 — Journal littéraire. De May & Juin, M.DCC.XIII – l'année M.DCC.XVII. Vols. 1–9i. 8º, La Haye, 1713–17.

H; M/G6 – 40 to 48; Tr/NQ.16.16 to 24; see *Correspondence*, VI, 79–80, 242–3 and pp. 67–8 above.

865 JOVIUS, Paulus, *Bp of Nocera dei Pagani*. Pauli Iovii Historiarum sui temporis. Vol. 1. 8º, Argentorati, 1556.

H; M/F2 – 32; Tr/NQ.8.128.

866 JULIANUS, *Roman Emperor*. Opera quæ quidem reperiri potuerunt omnia...Græce Latineque prodeunt cum notis. 4º, Parisiis, 1630.

H; M/A7 – 12; ?

867 JURIN, James. Dissertationis de motu aquarum fluentium contra nonnullas P. A. Michelotti animadversiones defensio...4º, Venetiis, 1724.

Not H; M (Tracts) D6 – 5; Tr/NQ.10.28³.

JUSTELLUS, Christophorus. Codex canonum Ecclesiæ Africanæ. C. Iustellus ex MSS. codicibus edidit, Græcam versionem adiunxit, & notis illustravit. 1614. *See* 408.

868 JUSTIN, *St, martyr*. Opera. Item Athenagoræ Atheniensis, Theophili Antiocheni, Tatiani Assyrii...tractatus aliquot...Quæ omnia Græcè & Latinè emendatiora prodeunt. Ed. nova...[Ed. by F. Sylburgius.] (3 pts.) Fº, Coloniæ, 1686.

H; M/B4 – 5; Tr/NQ.18.25¹ [bound with 893; a few signs of dog-earing].

869 JUSTINUS. Trogi Pompeii Historiarum Philippicarum epitoma... Accessit V. Strigelii commentarius...8º, Ursellis, 1602.

Not H; M/F2 – 14; Thame/977 (described as 'with the Autograph of Isaac Newton'); ?

870 JUVENALIS, Decimus Junius. Junii Juvenalis et Auli Persii Flacci Satyræ...16º, Amsterodami, 1651.

Not H; M/G1 – 39; ?

871 — D. Junii Juvenalis et A. Persii Flacci Satiræ. Interpretatione ac notis illustravit L. Prateus...in usum Serenissimi Delphini. Ed. 3ª, prioribus multò correctior. 8º, Londini, 1707.

Not H; M/F4 – 27; ?

872 — Satires. Translated into English verse. By Mr. Dryden. And several other eminent hands. Together with the Satires of Aulus Persius Flaccus. Made English by Mr. Dryden...3rd ed., adorn'd with sculptures. 8º, London, 1702.

Not H; M/J3 – 15; Tr/NQ.16.53.

873 KABBALA denudata seu Doctrina Hebræorum transcendentalis et metaphysica atque theologica...[Ed. by C. Knorr von Rosenroth.] (4 pts.) 2 vols. 4°, Sulzbaci, Francofurti, 1677–84.

H; M/B1 – 4 & 5; Tr/NQ.8.28 & 29 [vol. 1, pt 1 has pp. 151, 227, 241, 456, 570 turned up, pp. 152, 455, 677 down, pt 2 p. 241 down, pt 3 pp. 49, 54, 140 down, p. 147 up, vol. 2, pt 2 p. 194 up, p. 221 down, and several other signs of dog-earing esp. in vol. 1].

874 KALÎLAH WA-DIMNAH. Specimen sapientiæ Indorum veterum. Id est Liber ethico-politicus pervetustus...Nunc primum Græce ...cum versione nova Latina, opera S. G. Starkii. 8°, Berolini, 1697.

H; M/F1 – 20; Tr/NQ.10.76.

KEBLE, Joseph. The Statutes at large...from Magna Charta until this time...By J. Keble. 2 vols. 1695. *See* 1262.

875 KEILL, John. Epistola ad virum clarissimum J. Bernoulli (in qua Dominum Newtonum & seipsum defendit contra criminationes à Crusio quodam objectas, & in Actis Lipsiensibus publicatas...). 4°, Londini, 1720.

Not H; M (Tracts) J6 – 1; Tr/NQ.16.76[4].

876 — An examination of the reflections on the Theory of the earth [of T. Burnet]...8°, Oxford, 1699.

H; M/E4 – 15; Thame/977; ?

877 — Introductio ad veram astronomiam, seu Lectiones astronomicæ habitæ in Schola Astronomica Academiæ Oxoniensis. 8°, Oxoniæ, 1718.

H; M/E5 – 10; Tr/NQ.9.19 [p. 354 was formerly turned down].

878 — Introductio ad veram physicam. Seu Lectiones physicæ habitæ in Schola Naturalis Philosophiæ Academiæ Oxoniensis ...8°, Oxoniæ, 1702.

H; M/E5 – 11; Millikan Memorial Library, California Institute of Technology, Pasadena.

KEITH, George. Tracts on the longitude, by Hobbs, Keith, Rowe, &c. 1709. *See* 1636.

879 KEMPIS, Thomas à. De Imitatione Christi libri iv. 8°, 1672. *Edition not identified.*

Not H; M/G1 – 45; ?

880 KENDAL, John. Χρονομετρια or, The measure of time in directions ...Containing tables of the equation of arch's of direction...8°, London, 1684.

Not H; M/E3 – 13; Tr/NQ.16.140.

881 KENNEDY, Peter. An essay on external remedies. Wherein it is considered, whether all the curable distempers incident to human bodies, may not be cured by outward means...8°, London, 1715. H; M/E5 – 39; Tr/NQ.9.50¹ [bound with 1205].

— Ophthalmographia; or, A treatise of the eye, in 2 parts...[By P. Kennedy.] 1713. *See* 1205.

KENNETT, White, *Bp of Peterborough.* A complete history of England ...3 vols. (Vols. 1 & 2 ed. by J. Hughes; vol. 3 ed. by W. Kennett.) 1706. *See* 425.

KEPLER, Johann. Dioptrice. 1682. *See* 651.

882 KERCKRING, Theodor. Commentarius in Currum triumphalem Antimonii Basilii Valentini, a se Latinitate donatum. 8°, Amstelæ-dami, 1671. H; M/E1 – 25; ?

883 KERSEY, John. The elements of that mathematical art commonly called algebra, expounded in 4 books. 2 vols. in 1. F°, London, 1673–4. H; M/B2 – 14; ?

884 KETTLEWELL, John. An help and exhortation to worthy communicating. Or A treatise describing the meaning...of the Holy Sacrament...8th ed. 8°, London, 1717. Not H; M/H4 – 23; ?

885 KEYSLER, Johann Georg. Exercitatio historico-philologica, de Dea Nehalennia, numine veterum Walachrorum topico...4°, Cellæ, 1717. Not H; M/C1 – 32; Tr/NQ.9.172² [bound with 1148].

KIDD, William. A full account of the proceedings in relation to Capt. Kidd. In 2 letters...1701. *See* 643.

KING, Peter King, *Lord, 1st Baron of Ockham.* An enquiry into the constitution, discipline, unity & worship of the Primitive Church ...By an impartial hand [i.e. Peter King]. 2 pts. 1692 & 1712–13. *See* 559 & 560.

— The history of the Apostles Creed...[By Peter King.] 2nd ed. 1703. *See* 779.

886 KING, William, *Abp of Dublin.* De origine mali. 8°, Londini, 1702. H; M/H4 – 21; ?

887 KING, William, *LL.D.* Miscellanies in prose and verse. 8°, London, [1709]. Not H; M/J3 – 11; ?

888 KIRCHER, Konrad. Concordantiæ Veteris Testamenti Græcæ, Ebræis vocibus respondentes. 2 vols. 4°, Francofurti, 1607. H; M/C6 – 20 & 21; ?

889 KNATCHBULL, *Sir* Norton. Annotations upon some difficult texts in all the books of the New Testament. 8°, Cambridge, 1693.
Not H; M/H2 – 15; Tr/NQ.9.41 [title-page damaged].
'Error' in Newton's hand in margins of pp. 34, 37, 38; several signs of dog-earing.

890 KNIGHT, James. Eight sermons preached at the Cathedral Church of St. Paul...8°, London, 1721.
H; M/H6 – 38; ?

— The scripture doctrine of the most holy and undivided Trinity, vindicated from the misinterpretations of Dr. Clarke. [By J. Knight.]...1714. *See* 1476.

KNORR VON ROSENROTH, Christian. Kabbala denudata...
[Ed. by C. Knorr von Rosenroth.] (4 pts.) 2 vols. 1677–84. *See* 873.

891 KNOWLEDGE (The) of medals: or, Instructions for those who apply themselves to the study of medals, both ancient and modern...
Written by a Nobleman of France [L. Jobert]. Made English by an eminent hand. 2nd ed. To which is added, An essay concerning the error in distributing modern medals. By J. Addison. 12°, London, 1715.
Not H; M/J8 – 41; Tr/NQ.16.7.

892 KOENIG, Emanuel. Regnum minerale, physicè, medicè, anatomicè, chymicè, alchymicè, analogicè, theoreticè & practicè investigatum, perscrutatum & erutum...4°, Basileae Rauracorum, 1686.
Not H; M/B1 – 13; Tr/NQ.16.83.

893 KORTHOLT, Christian. In Iustinum Martyrem, Athenagoram, Theophilum Antiochenum, Tatianum Assyrium, commentarius. F°, Francofurti, 1686.
Not H; M/B4 – 5; Tr/NQ.18.25² [wants title-page; bound with 868].

894 KRAUSE, Johann Gottlieb. Nova litteraria anni MDCCXVIIII in supplementum Actorum eruditorum divulgata...8°, Lipsiae, [1720].
H; M/D1 – 11; Tr/NQ.16.164; *see also* 7.

895 LA BIGNE, Margarinus de. Bibliothecæ veterum Patrum et auctorum ecclesiasticorum. Ed. 4ª. 3 vols. [from a published set of 10]. F°, Parisiis, 1624.
H (lists '3 odd vol.'); M/E8 – 13 to 15; ?

896 LA BRUYÈRE, Jean de. The Characters, or The manners of the age ...Made English by several hands...3rd ed. corrected...8°, London, 1702.
Not H; M/E4 – 27; ?

897 LA CHASTRE, René de. Le prototype ou Tres parfait et analogique exemplaire de l'art chimicq;...8°, Paris, 1620.
H; M/G7 – 1; Tr/NQ.16.111.

898 LA COURT, Pierre Le change universel, ou Cours de change de toutes les villes de l'Europe. 8°, Bruxelles, 1695.
H; M/G6 – 7; Tr/NQ.7.8.

LA CROZE, Mathurin Veyssière de. Vindiciae veterum scriptorum, contra J. Harduinum...[By M. V. de la Croze.] 1708. *See* 1693.

899 LACTANTIUS FIRMIANUS, Lucius Coelius. De mortibus persecutorum liber. Accesserunt Passiones SS. Perpetuæ & Felicitatis. S. Maximiliani. S. Felicis. 12°, Oxonii, 1680.
Not H; M/G2 – 16; ?

900 — Opera quæ extant, ad fidem MSS. recognita et commentariis illustrata a T. Spark. 8°, Oxonii, 1684.
H; M/F4 – 17; ?

LACY, John. The general delusion of Christians, touching the ways of God's revealing Himself, to, and by the Prophets...[By J. Lacy.] 4 pts. 1713. *See* 663.

901 L'AGNEAU, David. Harmonie mystique, ou Accord des philosophes chymiques...Traduit par le Sr Veillutil...8°, Paris, 1636.
H; M/G8 – 5; Tr/NQ.16.129.

902 LAGNY, Thomas Fantet de. Méthodes nouvelles et abbregées pour l'extraction et l'approximation des racines...2e éd. 4°, Paris, 1692.
Not H; M/E6 – 34; Butler Library, Columbia University, New York [bound with 1368]; Sotheran **828** (1931) 3604.

903 LA HIRE, Philippe de. Nouveaux élémens des sections coniques, les lieux géometriques, la construction, ou effection des équations. 12°, Paris, 1679.
H; M/G2 – 39; Tr/NQ.9.164.

904 — Tabularum astronomicarum pars prior, de motibus solis et lunæ ...4°, Parisiis, 1687.
H; M/E6 – 18; Dibner Collection, Smithsonian Institution Libraries, Washington; Sotheran **786** (1923) 3395.
Mathematical calculations by Newton written on a slip of paper (attached to the end-paper) the obverse of which shows the top fragment of a letter addressed to Newton headed 'Crane-Court, [?] 11, 1722'.

905 LA HONTAN, Louis Armand de Lom d'Arce, *baron de*. Nouveaux voyages dans l'Amérique septentrionale...3 vols. 12°, La Haye, 1703.
H; M/G6 – 11 to 13; ?

LAMB, Thomas. A fresh suit against independency...[By T. Lamb.] 1677. *See* 640.

906 LA MOTTE, de. Le dénoüement de la quadrature du cercle. 4°, Utrecht, 1700.
Not H; M (Tracts) D5 – 25; Tr/NQ.8.51^4.

907 LAMPE, Friedrich Adolph. De cymbalis veterum libri III. . .12°, Trajecti ad Rhenum, 1703.
H; M/G2 – 14; ?

908 LAMY, Guillaume. Dissertation sur l'antimoine. 12°, Paris, 1682.
H; M/G2 – 26; Tr/NQ.16.107.

909 LANCASTER, Peter. A chronological essay on the Ninth Chapter of Daniel. . .4°, London, 1722.
H; M/J6 – 9; ?

910 LANCELOTTO, Giovanni Paolo. Institutiones juris canonici. . .12°, Parisiis, 1670.
Not H; M/A2 – 20; ?

911 LANDI, Costanzo, conte. Selectiorum numismatum, praecipue Romanorum expositiones. . .4°, Lugduni Batavorum, 1695.
H; M/F6 – 16; ?

912 LANGBAINE, Gerard. The lives and characters of the English dramatick poets. . .First begun by Mr. Langbain, improv'd and continued down to this time, by a careful hand. 8°, London, [1699].
Not H; M/J7 – 40; Tr/NQ.9.21.

913 LANGE, Johann Christian. Inventum novum quadrati logici universalis: in trianguli quoque formam commode redacti. . .(2 pts.) 8°, Gissae Hassorum, 1714.
H; M/F3 – 31; Tr/NQ.16.166.

914 LANGE, Johann Michael. Philologiæ Barbaro-Græce pars prior (& altera). 4°, Noribergæ & Altdorfi, 1708.
H; M/C1 – 19; Tr/NQ.8.76.

915 LANGIUS, Karl Nikolaus. Methodus nova & facilis Testacea marina . . .4°, Lucernæ, 1722.
H; M (Tracts) J6 – 15; Tr/NQ.8.52⁴.

916 LANSDOWNE, George Granville, 1st Baron. Poems upon several occasions. 4th ed. 12°, London, 1726.
Not H; M/J3 – 32; ?

917 LA QUINTINIE, Jean de. The compleat gard'ner. . .3rd ed. 8°, London, 1701.
Not H; M/J7 – 17; ?
LA RAMÉE, Pierre. See RAMUS, Petrus.

918 LA ROQUE, Jean de. Voyage [du Chevalier d'Arvieux] dans la Palestine, vers le Grand Emir. . .Fait par ordre du Roi Louis XIV . . .12°, Amsterdam, 1718.
H; M/G7 – 7; Tr/NQ.8.91.
— Voyage de l'Arabie Heureuse par l'Océan Oriental, & le Détroit de la Mer Rouge. . .[By J. de la Roque.] 1716. See 1699.

919 LARREY, Isaac de. Histoire de France, sous le règne de Louïs XIV. 4 vols. 8°, Rotterdam, 1718.
H; M/G8 – 22 to 25; Tr/NQ.10.110 to 113.

920 LAWRENCE, John. The clergy-man's recreation: shewing the pleasure and profit of the art of gardening. 5th ed. 8°, London, 1717.
Not H; M/J7 – 14; Tr/NQ.9.43¹ [bound with 921 & 922].

921 — The fruit-garden kalendar: or, A summary of the art of managing the fruit-garden...8°, London, 1718.
Not H; M/J7 – 14; Tr/NQ.9.43³ [bound with 920].

922 — The gentleman's recreation: or The second part of the art of gardening improved...2nd ed. 8°, London, 1717.
Not H; M/J7 – 14; Tr/NQ.9.43² [bound with 920].

923 LAZIUS, Wolfgang. De gentium aliquot migrationibus, sedibus fixis, reliquijs, linguarúmq; initijs & immutationibus ac dialectis, libri XII...F°, Basileæ, (1557).
H; M/E7 – 13; Tr/NQ.18.3 [a few signs of dog-earing].

924 LE BLANC, François. Traité historique des monnoyes de France, avec leurs figures...4°, Amsterdam, 1692.
H; M/F6 – 10; Tr/NQ.10.51.

925 LE BRUN, Charles. The conference of M. Le Brun, chief painter to the French King...director of the Academy of Painting and Sculpture; upon expression, general and particular. Transl. from the French...8°, London, 1701.
Not H; M/J1 – 57; ?

926 LE CLERC, Daniel. Histoire de la médecine...(3 pts.) 4°, Amsterdam, 1702.
H; M/J6 – 11; Tr/NQ.8.54 [a few signs of dog-earing].

927 LE CLERC, David. Quæstiones sacræ, in quibus multa Scripturæ loca...explicantur...8°, Amstelædami, 1685.
H; M/A1 – 1; ?

928 LE CLERC, Jean. Bibliothèque ancienne et moderne. Par J. Le Clerc. Pour l'année MDCCXIV–MDCCXXII. Vols. 1–18. 12°, Amsterdam, 1714–22.
H; M/G5 – 1 to 16, 19 & 20; Tr/NQ.7.10 to 27 [vol. 1 has pp. 380, 382, 429 turned down].

929 — Bibliothèque choisie, pour servir de suite à la Bibliothèque universelle. Année MDCCVIII–MDCCXIII. Vols. 16–27. 12°, Amsterdam, 1708–13.
H; M/G5 – 17 & 18, 21 to 30; Tr/NQ.7.28 to 39.

930 — Historia ecclesiastica duorum primorum a Christo Nato saeculorum, e veteribus monumentis depromta. 4°, Amstelodami, 1716.
H; M/C6 – 8; Tr/NQ.10.62.

931 — Joannis Clerici vita et opera ad annum MDCCXI., amici ejus opusculum. [By Le Clerc himself.] 12°, Amstelodami, 1711.
H; M/E2 – 26; Tr/NQ.10.91.

932 LE COMTE, Louis Daniel. Memoirs and observations...Made in a late journey through the Empire of China...Transl. from the Paris ed. 8°, London, 1698.
H; M/J4 – 21; ?

933 LEGAL provisions for the poor...By S. C[arter]. 4th ed....12°, London, 1718.
Not H; M/J8 – 42; Tr/NQ.10.72.

934 LEGUAT, François. Voyage et avantures de François Leguat, & de ses compagnons, en deux isles desertes des Indes Orientales... 2 vols. 12°, Londres, 1708.
H; M/G7 – 4 & 5; Tr/NQ.10.114 & 115.

LEIBNIZ, Gottfried Wilhelm von. Causa Dei asserta per justitiam ejus...[By G. W. von Leibniz.] 1710. See 354.

935 — A collection of papers, which passed between the late learned Mr. Leibnitz, and Dr. Clarke, in the years 1715 and 1716. Relating to the principles of natural philosophy and religion...(2 pts.) 8°, London, 1717.
H; M/H3 – 18; ?

— Essais de Théodicée sur la bonté de Dieu, la liberté de l'homme et l'origine du mal. [By G. W. von Leibniz.] 2 vols. in 1. 1710. See 571.

— Hypothesis physica nova...Autore G. G. L. L. [i.e. G. W. von Leibniz.] 1671. See 826.

— Recueil de diverses pièces...Par Mrs. Leibniz, Clarke, Newton, & autres autheurs célèbres. [Ed. by P. Des Maizeaux.] 2 vols. 1720. See 1379 & 1380.

— Scriptores rerum Brunsvicensium illustrationi inservientes...cura G. G. Leibnitii. 3 vols. 1707–11. See 1475.

936 LEIDEKKER, Melchior. De Republica Hebræorum libri XII...(2 pts.) F°, Amstelædami, 1704.
H; M/D8 – 6; Tr/NQ.11.47 [p. 32 turned down and a few other signs of dog-earing].

937 LEIGH, Edward. Critica sacra in 2 parts: or Observations on all the radices, or primitive Hebrew words of the Old Testament... 2nd ed., corrected and much enlarged. F°, London, 1650.
Not H; M/J9 – 23; Tr/NQ.10.58.

938 LEMERY, Nicolas. A course of chymistry...3rd ed., transl. from the 8th ed. in the French...8°, London, 1698.
H; M/E4 – 14; Tr/NQ.8.118 [pp. 165, 172, 210, 255, 265, 283, 296, 297 turned up, pp. 167, 176, 182, 254, 262, 266, 317 down, and many other signs of dog-earing].

939 — Traité de l'antimoine, contenant l'analyse chymique de ce minéral, & un recueil d'un grand nombre d'opérations. . .12°, Paris, 1707.
H; M/G7 – 9; Tr/NQ.16.128 [several signs of dog-earing].

940 LE MOINE, Pierre. Of the art both of writing & judging of history, with reflections upon ancient as well as modern historians. . .12°, London, 1695.
Not H; M/J8 – 49; Tr/NQ.9.64.

941 LE MORT, Jacob. Chymiæ veræ nobilitas & utilitas, in physica corpusculari, theoria medica. . .(4 pts.) 4°, Lugduni Batavorum, 1696.
H; M/B1 – 1; Tr/NQ.16.78.

942 LE NEVE, John. The lives and characters. . .of all the Protestant Bishops of the Church of England. . .Vol. 1, Pts 1 & 2. 8°, London, 1720.
H; M/G10 – 3 & 4; ?

943 — Monumenta Anglicana: being inscriptions on the monuments of several eminent persons. . .5 vols. 8°, London, 1717–19.
H; M/G10 – 5 to 9; ?

944 LEO, Joannes, *Africanus*. A geographical historie of Africa. . .Transl. and collected by J. Pory. F°, Londini, 1600.
H; not M; ?

945 LEON, *of Modena*. Cérémonies et coûtumes qui s'observent aujourd'huy parmy les Iuifs. Traduites de l'Italien de Léon de Modène par R. Simon. (*With* Comparaison des cérémonies des Iuifs, et de la discipline de l'Église. Par le Sieur de Simonville [i.e. R. Simon].) 12°, [Paris, *c.* 1681].
Not H; M/G2 – 29; Tr/NQ.16.4 [wants title-page].

946 LE PAYS, René. Amitiez, amours, et amourettes. 8°, Paris, 1678.
Not H; M/G2 – 45; ?

947 LE ROUX, Philibert Joseph. Dictionaire comique, satyrique, critique, burlesque, libre & proverbial. . .8°, Amsterdam, 1718.
H; M/G8 – 1; Tr/NQ.8.50.

948 LESLIE, Charles. The case stated, between the Church of Rome and the Church of England. . .8°, London, 1713.
H; M/H3 – 33; ?

L'ESTOILE, Pierre de. Mémoires pour servir à l'histoire de France . . .depuis 1515. jusqu'en 1611. . .[By P. de l'Estoile.] 2 vols. 1719. *See* 1063.

L'ESTRANGE, Hamon. The reign of King Charles. . .[By H. L'Estrange.] 1655. *See* 1386.

949 LETTER (A) to the Honourable A—r M—re [Arthur Moore], Com— ner [Commissioner] of trade and plantation. 8°, London, 1714.
Not H; M (Tracts) J2 – 38; Tr/NQ.16.194[7].

950 LETTRE d'un philosophe sur le secret du grand œuvre, écrite au sujet de ce qu'Aristée a laissé par écrit à son fils, touchant le magistère philosophique. Le nom de l'autheur est en Latin dans cett' anagramme. Dives Sicut Ardens. S. [i.e. A. T. de Limojon de Saint-Didier]. 12º, La Haye, 1686.
   Not H; M (Tracts) H8 – 14; Tr/NQ.16.95⁴.

951 LEUSDEN, Johannes. Compendium Græcum Novi Testamenti... Ed. 5ª. 8º, Londini, 1691.
   Not H; M/F1 – 9; Tr/NQ.10.135 [p. 44 turned down].

   L'HÔPITAL, Guillaume François Antoine de, *marquis de Sainte Mesme*. Analyse des infiniment petits, pour l'intelligence des lignes courbes. [By Le Marquis de L'Hôpital.] 1696. *See* 42.

952 — Traité analytique des sections coniques et de leur usage pour la résolution des équations...Ouvrage posthume. 4º, Paris, 1707.
   H; M/E6 – 8; Tr/NQ.10.50.

953 LHUYD, Edward. Lithophylacii Britannici ichnographia. Sive Lapidum aliorumque fossilium Britannicorum singulari figura insignium...8º, Londini, 1699.
   H; M/E5 – 3; Tr/NQ.16.116.

954 LIBAVIUS, Andreas. Alchymia, recognita, emendata, et aucta... tum commentario medico physico chymico...(3 pts.) Fº, Francofurti, 1606.
   Not H; M/H9 – 17; Thame/966; ?

955 LIBELLI seu decreta a Clodoveo et Childeberto, & Clothario...16º, [1550?].
   Not H; M/G1 – 44; ?

956 LIEBKNECHT, Johann Georg. Matheseos felix cum theologia nexus breviter delineatus...8º, Giessae, 1722.
   Not H; M (Tracts) H8 – 16; Tr/NQ.10.82¹.

957 LIGHTFOOT, John. Works. Revised and corrected by G. Bright. 2 vols. Fº, London, 1684.
   H; M/F8 – 16 & 17; ?

   LILY, William. A short introduction of grammar...[By W. Lily and J. Colet.] (2 pts.) 1692. *See* 1510.

958 LIMBORCH, Philippus van. Theologia Christiana ad praxin pietatis ac promotionem pacis Christianæ unicè directa. 4º, Amstelædami, 1686.
   Not H; M/C6 – 7; ?

   LIMIERS, Henri Philippe de. Histoire du règne de Louis XIV. Roi de France et de Navarre...par H. P. D. L. D. E. D. [i.e. H. P. de Limiers.] 7 vols. 1717. *See* 766.

LIMOJON DE SAINT-DIDIER, Alexandre Toussaint de. Lettre d'un philosophe sur le secret du grand œuvre, écrite au sujet de ce qu'Aristée a laissé par écrit à son fils, touchant le magistère philosophique... [By A. T. de Limojon de Saint-Didier.] 1686. *See* 950.

— Le triomphe hermetique, ou La pierre philosophale victorieuse... [By A. T. de Limojon de Saint-Didier.] 1689. *See* 1642.

959 LIPSIUS, Justus. Epistolarum selectarum chilias...8°, Genevæ, 1611. Not H; M/F3 – 23; Tr/NQ.8.135.

960 — Roma illustrata, sive Antiquitatum Romanarum breviarium...Ex nova recensione A. Thysii...Postrema ed. 12°, Amstelodami, 1657.
Not H; M/E1 – 35; ?

961 LIQUOR Alcahest, or A discourse of that immortal dissolvent of Paracelsus & Helmont... [By G. Starkey.] [2nd ed.?] 8°, London, 1684.
Not H; M/G9 – 27; Thame/976; ?

962 LISTER, Martin. Sex exercitationes medicinales de quibusdam morbis chronicis...8°, Londini, 1694.
H; M/A4 – 8; Tr/NQ.9.130.
LITURGIA, seu Liber precum communium...juxta usum Ecclesiæ Anglicanæ...1681. *See* 243.

— La liturgia ynglesa...Hispanizado por F. de Alvarado...Ed. 2ª corregida...1715. *See* 244.

963 LIVELY (The) oracles given to us, or The Christians birth-right and duty...By the author of The whole duty of man, &c. [R. Allestree]. 8°, Oxford, 1678.
Not H; M/H7 – 31; ?

964 LIVIUS, Titus. Historiæ Romanæ principis libri omnes superstites: post aliorum omnium emendationes nunc præterea castigati...à J. Grutero...8°, Francofurti, 1609.
Not H; M/F2 – 7; Tr/NQ.9.11 [imperfect: wants all after p. 828; extensive notes on fly-leaf, end-papers, and in text in the hand of Barnabas Smith].

965 — Historiarum quod exstat...Recensuit et notulis auxit J. Clericus. 10 vols. 8°, Amstelaedami, 1710.
Not H; M/F1 – 28 to 37; ?

966 LOCKE, John. A collection of several pieces...never before printed ...Publish'd by the Author of the Life of...John Hales [i.e. P. Des Maizeaux]. 8°, London, 1720.
H; M/E4 – 20; Dr H. F. Norman, Ross, Calif.

967 — An essay concerning humane understanding. In 4 books. F°, London, 1690.

H; M/B2 – 1; Dr H. F. Norman, Ross, Calif. [pp. 111, 133, 157, 349, 357, 359 turned up and a few other signs of dog-earing; two corrections in text in Locke's hand]; Thame/946.

968 — De intellectu humano. Ed. 4ª aucta & emendata, & nunc primum Latine reddita. F°, Londini, 1701.

H; M/B2 – 15; Tr/NQ.11.27.

969 — A letter to...Edward [Stillingfleet] Lᵈ Bishop of Worcester, concerning some passages relating to Mr. Locke's Essay of humane understanding...8°, London, 1697.

H; M/E4 – 23; ?

970 — Mʳ. Locke's reply to...the Lord Bishop of Worcester's [Stillingfleet] Answer to his letter concerning some passages relating to Mʳ. Locke's Essay...8°, London, 1697.

H; M/E4 – 22; ?

971 — Mʳ. Locke's reply to...the Lord Bishop of Worcester's [Stillingfleet] Answer to his Second letter...8°, London, 1699.

H; not M; ?

972 — A paraphrase and notes on the Epistles of St. Paul...4°, London, 1707.

H; M/C6 – 11; ?

973 — Posthumous works. 8°, London, 1706.

H; M/E4 – 21; ?

974 — A second letter concerning toleration. 4°, London, 1690.

H; not M; ?

975 — A third letter for toleration...4°, London, 1692.

H; not M; ?

976 — Some considerations of the consequences of the lowering of interest, and raising the value of money. 8°, London, 1692.

H; M/J8 – 34; Tr/NQ.16.159¹ [bound with 977 & 978; a few signs of dog-earing].

977 — Short observations on a printed paper, intituled, For encouraging the coining silver money in England, and after for keeping it here. 8°, London, 1695.

Not H; M/J8 – 34; Tr/NQ.16.159² [bound with 976].

978 — Further considerations concerning raising the value of money. Wherein Mr. Lowndes's Arguments...are particularly examined. 2nd ed. corrected. 8°, London, 1695.

Not H; M/J8 – 34; Tr/NQ.16.159³ [bound with 976].

LOCKHART, George. Memoirs concerning the affairs of Scotland, from Queen Anne's accession to the throne, to...1707...[By G. Lockhart.] 1714. *See* 1066.

979 LOGGAN, David. Cantabrigia illustrata. F°, Cantabrigiæ, [1690].
H; M/C5 – 1; Thame/948; ?

980 LOGICA, sive Ars cogitandi: in qua præter vulgares regulas plura
nova habentur ad rationem dirigendam utilia. [By A. Arnauld
and P. Nicole.] E tertia apud Gallos ed. recognita & aucta in
Latinum versa. 8°, Londini, 1687.
Not H; M/F1 – 7; Tr/NQ.10.71.

981 LONGINUS, Caesar. Trinum magicum, sive Secretorum magicorum
opus...12°, Francofurti, 1673.
Not H; M/E1 – 8; ?

982 LONGINUS, Dionysius. De sublimitate libellus, cum præfatione de
vita & scriptis Longini, notis, indicibus, & variis lectionibus. Ed.
2ª. (Greek & Latin.) 4°, Oxoniæ, 1718.
Not H; M/A4 – 10; Tr/NQ.9.10.

983 — An essay upon sublime style. Transl. from the Greek. 16°, Oxford,
1698.
Not H; M/J8 – 18; ?

984 LORD'S PRAYER. Oratio Dominica in diversas omnium fere gentium
linguas versa et propriis cujusque linguae characteribus expressa
...Editore J. Chamberlaynio. (2 pts.) 4°, Amstelædami, 1715.
H (2 copies, 1 of which is described as 'Spanish calf, gilt leaves');
M (1 only) H2 – 9; Tr/NQ.8.24 [not the specially bound
copy].

LOWNDES, William. A report containing an essay for the amendment
of the silver coins. [By W. Lowndes.] 1695. See 1393.

LOWTHORP, John. The Philosophical transactions and collections,
to the end of the year 1700. Abridg'd and dispos'd under general
heads. By J. Lowthorp. 3 vols. 1705. See 1307.

985 LUBIENIECKI, Stanisław. Theatrum cometicum...2 vols. F°,
Lugduni Batavorum, 1681.
H; M/A9 – 5 & 6; Tr/NQ.11.6 & 7 [a few signs of dog-earing in
vol. 1].

986 LUCANUS, Marcus Annaeus. Pharsalia, sive De bello civili Cæsaris
et Pompeii lib. x. Ex emendatione H. Grotii, cum eiusdem notis.
12°, Amsterodami, 1651.
Not H; M/G1 – 30; ?

987 LUCAS, Paul. Voyage du Sieur Paul Lucas au Levant...2 vols. 8°,
Paris, 1704.
H; M/G8 – 26 & 27; Tr/NQ.9.86 & 87.

LUCAS, Richard. An enquiry after happiness. [By R. Lucas.] Vol. 1.
[No more published.] 1685. See 558.

988 LUCIANUS, *Samosatensis*. Opera omnia quæ extant. Cum Latina doctiss. virorum interpretatione. I. Bourdelotius cum Regijs Codd. aliisque Mss. contulit, emendavit, supplevit...(2 pts.) (*Greek & Latin.*) F°, Lutetiæ Parisiorum, 1615.

H; M/C3 – 7; Tr/NQ.11.18 [p. 172 turned up, pp. 364, 394, 512 down, and a few other signs of dog-earing].

989 — Opera. Ex versione I. Benedicti. Cum notis integris I. Bourdelotii [etc.]. Accedunt inedita scholia in Lucianum, ex Bibliotheca I. Vossii. (*Greek & Latin.*) 2 vols. 8°, Amstelodami, 1687.

H; M/F4 – 6 & 7; Tr/NQ.8.38 & 39 [vol. 2 has pp. 297, 299, 470, 711 turned up, pp. 328, 356, 361, 476, 657, 684 down and a few other signs of dog-earing].

990 LUCRETIUS CARUS, Titus. De rerum natura libri vi. Quibus additæ sunt conjecturæ & emendationes T. Fabri cum notulis perpetuis ...12°, Cantabrigiæ, 1686.

H; M/F1 – 5; Tr/NQ.9.73.

Line numbers added in margins by Newton, in tens up to 380, then in larger intervals, ending in fifties at 1050 on p. 17; a few signs of dog-earing.

991 LUDOLFUS, Job. Iobi Ludolfi...ad suam Historiam Æthiopicam... commentarius...F°, Francofurti ad Moenum, 1691.

H; M/B2 – 11; ?

992 — A new history of Ethiopia...Made English, by J. P. 2nd ed. F°, London, 1684.

H; M/B2 – 10; Thame/965; ?

993 LUDUS mathematicus: or, The mathematical game...By E. W[ingate]. 12°, London, 1681.

Not H; M/G9 – 51; Thame/974; ?

LUIDIUS, Edvardus. *See* LHUYD, Edward.

994 LULL, Raymund. Ars magna, generalis et ultima...8°, Francofurti, 1596.

H; M/A3 – 4; Tr/NQ.10.70.

995 — De secretis naturæ sive Quinta essentia libri ii. His accesserunt, Alberti Magni Summi philosophi, De mineralibus & rebus metallicis libri v...8°, (Argentorati), 1541.

Not H; M/E3 – 18; Thame/976; ?

996 — De secretis naturæ, seu De quinta essentia liber unus...8°, Coloniæ, 1567.

H; M/E3 – 17; ?

997 — Liber, qui codicillus, seu vade mecum inscribitur, in quo fontes alchimicæ artis & reconditioris philosophiæ traduntur...8°, Coloniæ, 1563.

H; M/E3 – 15; Tr/NQ.16.133 [many signs of dog-earing].

998 — Mercuriorum liber iam tandem subsidio manuscripti exemplaris perfectè editus. Item eiusdem Apertorium, Repertorium, Artis intellectivæ theorica & practica, Magia naturalis...8°, Coloniæ Agrippinæ, 1567.
Not H; M/E3 – 19; ?

999 — Opera ea quæ ad adinventam ab ipso artem universalem, scientiarum artiumque omnium brevi compendio...8°, Argentorati, 1609.
H; M/E5 – 42; bound with 32; Sotheran **800** (1926) 11461, **804** (1927) 3806, **828** (1931) 3606, **843** (1935) 1498; ?

1000 — Raymundi Lullii Maioricani philosophi sui temporis doctissimi libelli aliquot chemici: nunc primùm, excepto Vade mecum, in lucem opera Doctoris Toxitæ editi...8°, Basileæ, 1572.
Not H; M/E3 – 16; Tr/NQ.16.37 [many signs of dog-earing].
References by Newton to other works by Lull: in margins of p. 7 'Ars magica p. 378, 379', p. 159 'Vide Lib. Merc. p. 156 l. 11 & p. 179 l. 16.'

1001 — Tractatus brevis et eruditus, De conservatione vitæ: item Liber secretorum seu quintæ essentiæ...8°, Argentorati, 1616.
Not H; M/E3 – 14; ?

1002 LUMEN de lumine: or A new magicall light discovered, and communicated to the world, by Eugenius Philalethes [i.e. T. Vaughan]...8°, London, 1651.
H; M/G9 – 38; ?

1003 LUMIÈRE (La) sortant par soy même des tenebres, ou Véritable théorie de la pierre des philosophes...[Variously attributed to M.-A. Crassellame and O. Tachenius.] 8°, Paris, 1687.
H; M/G2 – 41; Tr/NQ.16.117 [p. 206 turned down and many other signs of dog-earing].
LUPTON, William. Sermons. *See* 1494.
LUSITANUS, Amatus. *See* AMATUS, *Lusitanus.*

1004 LUTHER, Martin. A commentarie upon the Epistle of S. Paul to the Galathians...transl. into English for the unlearned...8°, [London], 1577.
Not H; M/H2 – 27; Tr/NQ.9.63¹ [bound with 385 & 1005; fols. 182v, 251r turned up].

1005 — Special and chosen sermons, collected out of his writings and preachings...Englished by W. G(ace). 8°, London, 1581.
Not H; M/H2 – 27; Tr/NQ.9.63² [bound with 1004].

1006 LUX Orientalis, or An enquiry into the opinion of the Eastern sages, concerning the præexistence of souls...[By J. Glanvill.] 8°, London, 1662.
Not H; M/H1 – 41; Tr/NQ.16.153 ['Isaac Newton' in Newton's hand on fly-leaf].

1007 LYDIAT, Thomas. Canones chronologici, nec non series summorum magistratuum et triumphorum Romanorum. Opus posthumum . . .8°, Oxonii, 1675.
H; M/F2 – 5; ?

1008 LYDIUS, Balthasar. Waldensia, id est, Conservatio veræ ecclesiæ, demonstrata ex confessionibus. . .8°, Roterodami, 1616.
H; M/A1 – 15; ?

1009 MABILLON, Jean. Librorum de re diplomatica supplementum. . . F°, Luteciæ–Parisiorum, 1704.
Not H; M/F8 – 4; Babson 408 [bound with 734]; Sotheran **804** (1927) 3804.

1010 MACLAURIN, Colin. Demonstration des loix du choc des corps. (Pièce qui a remporté le prix de l'Académie Royale des Sciences, 1724.) 4°, Paris, 1724.
Not H; M/F6 – 8; Queen's University, Belfast [p. 21 turned up to mark reference to 'Le celebre M. Newton' in text]; Sotheran **804** (1927) 3808.

1011 — — [Another copy.]
Not H; M (Tracts) D6 – 3; Tr/NQ.8.1².

1012 — Geometria organica: sive Descriptio linearum curvarum univer-salis. 4°, Londini, 1720. [Dedicated to Newton.]
H (2 copies); M (1 only) E6 – 19; Tr/NQ.10.56.

1013 MACROBIUS, Ambrosius Aurelius Theodosius. Opera. Ioh Isacius Pontanus secundò recensuit: adiectis ad libros singulos notis. Quibus accedunt I. Meursii breviores notæ. 8°, Lugduni Bata-vorum, 1628.
H; M/F3 – 1; Tr/NQ.8.70 [p. 259 turned down and several other signs of dog-earing].

MAGDEBURG CENTURIATORS. Ecclesiastica historia, integram Ecclesiæ Christi ideam. . .secundum singulas centurias, perspicuo ordine complectens. . . Per aliquot studiosos & pios viros in Urbe Magdeburgica. . .Centuriæ 1–13. 7 vols. (1560–74.) *See* 546.

1014 MAGNENUS, Johannes Chrysostomus. Democritus reviviscens: sive Vita & philosophia Democriti. Ed. ultima.12°, Hagæ-Comitis,1658.
Not H; M/G1 – 3; ?

1015 MAGNI philosophorum arcani revelator. Quo Hermetis discipuli, magnique scrutatores operis omnia ad suum laborem necessaria, clarissimè explicata invenient. . .8°, Genevæ, 1688.
Not H; M/E1 – 19; ?

1016 MAGNUS, Olaus, *Abp of Upsala*. Historia de gentibus septentrionali-bus. Sic in epitomen redacta. . .8°, Antverpiæ, 1558.
H; M/A2 – 14; Tr/NQ.7.5.

MAIER, Michael. *See* MAYER, Michael.

1017 MAIMBOURG, Louis. An historical treatise of the foundation and prerogatives of the Church of Rome, and of her bishops. Written originally in French, and transl. into English by A. Lovel. 8°, London, 1685.
Not H; M/H2 – 36; Tr/NQ.8.116.

1018 MAIMONIDES. De cultu divino ex R. Mosis Majemonidæ secunda lege, seu Manu forti liber vIII....Ex Hebræo Latinum fecit, & notis illustravit L. de C. de Veïl. 4°, Parisiis, 1678.
H; M/A7 – 10; Tr/NQ.10.124.

1019 — De idololatria liber, cum interpretatione Latina & notis D. Vossii. 4°, Amsterdami, 1641.
H; M/C1 – 12; Tr/NQ.8.46¹ [bound with 1697; a few signs of dog-earing].

1020 — De sacrificiis liber. Accesserunt Abarbanelis Exordium...Quæ ex Hebræo convertit in sermonem Latinum, & notis illustravit L. de C. de Veil. 4°, Londini, 1683.
H; M/F6 – 29; ?

1021 — Porta Mosis, sive Dissertationes aliquot...Nunc primùm Arabicè ...& Latinè editæ...Operâ & studio E. Pocockii. (2 pts.) 4°, Oxoniæ, (1654–)1655.
H; M/A6 – 20; ?

1022 — Tractatus de iuribus anni septimi et iubilaei. Textum Hebraeum addidit, in sermonem Latinum vertit, notisque illustravit I. H. Maius, Filius...4°, Francofurti ad Moenum, 1708.
Not H; M/A7 – 17; Tr/NQ.10.127.
MAITTAIRE, Michael. Miscellanea Græcorum aliquot scriptorum carmina, cum versione Latinâ et notis. [Ed. by M. Maittaire.] 1722. *See* 1087.

1023 MALEBRANCHE, Nicolas. De la recherche de la vérité...6ᵉ éd. 4 vols. 12°, Paris, 1712.
H; not M; ?

1024 MANETHO. Apotelesmaticorum libri vI. Nunc primum ex Bibliotheca Medicea editi curâ J. Gronovii qui etiam Latine vertit ac notas adjecit. (*Greek & Latin.*) 4°, Lugduni Batavorum, 1698.
H; M/C1 – 5; Tr/NQ.16.67.

1025 MANUCCI, Niccolao. Histoire générale de l'Empire du Mogol depuis sa fondation. Sur les mémoires portugais de M. Manouchi, Vénitien. Par F. Catrou. 2 vols. 8°, Paris, 1705.
H; M/G7 – 2 & 3; Tr/NQ.7.3 & 4.

1026 — — [Another ed., complete in 1 vol.] 12°, La Haye, 1708.
H; M/G6 – 18; Tr/NQ.10.95.

1027 MARBECKE, John. A concordāce, that is to saie, a worke wherein
. . .ye maie redely finde any worde conteigned in the whole
Bible. . .F°, [London], 1550.
Not H; M (where it is listed as 'Holford's [?] Concordance')
J9 – 26; ?

1028 MARBODUS, *Bp of Rennes.* De gemmarum lapidumǭ pretiosorum
formis, naturis, atǭ viribus eruditū cū primis opusculū. . .Cū
scholiis P. Villingeñ. 8°, Coloniæ, 1539.
H; M/E3 – 34; ?

1029 — De lapidibus pretiosis enchiridion cū scholiis Pictorii. . .8°,
Parisiis, 1531.
Not H; M/A1 – 21; ?

1030 MARCKIUS, Johannes. Sylloge dissertationum philologico-theo-
logicarum, ad selectos quosdam textus Novi Testamenti. . .4°,
Rotterodami, 1721.
H; M/C1 – 7; Tr/NQ.8.25.

1031 MARIOTTE, Edme. Œuvres. . .Revuës & corrigées de nouveau.
2 vols. in 1. 4°, Leide, 1717.
H; M/E6 – 14; Queen's University, Belfast.
Note in Newton's hand at foot of p. ** 2 *referring to publication date of 'Le*
traité des couleurs'; p. 226 turned up, see pp. 20, 25 above; Sotheran
804 (1927) 3809.

MARIUS MERCATOR. *See* MERCATOR, Marius.

1032 MARMORA Oxoniensia, ex Arundellianis, Seldenianis, aliisque
conflata. Recensuit, & perpetuo commentario explicavit,
H. Prideaux. . .F°, Oxonii, 1676.
H; M/B2 – 9; Tr/NQ.11.10 [a few signs of dog-earing].

MAROT, Clément. Pseaumes de David, mise en rime françoise par
C. Marot et T. de Bèze. . .1613. *See* 1356.

1033 MARPERGER, Paul Jakob. Dissertatio juris publici inauguralis, de
revocatione & amissione privilegiorum. . .4°, Trajecti ad
Rhenum, 1716.
Not H; M (Orationes) C6 – 19; Tr/NQ.10.35⁵.

1034 MARROW (The) of alchemy, being an experimental treatise, discovering the secret and most hidden mystery of the philosophers
elixer. . .By Eirenæus Philoponos Philalethes [i.e. G. Starkey].
8°, London, 1654.
Not H; M/G9 – 16; Memorial Library, University of Wisconsin –
Madison [many signs of dog-earing]; Thame/976; Duveen
p. 563.
— The marrow of alchymy (by Eirenæus Philoponos Philalethes
[i.e. G. Starkey]). . .1709. *See* 1644.

1035 MARSHALL, Benjamin. Chronological tables. (2 pts.) F°, Oxford, 1712–13.
H; M/B5 – 7; ?

1036 MARSHAM, *Sir* John. Canon chronicus Ægyptiacus, Ebraicus, Græcus, et disquisitiones. . .4°, Lipsiæ, 1676.
H; B1 – 28; ?

1037 MARTINEAU DU PLESSIS, Denis. Nouvelle géographie, ou Description exacte de l'univers: tirée des meilleurs auteurs tant anciens que modernes. . .3 vols. 12°, Amsterdam, 1700.
H; M/G6 – 4 to 6; Tr/NQ.9.132 to 134.

1038 MARTINEZ, Matthias. Dictionarium tetraglotton novum, in quo voces Latinæ omnes, & Græcæ his respondentes cum Gallica & Teutonica singularum interpretatione, ordine alphabetico proponuntur. Ed. novissima. 8°, Amstelodami, 1679.
Not H; M/F2 – 11; Tr/NQ.10.98¹ [bound with 514].

1039 MASSACHUSET (The) Psalter: or, Psalms of David with the Gospel according to John, in columns of Indian and English. . .
8°, Boston, N.E., 1709.
Not H; M/J8 – 36; Tr/NQ.16.160.

1040 MASSAEUS, Christianus. Chronicorum multiplicis historiæ utriusque testamenti. . .libri xx. F°, Antverpiæ, 1540.
Not H; M/G11 – 9; ?

1041 MAUNDRELL, Henry. A journey from Aleppo to Jerusalem, at Easter, A.D. 1697. 4th ed. . . .8°, Oxford, 1721.
H; M/J4 – 40; ?

1042 MAXIMES, traduites de l'Espagne. 8°, 1671. *Not identified.*
Not H; M/G2 – 23; ?

1043 MAXIMOS, *Peloponnesios*. Dissertatio, de Sacramento Eucharistiæ. Latinitate donata interprete J. T. Philips. (*Greek & Latin.*) 8°, Londini, 1715.
Not H; M/F3 – 24; Tr/NQ.9.56.

MAXWELL, John. The English atlas, by J. Senex and J. Maxwell. 1714. *See* 1491.

1044 MAYER, Michael. Lusus serius: or, Serious passe-time. A philosophicall discourse concerning the superiority of creatures under man. 12°, London, 1654.
Not H; M/G9 – 29; ?

1045 — Secretioris naturæ secretorum scrutinium chymicum, per oculis et intellectui accuratè accommodata. . .ingeniosissima emblemata. . .4°, Francofurti, 1687.
Not H; M/C1 – 8; Tr/NQ.16.88.
Newton added the references 'p. 8, 82' in margin of p. 111 and 'cap. 67' at p. 146; many signs of dog-earing.

1046 — Septimana philosophica, qua ænigmata aureola de omni naturæ genere...4°, Francofurti, 1620.
Not H; M/C1 – 18; ?

1047 — Silentium post clamores, hoc est, Tractatus apologeticus, quo causæ non solùm clamorum seu Reuelationum Fraternitatis Germanicæ de R.C. sed & silentii...8°, Francofurti, 1617.
Not H; M/E3 – 5; Tr/NQ.10.148² [bound with 1050; a few signs of dog-earing].

1048 — Symbola aureæ mensæ duodecim nationum...4°, Francofurti, 1617.
Not H; M/B1 – 22; S. M. Edelstein Collection, Jewish National & University Library, Jerusalem [a few signs of dog-earing].

1049 — Themis aurea; hoc est, De legibus Fraternitatis R.C. tractatus... 8°, Francofurti, 1618.
Not H; M/E3 – 5; Tr/NQ.10.148⁴ [bound with 1050].
In a note at the foot of p. 160 Newton deciphered the last 7 lines of text, printed in code, and added the key he used, see p. 20 and plate 3.

1050 — Tractatus de volucri arborea, absque patre et matre, in Insulis Orcadum, forma anserculorum proueniente...8°, Francofurti, 1619.
H; M/E3 – 5; Tr/NQ.10.148¹ [bound with 1047, 1049 & 1051; p. 136 turned down and a few other signs of dog-earing].

1051 — Verum inventum, hoc est, Munera Germaniæ; ab ipsa primitus reperta... & reliquo orbi communicata...8°, Francofurti, 1619.
Not H; M/E3 – 5; Tr/NQ.10.148³ [bound with 1050].

1052 — Viatorium, hoc est, De montibus planetarum septem seu metallorum ...8°, Rothomagi, 1651.
H; M/E3 – 4; ?

1053 MEAD, Joseph. Works. Corrected and enlarged...(3rd ed.) F°, London, 1672.
H; M/D7 – 14; ?

1054 MEAD, Richard. De imperio solis ac lunæ in corpora humana, et morbis inde oriundis. 4°, Londini, 1704.
H; M/A4 – 17; Tr/NQ.16.90.

1055 — A short discourse concerning pestilential contagion, and the methods to be used to prevent it. 8°, London, 1720.
H; not M; ?

1056 — — 8th ed., with...additions. 8°, London, 1722.
H; M/G10 – 10; ?

1057 MEAGER, Leonard. The new art of gardening. 2nd ed. 12°, London, 1697.
Not H; M/J8 – 50; ?

1058 MÉCHANIQUE (La) du feu...contenant Le traité de nouvelles cheminées...Par Mr. G*** [i.e. N. Gauger]. 8°, Amsterdam, 1714.
H; M/G2 – 37; ?

MÉDAILLES sur les principaux événements du règne de Louis Le Grand...1702. *See* 1248.

MEDE, Joseph. *See* MEAD, Joseph.

1059 MEDULLA historiæ Anglicanæ. Being a comprehensive history of the lives and reigns of the monarchs of England...[By W. Howell.] 3rd ed....8°, London, 1687.
H; not M; ?

1060 MEIBOMIUS, Marcus. Antiquæ musicæ auctores septem. Græce et Latine. M. Meibomius restituit ac notis explicavit. 2 vols. 4°, Amstelodami, 1652.
H; M/C6 – 29 & 30; Tr/NQ.8.36 & 37.

1061 MELA, Pomponius. De situ orbis libri III. Unà cum auctario P. I. Oliuarij Valentini, instauratione totius libelli & castigatione... 4°, Parisiis, 1557.
H; M/A7 – 14; Tr/NQ.8.134.

1062 — Pomponii Melæ libri III de situ orbis, nummis antiquis & notis illustrati ab J. Gronovio...8°, Lugd. Batavorum, 1696.
H; M/F1 – 2; Tr/NQ.7.75.

MÉMOIRES de mathématique et de physique tirez des registres de l'Académie Royale des Sciences. Année M.DCCVII–M.DCCX. 3 vols. 1708–13. *See* 1249. *And* Suite des Mémoires...M.DCCVII– M.DCCX. 4 vols. 1708–13. *See* 1251.

1063 — Mémoires pour servir à l'histoire de France. Contenant ce qui s'est passé de plus remarquable dans ce roiaume depuis 1515. jusqu'en 1611...[By P. de l'Estoile.] 2 vols. 8°, Cologne, 1719.
H; M/G8 – 6 & 7; Tr/NQ.8.83 & 84.

1064 — Mémoires sur le commerce des Hollandois, dans tous les états et empires du monde...[By P. D. Huet.] 12°, Amsterdam, 1717.
H; M/G8 – 13; Tr/NQ.10.120.

1065 — Mémoires sur les dernières révolutions de la Pologne, où on justifie le retour du Roy Auguste, par un gentilhomme polonnois [J. J. Przebendowski]. 8°, Rotterdam, 1710.
H; M/G8 – 2; Tr/NQ.8.103.

1066 MEMOIRS concerning the affairs of Scotland, from Queen Anne's accession to the throne, to...1707...[By G. Lockhart.] 8°, London, 1714.
H; M/J4 – 22; Tr/NQ.16.54.

1067 — Memoirs of the Dutch trade in all the states, kingdoms, and empires in the World...Done from the French [of P. D. Huet]. 8°, London, [*c.* 1700].
H; M/J7 – 46; Tr/NQ.8.107.

1068 MENANDER, *comicus.* Menandri et Philemonis reliquiæ, quotquot reperiri potuerunt; Græce et Latine, cum notis H. Grotii et J. Clerici. 8°, Amstelodami, 1709.
H; M/A4 – 2; ?

1069 MENESTRIER, Claude François. Histoire de roy Louis le Grand par les médailles, emblêmes, devises [etc.]...F°, Paris, 1691.
H; M/C2 – 6; ?

1070 MERCATOR improved, or, The description and use of a new instrument, with which may be solved all the problems of plain and Mercator's sailing...By B. G[oodday]. Philomath. 8°, London, 1725.
Not H; M (Tracts) J3 – 33; Tr/NQ.16.162³.

1071 MERCATOR, Marius. Opera quæcumque extant. Prodeunt nunc primum studio J. Garnerii, qui notas etiam ac dissertationes addidit. (2 pts.) F°, Parisiis, 1673.
H; M/E7 – 7; Tr/NQ.18.2.

1072 MERCATOR, Nicolaus. Institutionum astronomicarum libri II, de motu astrorum communi & proprio, secundum hypotheses veterum & recentiorum præcipuas...(2 pts.) 8°, Londini, 1676.
H; M/E5 – 34; Tr/NQ.10.152; *Math. papers,* I, 131 n. 48.
Extensive astronomical notes by Newton on pp. 213, 280, 281 and on endpaper, '1672' in margin of p. 151; see plate 2.

1073 — Logarithmo-technia: sive Methodus construendi logarithmos nova, accurata, & facilis...Huic etiam jungitur M. A. Riccii Exercitatio geometrica de maximis & minimis...(2 pts.) 4°, Londini, 1668.
H; M/E5 – 4; Tr/NQ.9.48⁴ [bound with 714].

1074 MERCURIAL (The) chronometer improv'd: or, A supplement to ...An essay, wherein a method is humbly propos'd for measuring equal time with the utmost exactness... By the author of that Essay [J. Clarke]. 8°, London, 1715.
Not H; M (Tracts) H8 – 14; Tr/NQ.16.95¹ᵃ.
MERCURY'S Caducean rod...By Cleidophorus Mystagogus [i.e. W. Yarworth]. 1702. *See* 1138.

1075 MEURSIUS, Johannes. De regno Laconico libri II. De Piræeo liber singularis. Et in Helladii Chrestomathiam animadversiones... 4°, Ultrajecti, 1687.
H; M/A6 – 9; Tr/NQ.8.77 [several signs of dog-earing].

1076 — Glossarium Græco-Barbarum...Ed. 2ª emendata. 4º, Lugduni
Batavorum, 1614.
H; M/C6 – 18; Tr/NQ.16.191 [wants title-page; 'pret. 9ˢ.' in
Newton's hand on fly-leaf].

1077 — Regnum Atticum. Sive, De regibus Atheniensium, eorumque rebus
gestis, libri iii. 4º, Amstelodami, 1633.
Not H; M/A6 – 7; Tr/NQ.8.93 [p. 53 turned up and many other
signs of dog-earing].

1078 — Reliqua Attica; sive, Ad librum de populis Atticae, paralipomena.
Liber singularis...(2 pts.) 4º, Ultrajecti, 1684.
H; M/A6 – 8; Tr/NQ.8.110³ [bound with 1079].

1079 — Themis Attica, sive De legibus Atticis libri ii. 4º, Trajecti ad
Rhenum, 1685.
H; M/A6 – 8; Tr/NQ.8.110¹ [bound with 1078 & 1080; a few
signs of dog-earing].

1080 — Theseus, sive De ejus vita rebusque gestis liber postumus...4º,
Ultrajecti, 1684.
H; M/A6 – 8; Tr/NQ.8.110² [bound with 1079; a few signs of
dog-earing].

1081 MICHELOTTI, Pietro Antonio. De separatione fluidorum in corpore
animali dissertatio physico-mechanico-medica. 4º, Venetiis,
1721.
Not H; M (Tracts) D6 – 5; Tr/NQ.10.28².

MIDDLETON, Conyers. A full and impartial account of all the late
proceedings in the University of Cambridge against Dr. Bentley
...[By C. Middleton.] 1719. *See* 644.

— Some remarks upon a pamphlet [by A. A. Sykes], entitled, The
case of Dr. Bentley farther stated and vindicated, &c....[By
C. Middleton.] 1719. *See* 1537.

MIÈGE, Guy. The new state of England, under our present monarch,
King William III...[By G. Miège.] 3rd ed....1700. *See*
1149.

MILLER, C. The Hig Dutch Minerva à-la-mode, or A perfect
grammar never extant before...[By C. Miller?] [1670?] *See*
763.

1082 MILTON, John. Poetical works. 2 vols. 4º, London, 1720.
H; M/C2 – 15 & 16; ?

1083 MINUCIUS FELIX, Marcus. Marci Minucii Felicis Octavius:
Cæcilius Cyprianus De idolorum vanitate, cum observationibus
omnibus N. Rigaltii...12º, Oxoniæ, 1678.
Not H; M/G2 – 10; ?

1084 MIRKHOND. The history of Persia...written in Arabick, by Mirkond...translated into Spanish, by A. Teixeira...and now render'd into English. By J. Stevens. 8°, London, 1715.

H; M/J4 – 10; Tr/NQ.9.117.

Newton appears to have equated the names of the 14th–16th Kings of Persia as given in the text with names found in classical sources, and thus in the margin of p. 60 he wrote 'Kai Axeres' for the printed 'Lorasph', at p. 65 'Darius Histaspis' for 'Gustasph', at p. 70 'Artaxerxes Longimanus' for 'Bahaman Daraz Daz', p. 33 turned up; Thame/915.

1085 MISCELLANEA Berolinensia ad incrementum scientiarum, ex scriptis Societati Regiæ Scientiarum exhibitis edita. Vol. 1. 4°, Berolini, 1710.

H; M/C1 – 31; Museum of the History of Science, Oxford University.

1086 — Miscellanea curiosa. Being a collection of some of the principal phænomena in nature, accounted for by the greatest philosophers of this age...[Ed. and in part written by E. Halley.] Vols. 1, 2. 8°, London, 1705–6.

H; M/E4 – 1 & 2; University of Chicago Library (vol. 1); Tr/NQ.16.55 (vol. 2); see p. 49 n. 1 above.

1087 — Miscellanea Græcorum aliquot scriptorum carmina, cum versione Latinâ et notis. [Ed. By M. Maittaire.] F°, Londini, 1722. [Newton was a subscriber to this work.]

H; M/C2 – 19; Tr/NQ.18.8.

1088 MODEST (A) censure on some mistakes concerning civil government, with relation to the late disputes about resistance and non-resistance. 8°, London, 1713.

H; M/J3 – 20; ?

1089 — A modest plea for the baptismal and scripture-notion of the Trinity ...[By A. A. Sykes.] 8°, London, 1719.

H; not M; ?

MŒSTISSIMÆ ac lætissimæ Academiæ Cantabrigiensis carmina funebria & triumphalia. Illis...Reginam Annam repentina morte abreptam deflet...1714. *See* 336.

1090 MOIVRE, Abraham de. Animadversiones in D. Georgii Cheynæi Tractatum de fluxionum methodo inversa. 8°, Londini, 1704.

H; M/E5 – 24; Tr/NQ.9.5 [a few signs of dog-earing].

1091 — Annuities upon lives; or, The valuation of annuities upon any number of lives...8°, London, 1725.

H; M/G10 – 13; Brown University Library, Providence, R.I.; Sotheran **804** (1927) 3810.

1092 — De fractionibus algebraicis radicalitate immunibus ad fractiones simpliciores reducendis...4°, [Paris?], 1722.
Not H; M (Tracts) D6 – 4; Tr/NQ.16.197².

1093 — De mensura sortis, seu, De probabilitate eventuum in ludis a casu fortuito pendentibus. (Philosophical Transactions, 329, Jan.–Feb., 1711.) 4°, London, 1712.
H; not M; ?

1094 — The doctrine of chances: or, A method of calculating the probability of events in play. 4°, London, 1718. [Dedicated to Newton.]
H; M/C2 – 31; Tr/NQ.10.30.

MOLL, Herman. Geographia classica: the geography of the ancients so far describ'd...[By H. Moll.] 2nd ed. 1717. See 666.

1095 — A new description of England and Wales, with the adjacent Islands...F°, London, 1724.
H; M/G3 – 11; ?

1096 — A system of geography: or, A new & accurate description of the earth...F°, London, 1701.
H; M/B2 – 8; Thame/966; ?

1097 MOLYNEUX, William. The case of Ireland's being bound by Acts of Parliament in England, stated. 8°, Dublin, 1698.
H; M/J7 – 43; Tr/NQ.8.69.

1098 — Dioptrica nova. A treatise of dioptricks, in 2 pts....4°, London, 1692.
H; M/E6 – 38; Tr/NQ.10.26 [a few signs of dog-earing including marking of Newton's name in text of p. 273, see p. 27 above].

MONMORT, Pierre Rémond de. Essay d'analyse sur les jeux de hazard. [By P. Rémond de Monmort.] 1708 & 2ᵉ éd. revûe... 1713. See 572 & 573.

1099 MONTAIGNE, Michel Eyquem de. Essais. Nouvelle éd....Par P. Coste. 3 vols. 4°, Londres, 1724.
Not H; M/C2 – 21 to 23; ?

1100 MONTANUS, Benedictus Arias. Antiquitatum Iudaicarum libri ix...4°, Lugduni Batavorum, 1593.
H; M/C6 – 10; ?

1101 — Elucidationes in omnia Sanctorum Apostolorum scripta. Eiusdem in S. Ioannis Apostoli et Evangelistæ Apocalypsin significationes. 4°, Antverpiæ, 1588.
Not H; M/J6 – 4; Tr/NQ.8.30² [bound with 1102].

1102 — Elucidationes in Quatuor Evangelia, Matthæi, Marci, Lucæ &

Iohannis. Quibus accedunt eiusdem elucidationes in Acta Apostolorum. 4°, Antverpiæ, 1575.
Not H; M/J6 – 4; Tr/NQ.8.30¹ [bound with 1101; 'Isaac Newton 1682. pret. 2ˢ 6ᵈ.' on fly-leaf].

MONTE-SNYDER, Johann de. Commentatio de pharmaco catholico... [By J. de Monte-Snyder.] 1666. *See* 1378.

1103 MONTFAUCON, Bernard de. Diarium Italicum. Sive Monumentorum veterum, bibliothecarum, musæorum, &c. Notitiæ singulares in itinerario Italico collectæ. 4°, Parisiis, 1702.
H; M/B6 – 11; Tr/NQ.10.27.

MOORE, Arthur. A letter to the Honourable A—r M—re [Arthur Moore], Com—ner [Commissioner] of trade and plantation. 1714. *See* 949.

1104 MOORE, *Sir* Jonas. A mathematical compendium: or Useful practices in arithmetick, geometry, and astronomy...3rd ed. ...16°, London, 1695.
Not H; M/G9 – 50; ?

1105 — Modern fortification: or, Elements of military architecture... 8°, London, 1689.
H; M/E4 – 37; Tr/NQ.10.101.

1106 — Moores arithmetick: discovering the secrets of that art, in numbers and species. In 2 bookes...8°, London, 1650.
H; M/G9 – 9; Tr/NQ.10.146 ['Isaac Newton' on end-paper, written by Newton above a scored-through note probably by previous owner.]

1107 — A new systeme of the mathematicks...4°, London, 1681.
H; not M; ?

1108 MORDEN, Robert. Geography rectified: or, A description of the world...3rd ed., enlarged...4°, London, 1693.
H; not M; ?

1109 — — 4th ed., enlarged...4°, London, 1700.
H; M/F5 – 15; Tr/NQ.9.29.

MORE, Henry. An answer to Several remarks upon Dʳ Henry More his Expositions of the Apocalypse and Daniel, as also upon his Apology. Written by S. E. Mennonite...1684. *See* 49.

1110 — An antidote against atheisme, or An appeal to the natural faculties of the minde of man, whether there be not a God. 8°, London, 1653.
Not H; M/H1 – 2; Tr/NQ.9.171² [bound with 1114].
4 short Latin notes by Newton in the margin of Preface, A1r, v.

1111 — Apocalypsis Apocalypseos: or The Revelation of St. John unveiled. 4°, London, 1680.
Not H; M/F5 – 7; ?

1112 — Discourses on several texts of scripture. 8°, London, 1692.
H; M/H3 – 34; ?

1113 — The Immortality of the Soul, so farre forth as it is demonstrable
from the knowledge of nature and the light of reason. 8°, London,
1659.
H; M/H1 – 1; ?

— Observations upon Anthroposophia theomagica, and Anima
magica abscondita [of Eugenius Philalethes, i.e. T. Vaughan].
By Alazonomastix Philalethes [i.e. H. More]. 1650. *See* 1199.

— Paralipomena prophetica, containing several supplements and
defences of Dr Henry More...(2 pts.) 1685. *See* 1244.

1114 — Philosophicall poems. 8°, Cambridge, 1647.
H; M/H1 – 2; Tr/NQ.9.171¹ [bound with 1110].

1115 — A plain and continued exposition of the several prophecies or divine
visions of the Prophet Daniel ...4°, London, 1681.
Not H; M/F5 – 5; Bancroft Library, University of California,
Berkeley [inscribed 'Is. Newton. Ex dono Reverendi Authoris'].
Marginal annotations by Newton on Revelation; see Popular Astronomy,
XXXIV (1926), 75–8; Thame/969.

— Remarks on D^r. Henry More's Expositions of the Apocalypse and
Daniel...1690. *See* 1391.

1116 — Tetractys anti-astrologica, or, The four chapters in the Explanation
of the grand mystery of Godliness, which contain a brief but solid
confutation of judiciary astrology...4°, London, 1681.
Not H; M/F5 – 8; Beinecke Library, Yale [inscribed 'Isaac
Newton Donum Reverendissimi Auctoris'].

1117 MORGAGNI, Giovanni Battista. Adversaria anatomica omnia...
4°, Patavii, 1719.
H; not M; ?

1118 MORGAN, Thomas. Philosophical principles of medicine...8°,
London, 1725.
H; M/G10 – 12; ?

1119 MORHOF, Daniel Georg. De metallorum transmutatione ad...
J. Langelottum...epistola. 8°, Hamburgi, 1673.
Not H; M/A2 – 12; Tr/NQ.16.126 [several signs of dog-earing].

1120 MORLAND, Joseph. Disquisitions concerning the force of the heart,
the dimensions of the coats of the arteries, and the circulation of
the blood. 8°, London, 1713.
H; M/J7 – 34; Tr/NQ.9.58.

1121 MORLEY, Christopher Love. Collectanea chymica Leydensia,

Maëtsiana, Margraviana, Le Mortiana...Nunc autem...per T. Muykens. 8°, Lugduni Batavorum, 1693.
H; M/E5 – 43; Tr/NQ.10.99.

MORZILIUS, Sebastianus Foxius. *See* FOX MORCILLO, Sebastian.

1122 MOXON, Joseph. Mechanick dyalling: teaching any man, though of an ordinary capacity and unlearned in the mathematicks, to draw a true sun-dial on any given plane...3rd ed. 4°, London, 1697.
Not H; M (Tracts) J3 – 24; Tr/NQ.9.125⁶.

1123 MOYLE, Walter. Works; none of which were ever before publish'd. [Ed. by T. Sergeant.] 2 vols. 8°, London, 1726.
H; M/E4 – 16 & 17; Tr/NQ.7.77 & 78.

1124 MUN, Thomas. England's treasure by foreign trade. 8°, [London, 1713?].
Not H; M (Tracts) J2 – 38; Tr/NQ.16.194² [wants title-page].

MUNCKER, Thomas. Mythographi Latini. C. Jul. Hyginus [etc.] ...T. Munckerus omnes ex libris MSS....emendavit, & commentariis...instruxit...1681. *See* 1139.

1125 MUNSTERUS, Sebastianus. Cosmographiæ universalis lib. vi. F°, Basileæ, 1550.
Not H; M/D8 – 18; ?

1126 — Dictionarium Chaldaicum, non tā ad Chaldaicos interpretes q̃ Rabbinorū intelligenda cōmentaria necessarium...4°, Basileae, 1527.
Not H; M/A6 – 13; Tr/NQ.8.18.

1127 — Kalendariũ Hebraicum...4°, Basileae, 1527.
Not H; M/J6 – 26; Tr/NQ.9.45 [a few signs of dog-earing].

1128 — Proverbia Salomonis, iam recens iuxta Hebraicā veritatē translata, & annotatiōibus illustrata, autore S. Munstero. 8°, [Basle, 1524].
Not H; M/F1 – 8; Tr/NQ.7.9.

1129 MURATORI, Lodovico Antonio. Delle antichità Estensi ed Italiane ...Pt 1. F°, Modena, 1717.
H; M/F7 – 16; Tr/NQ.18.24.

1130 MUSAEUM Hermeticum omnes sopho-spagyricae artis discipulos fidelissime erudiens. 4°, Francofurti, 1625.
Not H; M/F5 – 35; Tr/NQ.16.115 [imperfect, wants title-page and pp. 1–35, Lambspring's De lapide philosophorum; pp. 231, 353 turned down and several other signs of dog-earing].

1131 — Musæum Hermeticum reformatum et amplificatum, omnes

sopho-spagyricæ artis discipulos fidelissimè erudiens...Continens tractatos chimicos xxi....4°, Francofurti, (1677–)1678.
Not H; M/B1 – 18; Tr/NQ.16.148 [pp. 132, 150, 239 turned down, p. 278 up, and several other signs of dog-earing].

1132 MUSSCHENBROEK, Pieter van. Disputatio medica inauguralis de aëris præsentia in humoribus animalibus...4°, Lugd. Bat., 1715.
Not H; M (Orationes) C6 – 19; Tr/NQ.10.35².

1133 — Epitome Elementorum physico-mathematicorum, conscripta in usos academicos. 8°, Lugduni Batavorum, 1726.
H; not M; ?

1134 — Oratio de certa methodo philosophiæ experimentalis, dicta publice A.D. XIII. Septemb. 1723. 4°, Trajecti ad Rhenum, 1723.
Not H; M (Orationes) C6 – 19; Tr/NQ.10.35¹ [presentation inscription on fly-leaf 'Viro Amplissimo, Nobilissimo, Geometrarum Principi, Solidæ Philosophiæ Instauratori Isaco Newtono mittit Auctor'].

1135 MUYS, Wyer Wilhelm. Dissertatio & observationes de salis ammoniaci præclaro ad febres intermittentes usu, una cum epistola præfixa ad Regiam Societatem Londinensem missæ. 4°, Franequeræ, 1716.
Not H; M (Tracts) J6 – 15; Tr/NQ.8.52⁵.

1136 MYLIUS, Johannes Daniel. Opus medico-chymicum: continens tres tractatus sive basilicas...4°, Francofurti, 1618.
Not H; M/B1 – 16; ?

1137 MYNSICHT, Hadrianus. Thesaurus et armamentarium medico-chymicum...Cui in fine adiunctum est Testamentum Hadrianeum de aureo philosophorum lapide. 8°, Rothomagi, 1651.
Not H; M/E2 – 42; Tr/NQ.8.68 [a few signs of dog-earing].

1138 MYSTAGOGUS, Cleidophorus, *pseud.* Mercury's Caducean rod: or The great and wonderful office of the universal mercury, or God's Viceregent, displayed...By Cleidophorus Mystagogus [i.e. W. Yarworth]. 8°, London, 1702.
Not H; M/E2 – 37; Tr/NQ.16.131¹ [bound with 1302]; probably the book sent to Newton by the author, ?1702 (*Correspondence*, VII, 441, where the work referred to in Yarworth's undated letter is tentatively identified as his *The complete distiller*...2nd ed., 1705, though Newton is known to have owned only the 1st ed., 1692, no. 1760).
Note by Newton on fly-leaf: '*Willis his search of causes p. 3, 21. Sanguis naturæ p. 10 & Epistle p. 27. Philadelphia p. 13*' (*page references to works quoted in the text*); many signs of dog-earing.

— A philosophical epistle, discovering the unrevealed mystery of the three fires of the Sophi. [*Signed* Cloidophorus Mystagogus, i.e. W. Yarworth.] [*c.* 1702.] *See* 1302 & 1303.

1139 MYTHOGRAPHI Latini. C. Jul. Hyginus [etc.]...T. Munckerus omnes ex libris MSS....emendavit, & commentariis...instruxit...8°, Amstelodami, 1681.
H; M/F4 – 8; ?

1140 NATURAL (The) history of Oxford-shire, being an essay toward the natural history of England. By R. P[lot]. F°, Oxford, 1677.
Not H; M/G11 – 7; Tr/NQ.11.34 [bearing the signatures 'Anne Hedinton 1678 her Bucke' and 'Ann Neale of West Wycombe her Book 1721', the volume is very unlikely to have come from Newton's library].

1141 NEMOURS, Marie d'Orléans, *duchesse de.* Mémoires: contenant ce qui s'est passé de plus particulier en France pendant la Guerre de Paris, jusqu'à la prison du Cardinal de Retz en 1652...8°, Amsterdam, 1718.
H; M/G6 – 25; Tr/NQ.9.142.

1142 NEPOS, Cornelius. Cornelius Nepos per Rutgersium. 8°, 1688.
Edition not identified.
Not H; M/E1 – 38; ?

1143 — Vitæ excellentium Imperatorum & in eas J. Loccenii notæ politicæ. 12°, Hamburgi, 1673.
H; M/G1 – 23; ?

1144 — De vita excellentium Imperatorum. Interpretatione et notis illustravit N. Courtin...in usum Serenissimi Delphini. 4°, Parisiis, 1675.
H; M/F6 – 18; Tr/NQ.10.41 [a few signs of dog-earing].

1145 — The lives of illustrious men. Done into English from the original Latin...8°, London, 1713.
H; not M; ?

1146 NEPTUNE (Le) françois, ou Atlas nouveau des cartes marines... Reveu et mis en ordre par les Sieurs Pene, Cassini et autres. F°, Paris, 1693.
H; not M; ?

1147 NEUMANN, Caspar. Bigam difficultatum physico-sacrarum: de gemmis Urim & Tummim dictis, Exod. xxviii, 30. & de Cibo Samariæ obsessæ 2. Reg. vi, 25....Ed. 2ª. 4°, Lipsiæ, 1709.
Not H; M/C1 – 32; Tr/NQ.9.172³ [bound with 885 & 1148].

1148 — Clavis Domus Heber, reserans januam ad significationem hiero-

glyphicam literaturæ Hebraicæ perspiciendam...(2 pts.) 4°,
Wratislaviæ, 1712.
Not H; M/C1 – 32; Tr/NQ.9.172¹ [bound with 885 & 1147];
Newton thanked Neumann, ?1712, for presenting him with this
book (*Correspondence*, VII, 481).

1149 NEW (The) state of England, under our present monarch, King
William III...[By G. Miège.] 3rd ed....8°, London, 1700.
H; not M; ?

1150 — New Year's gift. 6 pts. 8°, 1656. *Not identified.*
Not H; M/G9 – 21; ?
NEWMAN, Samuel. A concordance to the Holy Scriptures...By
S. N. [i.e. Samuel Newman.] 2nd ed. corrected and enlarged.
1672. *See* 218. 4th ed. very much enlarged...1698. *See* 219.

1151 NEWTON, *Sir* Henry. Henrici Newton sive De Nova Villa, Societatis
Regiæ, Londini, Arcadiæ Romanæ, Academiæ Florentinæ, et ejus
quæ vulgò vocatur della Crusca, socii, epistolæ, orationes, et
carmina. (2 pts.) 4°, Lucæ, 1710.
H; M/F6 – 27; Tr/NQ.10.125¹ [bound with 1152].

1152 — Orationes quarum altera Florentiæ anno MDCCV. Altera vero
Genuæ anno MDCCVII. habita est. 4°, Amstelodami, 1710.
Not H; M/F6 – 27; Tr/NQ.10.125² [bound with 1151].

1153 NEWTON, Isaac. Analysis per quantitatum series, fluxiones, ac
differentias: cum Enumeratione linearum tertii ordinis. [Ed. by
W. Jones.] 4°, Londini, 1711.
H (3 copies); M (1 only) E6 – 32; Tr/NQ.8.26.
Corrections and clarifications by Newton in margins of pp. 42, 45, 50, 52;
the 2 sheets of Tabula intended to stand at 'De Quadratura, Prop. X,
Scholium', between pp. 62–3, are bound out of sequence between pp. 92–3.

1154 — Arithmetica universalis; sive De compositione et resolutione
arithmetica liber. Cui accessit Halleiana Æquationum radices
arithmetice inveniendi methodus. 8°, Cantabrigiæ, 1707.
H; M/E5 – 32; in private hands.
Corrections by Newton of minor misprints, insertions of more appropriate
running heads, transpositions marked, passages marked for deletion; see
Math. papers, v, 14.

1155 — Universal arithmetick: or, A treatise of arithmetical composition
and resolution. To which is added, Dr. Halley's Method of
finding the roots of æquations arithmetically. Transl. from the
Latin by the late Mr. Raphson, and revised and corrected by
Mr. Cunn. 8°, London, 1720.
H; M/E5 – 31; Tr/NQ.9.30.

Correction in Newton's hand in margin and in text of p. 4, line 4, altering the printed line to read 'makes a, whose Aggregate, or Sum is 2b + a, and so in others'.

1156 — Opticks: or, A treatise of the reflexions, refractions, inflexions and colours of light. Also Two treatises of the species and magnitude of curvilinear figures. (2 pts.) 4°, London, 1704.

H; M/E6 – 28; Sotheran, *Bibl.*, 2 (1921) 12535, where it is described as 'With auto. "Isaac Newton" on endpaper'; ?

1157 — — [Another copy.]

H; M/E6 – 30; Tr/NQ.16.198.

*Lacks pp. 1–17 and the two appended treatises; numerous notes and corrections by Newton; several signs of dog-earing; Sotheran **828** (1931) 3602.*

1158 — — [Another copy.]

Not H; not M; Portsmouth (1888); ULC/Adv.b.39.3.

Lacks title-page, plates, and all after Pt 2, p. 137; printer's proofs with numerous additions and corrections by Newton for the 2nd ed., 1717.

1159 — Opticks: or, A treatise of the reflections, refractions, inflections and colours of light. 2nd ed., with additions. 8°, London, 1717.

H; M/E6 – 30; Babson 133.

Many corrections by Newton, some adopted in later eds., with 15 new lines, giving the Seven Precepts, added at end of text, p. 382; Sotheran, Bibl., 2 (1921) 12539: 'Sir Isaac Newton's own corrected copy'.

1160 — — [Another copy.]

H; M/E5 – 7; ?

1161 — Opticks: or, A treatise of the reflections, refractions, inflections and colours of light. 3rd ed., corrected. 8°, London, 1721.

H (2 copies); not M; ?

1162 — Optice: sive De reflexionibus, refractionibus, inflexionibus & coloribus lucis libri III. Authore Isaaco Newton. Latine reddidit S. Clarke. Accedunt Tractatus duo ejusdem authoris de speciebus & magnitudine figurarum curvilinearum, Latine scripti. (2 pts.) 4°, Londini, 1706.

H; M/E6 – 27; ULC/Adv.b.39.4.

A number of notes, loose inserts, and marginal corrections by Newton; significant signs of dog-earing of pp. 29, 269, 293, 311, 313, 319, 329, 335, and esp. p. 346; see pp. 25–6 above.

1163 — — [Another copy.]

H ('Optics lat. Interleavd'; one of the six 'Books that has Notes of Sir Is. Newtons'); not subsequently recorded; ?

1164 — Optice: sive De reflexionibus, refractionibus, inflexionibus & coloribus lucis, libri III. Latine reddidit S. Clarke. Ed. 2ª, auctior. 8°, Londini, 1719.

H; not M; ?

1165 — Traité d'optique sur les réflexions, réfractions, inflexions, et les couleurs de la lumière. Traduit de l'anglois par M. Coste sur la 2ᵉ éd. augmentée par l'auteur. 2 vols. 12°, Amsterdam, 1720.
H; M/G7 – 39 & 40; ?

1166 — Traité d'optique sur les réflexions, réfractions, inflexions, et les couleurs de la lumière. Traduit par M. Coste sur la 2ᵉ éd. angloise, augmentée par l'auteur. 2ᵉ éd. françoise, beaucoup plus correcte que la première. 4°, Paris, 1722.
H (3 copies); M (1 only) E6 – 29; Turner Collection, University of Keele Library.

1167 — Philosophiæ naturalis principia mathematica. 4°, Londini, 1687. ('Prostat apud plures Bibliopolas' title-page.)
H ('Interleavd imperf sewed'; one of the six 'Books that has Notes of Sir Is. Newtons'); Portsmouth (1888); ULC/Adv.b.39.1.
Numerous notes and corrections for 2nd ed. by Newton in printed text and on interleaves; Cohen, *Introd. to Newton's 'Principia'* (Cambridge, 1971), pp. 25–6, pls. 2–3, Koyré & Cohen (eds.), *'Principia'*, variorum ed. (2 vols., Cambridge, 1972), I, where it is coded E_1i; see p. 22 above.

1168 — — [Another copy.]
H; M/E6 – 25; Tr/NQ.16.200.
Numerous notes and corrections by Newton for 2nd ed.; Sotheran **828** (1931) 3602; Cohen, *Introd. to Newton's 'Principia'* (Cambridge, 1971), p. 25, pls. 1, 13, where it is coded E_1a; see pp. 22, 53–4 above.

1169 — Philosophiæ naturalis principia mathematica. Ed. 2ª auctior et emendatior. 4°, Cantabrigiæ, 1713.
Not H; not M; Portsmouth (1888); ULC/Adv.b.39.2.
Interleaved with numerous interpolations and corrections by Newton to printed text, on interleaves and on inserted paper slips; Cohen, *Introd. to Newton's 'Principia'* (Cambridge, 1971), pl. 4; Koyré & Cohen (eds.), *'Principia'*, variorum ed. (2 vols., Cambridge, 1972), I, where it is coded E_2i; see p. 22 above.

1170 — — [Another copy.]
H; M/E6 – 26; Tr/NQ.16.196.
Numerous notes and corrections by Newton; Sotheran **828** (1931) 3602; Koyré & Cohen (eds.), *'Principia'*, variorum ed. (2 vols., Cambridge, 1972), I, where it is coded E_2a; see pp. 22, 53–4 above.

1171 — Philosophiæ naturalis principia mathematica. Ed. ultima. Cui accedit Analysis per quantitatum series, fluxiones ac differentias cum Enumeratione linearum tertii ordinis. 4°, Amstælodami, 1723.
H; not M; ?

1172 — Philosophiæ naturalis principia mathematica. Ed. 3ª aucta &
emendata. 4°, Londini, 1726.
H (2 copies); M (1 only) C2 – 17; Tr/NQ.17.34.
— Recueil de diverses pièces...Par Mrs. Leibniz, Clarke, Newton,
& autres autheurs célèbres. [Ed. by P. Des Maizeaux.] 2 vols.
1720. *See* 1379 & 1380.
See also 122, 422–4, 1370, 1371 & 1667.

1173 NICEPHORUS CALLISTUS. Ecclesiasticæ historiæ libri xviii. In
2 tomos distincti, ac Græcè nunc primùm editi. Adiecta est Latina
interpretatio I. Langi, à F. Ducæo cum Græcis collata & re-
cognita. 2 vols. F°, Lutetiæ Parisiorum, 1630.
H; M/D7 – 5 & 6; Tr/NQ.11.45 [vol. 2 only].
NICOLE, Pierre. Logica, sive Ars cogitandi...[By A. Arnauld and
P. Nicole.] E tertia apud Gallos ed. recognita & aucta in
Latinum versa. 1687. *See* 980.

1174 NIEUWENTIJT, Bernard. Analysis infinitorum, seu Curvili-
neorum proprietates ex polygonorum natura deductæ. 8°,
Amstelædami, 1695.
H; M/E4 – 35; Tr/NQ.8.104.

1175 — Considerationes secundæ circa calculi differentialis principia; et
Responsio ad...G. G. Leibnitium. 8°, Amstelædami, 1696.
Not H; M (Tracts) H8 – 13; Tr/NQ.16.99⁴.

1176 — The religious philosopher: or The right use of contemplating the
works of the Creator...3 vols. 8°, London, 1718–19.
H; M/E5 – 26 to 28; ?

1177 NODOT, François. Nouveaux mémoires de Mʳ. Nodot; ou Obser-
vations qu'il a faites pendant son voyage d'Italie...2 vols. 12°,
Amsterdam, 1706.
H; M/G7 – 10 & 11; Tr/NQ.9.135 & 136.

1178 — Relation de la cour de Rome, où l'on voit le vray caractère de
cette cour...Pt 1. 8°, Paris, 1701.
H; M/G8 – 14; Tr/NQ.8.112.

1179 NONNUS, *Panopolitanus*. Dionysiaca, nunc primum in lucem edita,
ex Bibliotheca Ioannis Sambuci Pannonij. Cum lectionibus, &
coniecturis G. Falkenburgij...(*Greek.*) 8°, Antverpiæ, 1569.
H; M/A7 – 11; Tr/NQ.8.31.
NORMAN, Robert. The newe attractive, shewing the nature,
propertie, and manifold vertues of the loadstone...1720. *See*
1731.

1180 NORTON, Samuel. Alchymiæ complementum, et perfectio, seu
Modus et processus argumentandi...ab E. Deano auctior &
perfectior editus...4°, Francofurti, 1630.
Not H; M/B1 – 17; Tr/NQ.16.101⁷.

1181 — Catholicon physicorum, seu Modus conficiendi tincturam physicam & alchymicam...editus labore & industriâ E. Deani...4°, Francofurti, 1630.
Not H; M/B1 – 17; Tr/NQ.16.101² [a few signs of dog-earing].

1182 — Elixer, seu Medicina vitae, seu Modus conficiendi verum aurum et argentum...editus industriâ, & operâ E. Deani...4°, Francofurti, 1630.
Not H; M/B1 – 17; Tr/NQ.16.101³.

1183 — Mercurius redivivus, seu Modus conficiendi lapidem philosophicum ...editus opera & studio E. Deani...auctior & perfectior. 4°, Francofurti, 1630.
Not H; M/B1 – 17; Tr/NQ.16.101¹ [bound with 1180–82, 1184–7].

1184 — Metamorphosis lapidum ignobilium in gemmas quasdam pretiosas, seu Modus transformandi perlas parvas, et minutulas, in magnas & nobiles...editus diligentia E. Deani...4°, Francofurti, 1630.
Not H; M/B1 – 17; Tr/NQ.16.101⁵.

1185 — Saturnus saturatus dissolutus, et coelo restitutus, seu Modus componendi lapidem philosophicum...vero edente E. Deano, ...4°, Francofurti, 1630.
Not H; M/B1 – 17; Tr/NQ.16.101⁶.

1186 — Tractatulus de antiquorum scriptorum considerationibus in alchymia...editus studio, labore & industriâ E. Deani...4°, Francofurti, 1630.
Not H; M/B1 – 17; Tr/NQ.16.101⁸.

1187 — Venus vitriolata, in elixer conversa...sivè Modus conficiendi lapidem philosophicum...editus studiis, & diligentiâ E. Deani ...4°, Francofurti, 1630.
Not H; M/B1 – 17; Tr/NQ.16.101⁴.

1188 NORWOOD, Matthew. The seaman's companion, being a plain guide to the understanding of arithmetick, geometry, trigonometry, navigation, and astronomy. Applied chiefly to navigation ...3rd ed. corrected and amended. 4°, London, [1678].
Not H; M (Tracts) J3 – 41; Tr/NQ.8.111⁴.

1189 NORWOOD, Richard. Norwood's epitomie: or The application of the doctrine of triangles...8°, London, 1645.
Not H; M/G9 – 12; ULC/White.d.178; Thame/979.
Correction in Newton's hand in p. 14, l. 22, making the misprinted number '46' read '45'.

1190 — Trigonometrie: or, The doctrine of triangles...8th ed. being diligently corrected...4°, London, 1685.
H; M/J3 – 38; Tr/NQ.9.33.

1191 NOUVELLE bibliothèque choisie, où l'on fait connoître les bons livres en divers genres de litérature, & l'usage qu'on en doit faire. [By N. Barat.] 2 vols. 12°, Amsterdam, 1714.
H; M/G6 – 2 & 3; Tr/NQ.9.144 & 145.

1192 NOVUM lumen chymicum. E naturæ fonte & manuali experientia depromptum: cui accessit Tractatus de sulphure...[By M. Sendivogius.] 8°, Genevæ, 1639.
Not H; M/E3 – 42; British Library/C.112.aa.3.(1.)
Annotations by Newton on pp. 36, 97, 116; pp. 45, 47, 67, 69, 75, 79, 88 turned down and a few other signs of dog-earing; *see also* 445 & 1485.

1193 NUBES testium: or A collection of the Primitive Fathers...[By J. Gother.] 4°, London, 1686.
H; M/F5 – 20; Tr/NQ.8.131¹ [bound with 1234; a few signs of dog-earing].

1194 NUMISMATUM antiquorum sylloge populis Græcis, municipiis, & Coloniis Romanis cusorum, ex Cimeliarchio editoris. 4°, Londini, 1708.
Not H; M (Tracts) E6 – 35; Tr/NQ.10.34¹ [a few signs of dog-earing].

1195 NUNNS (The) complaint against the Fryars...[By A. L. Varet.] 8°, London, 1676.
Not H; M/H1 – 44; Tr/NQ.16.29.

1196 NUYSEMENT, Jacques. Tractatus de vero sale secreto philoso-phorum, & de universali mundi spiritu...nunc Latine versus a L. Combachio...12°, Lugduni Batavorum, 1672.
Not H; M/E1 – 24; ?

1197 OBSERVATIONES on scripture, MSS. by R. W. F°. *Not identified.*
Not H; M/J9 – 24; ?

1198 OBSERVATIONS on Dr. Waterland's second defense of his queries ...[By S. Clarke.] 8°, London, 1724.
H; not M; ?

1199 — Observations upon Anthroposophia theomagica, and Anima magica abscondita [of Eugenius Philalethes, i.e. T. Vaughan]. By Alazonomastix Philalethes [i.e. H. More]. 8°, [London], 1650.
Not H; M/G9 – 30: Tr/NQ.16.134³ [bound with 50].

1200 OCKLEY, Simon. The history of the Saracens...2nd ed. 2 vols. 8°, London, 1718.
H; M/J4 – 12 & 13; Tr/NQ.8.99 & 100.

1201 OF trade. 1. In general. 2. In particular. 3. Domestick [etc.]...By J. P[ollexfen]. Esq; To which is annex'd, The argument of the late Lord Chief Justice Pollexphen, upon an action of the case, brought by the East-India Company against Mr. Sands an interloper. (2 pts.) 8°, London, 1700.

H; M/J1 – 44; Tr/NQ.8.125² [bound with 299; a few signs of dog-earing].

1202 OISELIUS, Jacobus. Thesaurus selectorum numismatum anti-quorum...4°, Amstelodami, 1677.

H; M/F6 – 22; ?

1203 OLDHAM, John. The works of Mr. J. Oldham together with his remains. 8°, London, 1686.

Not H; M/J3 – 29; ?

OLDMIXON, John. Arcana Gallica: or, The secret history of France, for the last century...[By J. Oldmixon.] 1714. *See* 73.

— The secret history of Europe...[By J. Oldmixon.] 2 vols. 1712. *See* 1477.

1204 OMERIQUE, Antonio Hugo de. Analysis geometrica, sive Nova, et vera methodus resolvendi tam problemata geometrica, quam arithmeticas quæstiones....4°, Gadibus, 1698.

H; M/B1 – 12; Tr/NQ.8.115; Newton found this 'a judicious & valuable piece' (*Correspondence*, VII, 412).

1205 OPHTHALMOGRAPHIA: or, A treatise of the eye, in 2 parts... [By P. Kennedy.] 8°, London, 1713.

H; M/E5 – 39; Tr/NQ.9.50² [bound with 881].

1206 OPTATUS, *St, Bp of Milevi*. Opera cum observationibus et notis G. Albaspinæi...(2 pts.) (*Greek & Latin*.) F°, Parisiis, 1631.

H; M/D8 – 4; D. J. McKitterick, Cambridge ['pret 15ˢ' in Newton's hand on fly-leaf]; see p. 57 above.

1207 OPUSCULA mythologica, ethica et physica. Græce & Latine... [Ed. by T. Gale.] 8°, Cantabrigiæ, 1671.

H; M/F2 – 16; ?

1208 — Opuscula quædam chemica. G. Riplei Medulla philosophiæ chemicæ [etc.]...Omnia partim ex veteribus manuscriptis eruta partim restituta...8°, Francofurti, 1614.

H; M/E3 – 24; ?

ORATIO DOMINICA. *See* LORD'S PRAYER.

1209 ORIGENES, *Adamantius*. Contra Celsum libri VIII. Ejusdem Philo-calia. G. Spencerus, utriusque operis versionem recognovit, & annotationes adjecit. (*Greek & Latin*.) 4°, Cantabrigiæ, 1658.

H; M/C6 – 13; Tr/NQ.10.39.

On front paste-down in probable early Newton hand 'Origenes est bonus Scripturarũ Interpres, malus dogmatistes. Hieron. Epist. ad Pañach & Ocean' [i.e. Epistulae LXXXIV.2]; on fly-leaf 'pret 12ˢ' by Newton; several signs of dog-earing.

1210 — Dialogus contra Marcionitas, sive De rectâ in Deum fide: Exhortatio ad Martyrium: Responsum ad Africani epistolam de historiâ Susannæ...Operâ & studio J. R. Wetstenii. (*Greek &* *Latin.*) 4°, Basileæ, 1674.
H; M/C1 – 4; Tr/NQ.16.142 ['pret 7ˢ.' in Newton's hand on fly-leaf].

1211 — Hexaplorum Origenis quæ supersunt, multis partibus auctiora, quàm a F. Nobilio & J. Drusio edita fuerint. Ex manuscriptis & ex libris editis eruit & notis illustravit B. de Montfaucon... (*Greek & Latin.*) 2 vols. F°, Parisiis, 1713.
H; M/C4 – 5 & 6; Tr/NQ.17.30 & 31.

1212 — Opera, quae quidem extant omnia, per D. Erasmum partim versa, partim vigilanter recognita...2 vols. F°, Apud Basileam, 1536.
H; M/D8 – 1 & 2; Tr/NQ.17.19 & 20¹ [bound with 705].

1213 — Ὀριγένους περὶ εὐχῆς σύνταγμα μέχρι τοῦδε τοῦ χρόνου ἀνέκδοτον. (*Greek & Latin.*) 12°, Ὀξονία, [1686].
H; M/A1 – 16; ?

1214 ORPHEUS. Argonautica, Hymni, et De lapidibus curante A. C. Eschenbachio...Accedunt H. Stephani in omnia & J. Scaligeri in Hymnos notæ. (*Greek & Latin.*) 12°, Trajecti ad Rhenum, 1689.
H; M/F1 – 18; Tr/NQ.16.28 [a few signs of dog-earing].

1215 ORTELIUS, Abraham. Thesaurus geographicus recognitus et auctus ...4°, Hanoviæ (1611).
Not H; M/C1 – 23; Tr/NQ.8.43 [20 pages still turned and a few other signs of dog-earing].

1216 OTT, Johann. Cogitationes physico-mechanicæ de natura visionis... 4°, Heidelbergæ, 1670.
Not H; M (Tracts) D5 – 25; Tr/NQ.8.51⁷.

1217 OUDIN, Antoine. Curiositez françoises, pour supplément aux dictionnaires...8°, Paris, 1656.
H; M/G8 – 12; Tr/NQ.10.117.

1218 OUGHTRED, William. Clavis mathematicæ denuo limata, sive potius fabricata...Ed. 3ᵃ auctior & emendatior. 8°, Oxoniæ, 1652.
Not H; M/E3 – 2; Tr/NQ.8.59 [pp. 1–108 only of the printed text, the rest of the volume consisting of handwritten copies of mathematical texts and of the text of p. 109, in an unidentified hand]; *Math. papers*, I, 22 n. 18.

1219 — Clavis mathematicæ denuo limata, sive potius fabricata...Ed. 4ᵃ auctior & emendatior. 8°, Oxoniæ, 1667.
H; M/E3 – 1; Tr/NQ.10.142.

1220 — Trigonometria: hoc est, Modus computandi triangulorum latera & angulos, ex canone mathematico traditus & demonstratus. 4°, Londini, 1657.

H; M/J3 – 35; Tr/NQ.16.183.

2 illustrative diagrams in an early Newton hand added at the top left-hand corner of p. [2].

1221 OVIDIUS NASO, Publius. Opera omnia, ex recensione G. Bersmani, cum ejusdem, aliorumque virorum doctissimorum notationibus. Ed. nova. (4 pts.) 12°, Londini, (1655–)1656.

H; M/E1 – 5; Tr/NQ.9.168 [pt 3, p. 89 turned up and a few other signs of dog-earing].

1222 — Publii Ovidii Nasonis operum tomus primus(–tertius)...N. Heinsius...castigavit...3 vols. 12°, Amstelodami, 1664.

Not H; M/G1 – 20 to 22; ?

1223 — P. Ovidii Nasonis operum tomus primus(–tertius). [Ed. by M. Maittaire.] 3 vols. 12°, Londini, 1715.

Not H; M/F1 – 38 to 40: ?

1224 — P. Ovidii Metamorphosis, seu Fabulæ poeticæ: earumque interpretatio ethica, physica et historica Georgii Sabini...Ultima ed. 16°, Francofurdi, 1593.

Not H; M/G1 – 35; Babson 409 [inscribed 'Isaci Newtoni liber Octobris 15 1659. prætium -0-1-6' on fly-leaf].

1225 — Ovids Metamorphosis Englished. By G. Sandys. 4th ed. 12°, London, 1656.

Not H; M/G9 – 2; ?

1226 OWEN, John. Epigrammatum libri XII. 24°, Londini, 1668.

Not H; M/A1 – 24; ?

1227 OWEN, Robert. Hypermnestra: or, Love in tears. A tragedy. 2nd ed. 12°, London, 1722.

Not H; M (Plays) H8 – 19; Tr/NQ.10.92².

OXFORD. *University.* Parecbolæ sive excerpta è Corpore Statutorum Universitatis Oxoniensis...1721. *See* 1246.

1228 — *University College.* The proceedings of the visitors of University College, with regard to the late election of a Master, vindicated. 2nd ed. F°, Oxford, 1723.

Not H; M/F8 – 4; Babson 408 [bound with 1009]; Sotheran **804** (1927) 3804.

1229 PACCHIONI, Antonio. Dissertationes binæ ad...D. J. Fantonum datæ, cum ejusdem responsione illustrandis duræ meningis, ejusque glandularum structuræ, atque usibus concinnatæ...8°, Romæ, 1713.

Not H; M (Tracts) H8 – 16; Tr/NQ.10.82².

1230 PAGNINUS, Sanctes. קצר אוצר לשון הקדש. Hoc est, Epitome thesauri linguæ sanctæ. 3ª ed. 8°, Antverpiæ, 1578.
Not H; M/E2 – 27; Tr/NQ.8.127.

1231 PALFYN, Jean. Description anatomique des parties de la femme, qui servent à la géneration...4°, Leide, 1708.
H; M/J3 – 45; ?

1232 PALLADIUS, Bp of Aspona. De vita S. Johannis Chrysostomi dialogus ...Omnia nunc primùm Græco–Latina prodeunt curâ & studio E. Bigotii. 4°, Luteciæ Parisiorum, 1680.
H; M/F6 – 21; ?

1233 — Palladii Divi Evagrii Discipuli Lausiaca quæ dicitur historia, et Theodoreti Episcopi Cyri Θεοφιλῆς, id est religiosa historia... 4°, Parisiis, 1555.
Not H; M/A7 – 16; ?

1234 PAMPHLET (The) entituled, Speculum ecclesiasticum [of T. Ward], or An ecclesiastical prospective-glass, considered in its false reasonings and quotations. [By H. Wharton.] 4°, London, 1688.
Not H; M/F5 – 20; Tr/NQ.8.131² [bound with 1193].

1235 PANCIROLI, Guido. Notitia utraque dignitatum cum Orientis, tum Occidentis, ultra Arcadii Honoriique tempora. Et in eam G. Panciroli...commentarium...Ultima ed., auctior, et correctior. F°, Lugduni, 1608.
H; M/D8 – 7; ?

1236 — Rerum memorabilium sive Deperditarum pars prior (& Liber secundus). Commentarijs illustrata...Ab H. Salmuth. 4°, Francofurti, 1660.
H; M/C1 – 3; Tr/NQ.8.40 [pt 1, p. 104 and pt 2, p. 28 turned up].

1237 PANEGYRICI. xii. Panegyrici veteres...emendati, aucti; nuper quidem ope I. Livineii: nunc verò operâ I. Gruteri...16°, Francofurti, 1607.
Not H; M/G1 – 19; ?

PANTALEON. La refutation de l'anonyme Pantaleon, soy disant Disciple d'Hermés. 1689. See 511.

1238 PARACELSUS. Aurora thesaurusque philosophorum, Theophrasti Paracelsi, Germani philosophi, & Medici præ cunctis omnibus accuratissimi. Accessit Monarchia physica per G. Dorneum... 8°, Basileæ, 1577.
Not H; M/F3 – 29; Tr/NQ.9.170² [bound with 1239; a few signs of dog-earing].

1239 — Congeries Paracelsicæ chemiæ de transmutationibus metallorum, ex omnibus quæ de his ab ipso scripta reperire licuit hactenus.

Accessit genealogia mineralium, atcg metallorum omnium, eiusdem autoris. G. Dorneo interprete. 8°, Francofurti, 1581.
H; M/F3 – 29; Tr/NQ.9.170¹ [bound with 1238 & 1241; several signs of dog-earing].

1240 — De summis naturæ mysteriis commentarii iii, à G. Dorn conversi ...8°, Basileæ, 1584.
Not H; M/D1 – 26; ?

1241 — Libri v. de vita longa, incognitarum rerum, & hucusque à nemine tractatarum refertissimi...8°, Basileæ, [1562].
Not H; M/F3 – 29; Tr/NQ.9.170³ [bound with 1239].

1242 — Opera omnia medico-chemico-chirurgica...Ed. novissima... 3 vols. in 2. F°, Genevæ, 1658.
H; M/C3 – 18 & 19; Ekins 1 – 5 – 0; ?

1243 — Tract. varii. 4°, 1600. *Not identified.*
Not H; M/E3 – 31; ?

1244 PARALIPOMENA prophetica, containing several supplements and defences of Dʳ Henry More his Expositions of the Prophet Daniel and the Apocalypse...(2 pts.) 4°, London, 1685.
Not H; M/F5 – 6; Tr/NQ.9.37; Thame/969.
PARDIES, Ignace Gaston. A discourse of local motion...By A. M. [i.e. I. G. Pardies]. Englished out of French. 1670. *See* 526.

1245 — La statique ou La science des forces mouvantes. 12°, Paris, 1673.
H; M/G2 – 46; ?; Newton thanked Collins for 'ye little but ingenious tract', 17 Sept. 1673 (*Correspondence*, i, 307).

1246 PARECBOLÆ sive excerpta è Corpore Statutorum Universitatis Oxoniensis. Accedunt Articuli religionis xxxix. in Ecclesia Anglicana recepti: nec non juramenta fidelitatis & suprematus. 8°, Oxoniæ, 1721.
Not H; M/F1 – 10; Tr/NQ.7.7.

1247 PARIS. *Académie Royale des Inscriptions et Belles Lettres.* Histoire de l'Académie Royale des Inscriptions et Belles Lettres, depuis son établissement jusqu'à présent...Vols. 1–4. 4°, Paris, 1717–23.
H; M/D6 – 22 to 25; ?

1248 — *Académie Royale des Médailles et des Inscriptions.* Médailles sur les principaux événements du règne de Louis Le Grand, avec des explications historiques. Par l'Académie Royale des Médailles & des Inscriptions. F°, Paris, 1702.
H; M/C2 – 24; Tr/NQ.18.9.
Loose slip of paper now attached opposite p. 89 contains draft by Newton for exergue of medal to commemorate Marlborough's Blenheim campaign: 'Expugnata Host. castra ad Donaverdam Iun XXI. Deletus exercitus ad Hochstet Aug. II. Captis XIII. M. Sign. relat CLXXI. Et Bavaria

recepta MDCCIV'. (Fortescue's *History of the British Army*, I (1899), 443, gives totals of 11,000 prisoners and 171 standards.)

1249 — *Académie Royale des Sciences*. Histoire de l'Académie Royale des Sciences. Année M.DCCVII, M.DCCVIII, M.DCCX. Avec les Mémoires de mathématique & de physique, pour la même année. Tirez des registres de cette Académie. 3 vols. 12°, Amsterdam, 1708–13.

H; M/G7 – 30, 32, 35; Tr/NQ.16.38 to 40.

1250 — Histoire de l'Académie Royale des Sciences. Année M.DCCX – Année M.DCCXXII. Avec les Mémoires de mathématique & de physique, pour la même année. Tirés des registres de cette Académie. 13 vols. 4°, Paris, 1712–24.

H; M/D6 – 7 to 15, 17 to 20; Tr/NQ.10.9 to 21 [vols. for 1712, 1713, 1719 have a few signs of dog-earing].

1251 — Suite des Mémoires de mathématique et de physique tirez des registres de l'Académie Royale des Sciences, de l'année M.DCCVII, M.DCCVIII, M.DCCIX, M.DCCX. 4 vols. 12°, Amsterdam, 1708–13.

H; M/G7 – 31, 33, 34, 36; Tr/NQ.16.32 to 35.

1252 — Suite des Mémoires de l'Académie Royale des Sciences. Année M.DCCXVIII. (De la grandeur et de la figure de la terre. [By J. Cassini.]) 4°, Paris, 1720.

H; M/D6 – 16; Tr/NQ.10.22 [pp. 148, 154 turned down and a few other signs of dog-earing]; sent to Newton by Varignon as a gift from Cassini, August 1722 (*Correspondence*, VII, 206–10).

1253 — Recueil d'observations faites en plusieurs voyages par ordre de sa Majesté pour perfectionner l'astronomie et la géographie... Par Messieurs de l'Académie Royale des Sciences. F°, Paris, 1693.

H; M/F8 – 18; ?

1254 — Reglement ordonné par le Roy pour l'Académie Royale des Sciences. Du 26. de Janvier 1699. 4°, Paris, 1699.

H; M/E6 – 34; Butler Library, Columbia University, New York [bound with 1368]; Sotheran **828** (1931) 3604.

1255 — — [Another copy.]

Not H; M [shelf-mark not known]; William Andrews Clark Memorial Library, University of California, Los Angeles; Sotheran **804** (1927) 3798.

1256 PARIS, Matthew. Historia major. Juxta exemplar Londinense 1640, verbatim recusa...Editore W. Wats. F°, Londini, 1684.

H; M/D7 – 1; ?

1257 PARKINSON, John. Theatrum botanicum: the theater of plants, or, An herball of large extent...F°, London, 1640.

Not H; M/J9 – 1; ?

1258 PARKYNS, *Sir* Thomas. The Inn-play, or Cornish-hugg wrestler: digested in a method which teacheth to break all holds, and throw most falls mathematically...4°, Nottingham, 1713.
H; M/F5 – 24; Tr/NQ.9.61; see pp. 73–4 above.

1259 PARLIAMENT. Acts of Parliament. 22 vols. F°, [London, 1697–1722?].
H (described as 'Begun 7ᵐᵒ Gulielmi & end with 9° Georgii'); not M; ?

1260 — The Acts of Parliament relating to the building fifty new churches in and about the cities of London and Westminster...8°, London, 1721.
H; M/J4 – 4; Tr/NQ.9.22; Newton received printed notices dated 10 August 1717 and 2 July 1720 summoning him to meetings of the Commission for Building 50 New Churches (*Correspondence*, VI, 406–7, and VII, 484).

1261 — A collection of all the Statutes now in use in the Kingdom of Ireland, with notes...F°, Dublin, 1678.
H; M/H9 – 18; ?

1262 — The Statutes at large...from Magna Charta until this time. By J. Keble. 2 vols. F°, London, 1695.
H; M/F8 – 13 & 14; ?

1263 PARNASSE (Le) assiegé ou La guerre declarée entre les philosophes anciens & modernes. 12°, Lyon, 1697.
H; M/G2 – 38; Tr/NQ.16.5; Ekins 2 – 6.

1264 PASOR, Georg. Lexicon Græco-Latinum in Novum [Domini Nostri Jesu Christi Testamentum...(3 pts.) 8°, Londini, (1649–)1650.
H; M/D1 – 22; Tr/NQ.7.81 ['Isaac Newton Trin: Coll: Cant: pret: 6ᵈ. 1661' on fly-leaf, 'Isaac Newton hunc librum possidet. pret: 6ᵈ. Martij 29 1661.' on verso of title-page; p. 595 turned down].

1265 PATERCULUS, Gaius Velleius. C. Velleius Paterculus cum selectis variorum notis. A. Thysius edidit & accuratè recensuit. 8°, Lugd. Bat., 1653.
H; M/F2 – 13; Tr/NQ.10.106.

1266 — C. Velleius Paterculus, cum selectis variorum notis. A. Thysius edidit & accurate recensuit. 8°, Lugd. Batavorum, 1659.
H; not M; ?

1267 PATRICIUS, Franciscus. Magia philosophica, hoc est F. Patricii summi philosophi Zoroaster & eius 320. Oracula Chaldaica. Asclepii Dialogus. & Philosophia magna Hermetis Trismegisti ...Latine reddita. 8°, Hamburgi, 1593.
Not H; M/E1 – 37; ?

1268 PATRICK, Simon, *Bp of Ely*. The glorious Epiphany, with the devout Christian's love to it. 8°, London, 1678.
Not H; M/H3 – 40; ?

1269 — The parable of the pilgrim: written to a friend. 5th ed. 4°, London, 1678.
Not H; M/H2 – 3; ?
PAULUS, *Diaconus*. *See* WARNEFRIDUS, Paulus, *Diaconus*.

1270 PAUSANIAS. Graeciae descriptio accurata...cum Latina R. Amasaei interpretatione. Accesserunt G. Xylandri & F. Sylburgii annotationes, ac novae notae I. Kuhnii. F°, Lipsiæ, 1696.
H; M/C3 – 21; Tr/NQ.17.17 [pp. 13, 421, 603, 668, 815 turned down, p. 408 up, and a few other signs of dog-earing].

1271 PEARSON, John, *Bp of Chester*. An exposition of the Creed. 7th ed. revised and corrected. F°, London, 1701.
Not H; M/B2 – 19; ?

1272 — Opera posthuma chronologica, &c....Prælo tradidit, edenda curavit & dissertationis novis additionibus auxit H. Dodwellus ...4°, Londini, 1688.
H; M/A7 – 3; Tr/NQ.10.7.

1273 PELEGROMIUS, Simon. Synonymorum sylva...Nunc autem è Belgarum sermone in Anglicanum transfusa...per H. F....8°, Londini, 1668.
Not H; M/E2 – 24; Tr/NQ.16.157.

1274 PEMBERTON, Henry. Epistola [of H. Pemberton] ad amicum [J. Wilson] de Cotesii inventis, curvarum ratione, quæ cum circulo & hyperbola comparationem admittunt. 4°, Londini, 1722.
H; M/E6 – 33; Tr/NQ.18.6.

1275 PEMBLE, William. Workes. Containing sundry treatises and expositions...3rd ed. F°, London, 1635.
Not H; M/J9 – 21; ?

1276 PEPYS, Samuel. Memoires relating to the state of the Royal Navy of England, for ten years, determin'd December 1688. 8°, [London?], 1690.
Not H; M/J1 – 41; Babson 881.

1277 PERIZONIUS, Jacobus. Origines Babylonicæ et Ægyptiacæ. 2 vols. 8°, Lugduni Batavorum, 1711.
H; M/A2 – 2 & 3; ?

1278 — Rerum per Europam maxime gestarum ab ineunte Saeculo Sexto-decimo usque ad Caroli V. mortem &c. Commentarii historici. 8°, Lugduni Batavorum, 1710.
H; M/A2 – 4; Tr/NQ.8.89.

1279 PERKINS, William. A warning against the idolatrie of the last times. And an instruction touching religious or Divine worship. 8°, Cambridge, 1601.
Not H; M/G9 – 5; ?

1280 PERRY, John. An account of the stopping of the Daggenham Breach...8°, London, 1721.
H; M/J4 – 17; ?

1281 — Maps of Ireland. F°, [c. 1720].
Not H; M/B5 – 3 [bound with 1593]; ?

1282 — The state of Russia, under the present Czar...8°, London, 1716.
H; M/J4 – 16; Tr/NQ.9.131.

PERSIUS FLACCUS, Aulus. *For the Satires of Persius Flaccus published with those of Juvenal, see* JUVENALIS, Decimus Junius.

1283 PETAU, Denis. Abrégé chronologique de l'Histoire universelle sacrée et profane. Traduction nouvelle, suivant la dernière éd. latine. Nouvelle éd. continuée jusqu'à présent. 5 vols. 12°, Paris, 1715.
H; M/G8 – 29 to 33; Tr/NQ.7.59 to 63.

1284 — Opus de doctrina temporum...cum praefatione et dissertatione ...J. Harduini...3 vols. F°, Antwerpiæ, 1703.
H; M/C5 – 15 to 17; ?

1285 — Opus de theologicis dogmatibus, auctius in hac nova editione... notulis T. Alethini [i.e. J. Le Clerc]. 6 vols. in 3. F°, Antwerpiæ, 1700.
H; M/C5 – 18 to 20; Tr/NQ.17.26 to 28 [vol. 1 has p. 101 turned down and a few other signs of dog-earing].

1286 — Rationarium temporum, in partes duas, libros XIII, distributum... Ed. ultima...8°, Franequeræ, 1694.
H; M/F1 – 19; Tr/NQ.16.8 [a few signs of dog-earing].

1287 PETIVER, James. A catalogue of Mʳ Rayˢ English Herbal illustrated with figures on (50) folio copper plates. [Plates 1 and 11 dedicated to Newton.] F°, London, [c. 1700].
Not H; M/J9 – 6; Tr/NQ.18.26² [bound with 1289].

1288 — Gazophylacii naturæ & artis decas prima(–quinta)...8°, Londini, 1702-6.
Not H; M (Tracts) H8 – 13; Tr/NQ.16.99⁵ [without the plates].

1289 — Gazophylacii naturæ & artis decas prima(–decima)...[The leaves of the text of the 8° ed., 1702-6, mounted on sheets facing the folio plates. Plates 39 and 61 dedicated to Newton.] F°, [London, 1711].
Not H; M/J9 – 6; Tr/NQ.18.26¹ [bound with 1287 & 1290].

1290 — Plants already engraved in Mr. Petiver's English Herbal. F°, London, [1715].
Not H; M/J9 – 6; Tr/NQ.18.26³ [bound with 1289].

1291 PETTUS, *Sir* John. Fleta minor. The laws of art and nature, in knowing, judging, assaying, fining, refining and inlarging the bodies of confin'd metals. In 2 pts. The first contains Assays of L. Erckern...The second contains Essays on metallick words, as a dictionary...By Sir John Pettus. F°, London, 1683.
H; M/B3 – 8; Tr/NQ.18.7.

1292 PETTY, *Sir* William. The discourse made before the Royal Society the 26. of November 1674. Concerning the use of duplicate proportion...12°, London, 1674.
H; M/G9 – 48; ?

PETYT, William. Britannia languens, or A discourse of trade...[By W. Petyt.] 1680. *See* 299.

PEURBACH, Georg von. *See* PURBACH, Georg.

1293 PEZELIUS, Christophorus. Mellificium historicum integrum...4°, Marpurgi, 1631.
Not H; M/J6 – 27; ?

1294 PFAFF, Christoph Matthaeus. Dissertatio critica de genuinis librorum Novi Testamenti lectionibus...8°, Amstelodami, 1709.
H; M/A2 – 13; ?

1295 PFEIFFER, August. Theologiæ, sive potius Ματαιολογίας Judaicæ atque Mohammedicæ seu Turcico-Persicæ principia sublesta et fructus pestilentes...8°, Lipsiæ, 1687.
Not H; M/F3 – 16; ?

1296 PHILADELPHIA, or Brotherly love to the studious in the hermetick art...Written by Eyreneus Philoctetes [i.e. G. Starkey?]. 12°, London, 1694.
Not H; M/G9 – 5; Thame/974; ?

PHILALETHES, Æyrenæus, *pseud. See* STARKEY, George.

PHILALETHES, Alazonomastix, *pseud. See* MORE, Henry.

PHILALETHES, Eirenæus, *pseud. See* STARKEY, George.

PHILALETHES, Eirenæus Philoponos, *pseud. See* STARKEY, George.

PHILALETHES, Eugenius, *pseud. See* VAUGHAN, Thomas.

PHILELEUTHERUS, *Lipsiensis, pseud. See* BENTLEY, Richard.

PHILEMON, *comicus.* Menandri et Philemonis reliquiæ, quotquot reperiri potuerunt; Græce et Latine, cum notis H. Grotii et J. Clerici. 1709. *See* 1068.

1297 PHILIPPS, Jenkin Thomas. Dissertatio historico-philosophica de atheismo. Sive Historia atheismi...8°, Londini, 1716.
H; M/D1 – 16; Tr/NQ.9.26.

1298 — Dissertationes varii argumenti...Ed. 2ᵃ...(2 pts.) 8°, Londini, 1715.
Not H; M/E2 – 23; Tr/NQ.16.61.

— Thirty four conferences between the Danish missionaries and the Malabarian Bramans...Transl....by Mr. Philipps. 1719. *See* 1612.

PHILIPS, John. Cyder. A poem. In 2 books. [By J. Philips.] 1720. *See* 473.

1299 — Poems on several occasions. 3rd ed. 12°, London, 1720.
Not H; M/J8 – 25; Tr/NQ.8.109² [bound with 1501].

1300 PHILO, *Judaeus.* Omnia quæ extant opera. Ex accuratissima S. Gelenii, & aliorum interpretatione...(*Greek & Latin.*) F°, Lutetiæ Parisiorum, 1640.
H; M/B3 – 10; ?

PHILOCTETES, Eyreneus, *pseud. See* STARKEY, George.

1301 PHILOSOPHIÆ chymicæ IV. vetustissima scripta, I. Senioris Zadith F. Hamuellis Tabula chymica. II. Innominati philosophi Expositio tabulæ chymicæ. III. Hermetis Trismegisti Liber de compositione. IV. Anonymi veteris philosophi Consilium coniugii ...Omnia ex Arabico sermone Latina facta, & nunc primum in lucem producta. 8°, Francofurti, 1605.
Not H; M/E3 – 46; Tr/NQ.16.136.
Long index-note by Newton on fly-leaf headed 'Authores a Seniore citati' and listing 18 authors with page numbers for each ranging from 1 to 20 (for 'Hermes'), also short references to other writers at pp. 3–5, 47, 98, 114, 117–19, 124, 133, 207; several signs of dog-earing.

1302 PHILOSOPHICAL (A) epistle, discovering the unrevealed mystery of the three fires of the Sophi. [*Signed* Cloidophorus Mystagogus, i.e. W. Yarworth.] 8°, [*c.* 1702].
Not H; M/E2 – 37; Tr/NQ.16.131² [bound with 1138; many signs of dog-earing].

1303 — — [Another copy.]
Not H; M (Tracts) H8 – 14; Tr/NQ.16.95³.

1304 — The philosophical transactions (of the Royal Society)...Nos. 1–380. 16 vols. 4°, London, 1665–1723.
H; M/D2 – 1 to 16; Thame/981; ?

1305 — Philosophical transactions. No. 38, Monday, August 17. 1668. [Includes a review of Mercator's Logarithmo-technia...1668. *See* 1073.] 4°, [London, 1668].
Not H; M/E5 – 4; Tr/NQ.9.48⁵ [bound with 714].

1306 — Philosophical transactions. No. 43, Monday, Januar. 11. 166$\frac{8}{9}$. [Includes a Summary account by John Wallis on the general laws of motion.] 4°, [London, 1669].
Not H; M/E5 – 4; Tr/NQ.9.48⁶ [bound with 714].

1307 — The philosophical transactions and collections, to the end of the year 1700. Abridg'd and dispos'd under general heads. By J. Lowthorp. 3 vols. 4°, London, 1705.

H; M/B6 – 16 to 18; ?

1308 — The philosophical transactions, from the year 1700 to the year 1720. Abridg'd and dispos'd under general heads. By H. Jones. Vols. 4, 5. 4°, London, 1721.

H; M/B6 – 19 & 20; Tr/NQ.10.47 & 48.

1309 PHILOSOPHIE naturelle de trois anciens philosophes renommez: Artephius, Flamel, & Synesius, traitant de l'art occulte, & de la transmutation metallique. Dernière éd. [of Trois traitez de la philosophie naturelle...]. 4°, Paris, 1682.

Not H; M/F6 – 31; Tr/NQ.16.93 [pp. 21, 31 turned down and several other signs of dog-earing].

1310 — — [Another copy.]

Not H; M (Tracts) J6 – 2; Tr/NQ.16.77⁵.

1311 — La philosophie naturelle restablie en sa pureté. Où l'on void à découvert toute l'œconomie de la nature, & où se manifestent quantité d'erreurs de la philosophie ancienne...[By J. d'Espagnet.] 8°, Paris, 1651.

H; M/G6 – 15; Tr/NQ.10.81 [a few signs of dog-earing].

1312 PHILOSTRATUS. Philostratorum quæ supersunt omnia, Vita Apollonii libris VIII, Vitae Sophistarum libris II...Accessere Apollonii Tyanensis Epistolae...Omnia ex Mss. codd. recensuit notis perpetuis illustravit...G. Olearius. (*Greek & Latin.*) F°, Lipsiæ, 1709.

H; M/B4 – 14; Tr/NQ.18.13 [pp. 699, 890 turned down].

1313 PHOTIUS, *Patriarch of Constantinople.* Epistolæ. Per R. Montacutium Latinè redditæ, & notis subinde illustratæ. (*Greek & Latin.*) F°, Londini, 1651.

H; M/J9 – 7; Tr/NQ.11.29 [pp. 53, 260 turned down].

1314 — Myriobiblon, sive Bibliotheca librorum quos legit et censuit Photius. Græcè edidit D. Hoeschelius...Latinè verò reddidit...A. Schottus. F°, Rothomagi, 1653.

H; M/E8 – 18; ?

1315 PIAZZA, Girolamo Bartolomeo. A short and true account of the Inquisition and its proceeding, as it is practis'd in Italy...4°, London, 1722.

H; M/F5 – 12; ?

1316 PILOTE (Le) de l'onde vive, ou Le secret du flux et reflux de la mer ...2ᵉ éd. reveuë & augmentée de deux traitez nouveaux sur la philosophie naturelle. [By M. Eyquem du Martineau.] 12°, Paris, 1689.

H; M/G6 – 16; Tr/NQ.16.163; Ekins 3 – 0.

1317 PINDARUS. Olympia, Pythia, Nemea, Isthmia. Cæterorum octo
lyricorum carmina...nonnulla etiam aliorum. Omnia Græcè
et Latinè...[Ed. by H. Stephanus.] 8°, [Paris], 1560.
H; M/E1 – 44; Keynes Collection, King's College, Cambridge
['Isaacus Newton hunc librum possidet. Pret. 8ᵈ. 1659' in
Newton's hand on verso of title-page; p. 215 turned down,
pp. 221, 235, 251 up, and a few other signs of dog-earing];
bought by J. M. Keynes from Thorp of Guildford, March 1921,
for 10s.

1318 — Olympia, Pythia, Nemea, Isthmia. I. Benedictus ad metri rationem
...totum authorem innumeris mendis repurgavit...Ed. puris-
sima...(*Greek & Latin*.) 4°, Salmurii, 1620.
H; M/F6 – 13; ?

1319 — Olympia, Nemea, Pythia, Isthmia. Una cum Latina omnium
versione carmine lyrico per N. Sudorium. (*Greek & Latin*.) F°,
Oxonii, 1697.
H; M/B2 – 23; Tr/NQ.11.1 [pp. 121, 127 turned down and a
few other signs of dog-earing].

PITCAIRNE, Archibald. Epistola Archimedis ad Regem Gelonem,
Albæ Græcæ reperta...1688. [By A. Pitcairne.] [*c.* 1710.] *See*
566.

1320 PITISCUS, Samuel. Lexicon antiquitatum Romanarum: in quo ritus
et antiquitates...exponuntur...2 vols. F°, Leovardiae, 1713.
H; M/B4 – 16 & 17; Tr/NQ.18.29 & 30.

1321 PITOT, Allain. L'automate de longitude. Nouveau systéme d'hydro-
métrie...8°, Londres, 1716.
Not H; M (Tracts) J8 – 11; Tr/NQ.10.119⁵.

1322 PITTON DE TOURNEFORT, Joseph. Relation d'un voyage du
Levant, fait par ordre du Roy...3 vols. 8°, Lyon, 1717.
H; M/G10 – 32 to 34; Tr/NQ.10.121 to 123.

PLACE, E. An essay towards a new method to shew the longitude at
sea; especially near the dangerous shores. [By E. Place.] 1714.
See 578.

1323 PLATINA, Bartolomeo. Historia de vitis Pontificum Romanorum...
4°, Coloniæ, 1600.
Not H; M/F6 – 25; ?

1324 PLATO. De rebuspub. sive De iusto, libri x, a I. Sozomeno è Græco
in Latinum, & ex dialogo in perpetuum sermonem redacti,
additis notis, & argumentis. 8°, Venetiis, 1626.
Not H; M/J6 – 29; Tr/NQ.9.31.

1325 — Opera omnia quæ exstant. M. Ficino interprete...(*Greek & Latin*.)
F°, Francofurti, 1602.
H; M/B4 – 13; ?

1326 PLEAS of the Crown. Or A brief, but full account of whatsoever can be found relating to that subject. [By Sir Matthew Hale.] 8°, London, 1678.

Not H; M/J1 – 42; Tr/NQ.8.86.

1327 PLINIUS CAECILIUS SECUNDUS, Gaius. Epistolarum libri x. & Panegyricus. Accedunt variantes lectiones. 12°, Lugd. Batavorum, 1640.

Not H; M/G1 – 15; ?

1328 — C. Plinii Cæcilii Secundi Epistolæ et Panegyricus, notis illustrata. 4°, Oxonii, 1677.

H; M/F4 – 18; Tr/NQ.7.48 [p. 117 turned down].

1329 PLINIUS SECUNDUS, Gaius. Histoire de la peinture ancienne, extraite de l'Hist. naturelle de Pline, liv. xxxv. Avec le texte latin, corrigé sur les Mss. de Vossius & sur la I. éd. de Venise...F°, Londres, 1725.

H; M/F7 – 8; Tr/NQ.18.19.

PLOT, Robert. The natural history of Oxford-shire, being an essay toward the natural history of England. By R. P[lot]. 1677. *See* 1140.

1330 PLUTARCHUS. Quæ extant opera, cum Latina interpretatione. Ex vetustis codicibus plurima nunc primùm emendata sunt, ut ex H. Stephani annotationibus intelliges...(*Greek & Latin.*) 13 vols. 8°, [Geneva], 1572.

H; M/F3 – 2 to 14; Tr/NQ.9.96 to 108 [vol. 9 has p. 11 of the Index turned up, vol. 10 has several signs of dog-earing].

1331 — Plutarchi Chæronensis quæ exstant omnia, cum Latina interpretatione H. Cruserii: G. Xylandri...2 vols. F°, Francofurti, 1599.

H; M/B4 – 10 & 11; ?

1332 POCOCKE, Edward. A commentary on the Prophecy of Hosea. F°, Oxford, 1685.

H; not M; ?

— Lamiato'l Ajam, carmen Tograi, poetæ Arabis doctissimi; unà cum versione Latina, & notis...operâ E. Pocockii...(2 pts.) 1661. *See* 1648.

— Porta Mosis, sive Dissertationes aliquot à R. Mose Maimonide... Nunc primùm Arabicè...& Latinè editæ...Operâ & studio E. Pocockii. (2 pts.) (1654–)1655. *See* 1021.

1333 POETÆ minores Græci. Hesiodus, Pythagoras, Mimnermus [etc.] ...Accedunt etiam Observationes R. Wintertoni in Hesiodum. (*Greek & Latin.*) 8°, Cantabrigiæ, 1684.

H; M/F2 – 25; Tr/NQ.9.129 [pp. 26, 204, 384 turned down, p. 296 turned up, and several other signs of dog-earing].

1334 POETICAL recreations: consisting of original poems, songs, odes, &c. With several new translations. In 2 parts. Part I. Occasionally written by Mrs. Jane Barker. Part II. By several gentlemen of the Universities, and others. 8°, London, 1688.
Not H; M/J3 – 27; Tr/16.158.

1335 POLE, Matthew. Synopsis criticorum aliorumque S. Scripturæ interpretum. 4 vols. in 5. F°, Londini, 1669–76.
H; M/F8 – 7 to 11; ?

1336 POLENI, Giovanni. De motu aquæ mixto libri II. Quibus multa nova pertinentia ad aestuaria, ad portus, atque ad flumina continentur. 4°, Patavii, 1717.
H; M (Tracts) D6 – 3; Tr/NQ.8.1⁵.

POLLEXFEN, John. Of trade. 1. In general. 2. In particular. 3. Domestick [etc.]...By J. P[ollexfen]. Esq....(2 pts.) 1700. *See* 1201.

1337 POLLUX, Julius. Onomasticon Græce & Latine...Omnia contulerunt...notas adjecerunt, editionemque curaverunt...J. H. Lederlinus et T. Hemsterhuis. 2 vols. F°, Amstelædami, 1706.
H; M/B2 – 21 & 22; ?

1338 POLYBIUS. Historiarum libri qui supersunt. I. Casaubonus ex antiquis libris emendavit, Latine vertit, & commentariis illustravit ...(*Greek & Latin.*) (2 pts.) F°, [Hanau], 1609.
H; M/C3 – 4; Tr/NQ.18.5.

POPE, Alexander. The Iliad. Translated by Mr Pope. 6 vols. 1715–20 & 6 vols. 1720. *See* 790 & 791.

1339 POPISH prayers. 8°. *Not identified.*
Not H; M/C9 – 5; ?

1340 PORTA, Giambattista della. Magiæ naturalis libri xx...12°, Lugd. Batavorum, 1651.
Not H; M/E1 – 18; ?

1341 POTIER, Michael. Veredarius hermetico-philosophicus lætum et inauditum nuncium adferens...8°, Francofurti, 1622.
Not H; M/E3 – 30; ?

1342 POTTER, Christopher. Want of charitie justly charged, on all such Romanists, as dare, without truth or modesty, affirme, that Protestancie destroyeth Salvation...2nd ed., revised and enlarged. 8°, London, 1634.
Not H; M/H1 – 20; Tr/NQ.16.59 [p. 184 turned down].

1343 POTTER, John, *Abp of Canterbury.* Archæologiæ Græcæ: or, The antiquities of Greece. 2 vols. (Vol. 2, 2nd ed.) 4°, Oxford, 1697; London, 1706.
H; M/J4 – 18 & 19; Tr/NQ.16.47 & 48 [vol. 2 has a few signs of dog-earing].

1344 POWER, Henry. Experimental philosophy, in three books...4°, London, 1664.
H; M/F5 – 11; ?

1345 PREPARATION (A) for death, in consideration of the future judgement; with a panegyrick on his late Royal Highness, William Henry, Duke of Gloucester. 8°, London, 1700.
Not H; M/H1 – 28; ?

1346 PRESENT (The) condition of the English Navy set forth in a dialogue betwixt young Fudg of the Admiralty, and Capt. Steerwell, an Oliverian Commander. 4°, London, 1702.
Not H; M (Tracts) J3 – 41; Tr/NQ.8.111⁶.

1347 — The present state of Russia...from the year 1714, to 1720...[By F. C. Weber.] Translated from the High-Dutch. 2 vols. 8°, London, 1722–3.
H; M/J4 – 14 & 15; Tr/NQ.8.101 & 102.

PRIDEAUX, Humphrey. Marmora Oxoniensia, ex Arundellianis, Seldenianis, aliisque conflata. Recensuit, & perpetuo commentario explicavit, H. Prideaux...1676. *See* 1032.

1348 — The Old and the New Testament connected in the history of the Jews and neighbouring nations...2 vols. in 3. 8°, London, 1716–18.
H; M/H5 – 20 to 22; ?

1349 PRIOR, Matthew. Poems on several occasions. F°, London, 1718.
H; M/B5 – 10; ?

1350 PRIVILEGES (The) of the House of Lords and Commons argued and stated, in two conferences between both houses, April 19, and 22, 1671. To which is added a discourse, wherein the rights of the House of Lords are truly asserted...Written by Arthur [Annesley], Earl of Anglesey. 8°, London, 1702.
Not H; M/J8 – 55; Tr/NQ.16.71 [title-page defaced and mounted].

1351 PROCLUS, *Diadochus*. Procli De sphæra liber I. Cleomedis De mundo, sive circularis inspectionis meteororum libri II...Omnia Græcè & Latinè...coniuncta...8°, Basileæ (1585).
Not H; M/F2 – 30; ?

1352 PROCOPIUS, *Caesariensis*. Ἀνέκδοτα. Arcana historia, qui est liber nonus Historiarum. Ex Bibliotheca Vaticana N. Alemannus protulit, Latinè reddidit. Notis illustravit. (*Greek & Latin*.) F°, Lugduni, 1623.
H; M/C2 – 1; Tr/NQ.11.37.

1353 PROPOSALS to supply His Majesty with twelve or fourteen millions of money, or more if requir'd, for the year 1697 without subscriptions, or advancing the present taxes. By A. D....and some

others his friends. [*With* A supplement to the Proposals...]
(2 pts.) 4°, London, 1697.
Not H; M (Tracts) J3 – 41; Tr/NQ.8.111[9].

1354 PRUDENTIUS, Aurelius Clemens. Opera. Interpretatione et notis
illustravit S. Chamillard...4°, Parisiis, 1687.
H; M/F6 – 19; ?

PRZEBENDOWSKI, Jan Jerzy. Mémoires sur les dernières révolutions de la Pologne...par un gentilhomme polonnois [J. J.
Przebendowski]. 1710. *See* 1065.

1355 PSALMS. The whole Booke of Psalmes: with their wonted tunes, as they
are sung in Churches, composed into 4 parts...8°, London, 1594.
Not H; M/A1 – 11; Tr/NQ.16.171[2] [bound with 198].

— The whole Book of Psalms collected into English metre by T.
Sternhold, J. Hopkins, and others...1661. *See* 1560.

1356 PSEAUMES de David, mise en rime françoise par C. Marot et T. de
Bèze. 8°, [Paris?], 1613.
Not H; M/G1 – 48; ?

1357 PTOLEMAEUS, Claudius. Harmonicorum libri III. Ex codd. MSS.
undecim, nunc primum Græce editus. J. Wallis recensuit, edidit,
versione & notis illustravit, & auctarium adjecit. (*Greek & Latin.*)
4°, Oxonii, 1682.
H; M/A7 – 4; Tr/NQ.10.36.

1358 — Theatri geographiæ veteris tomus prior in quo Cl. Ptol. Alexandrinii
geographiæ libri VIII Græcé et Latiné...Opera P. Bertio.
(*With* Tomus posterior in quo Itinerarium Antonini Imperatoris
...Edente P. Bertio.) F°, Amstelodami, 1619.
H; M/F8 – 12; ?

1359 PUFENDORF, Samuel von. De officio hominis et civis...Ed. 6[a]...
curante I. Webero...12°, Francofurti, 1700.
Not H; M/G2 – 19; ?

1360 — Introductio ad historiam Europaeam, Latine reddita a J. F.
Cramero. Ed. 3[a]...8°, Ultrajecti, 1702.
H; M/F1 – 1; ?

1361 — An introduction to the history of the principal kingdoms and
states of Europe. Made English from the original, the High-
Dutch: the 3rd ed., with additions. 8°, London, 1699.
Not H; M/J4 – 8; Tr/NQ.16.49.

1362 PURBACH, Georg. Theoricæ novæ planetarum. Quibus accesserunt:
I. de Monteregio Disputationes...Item, I. Essler Maguntini
Tractatus utilis ante LX annos conscriptus, cui titulum fecit,
Speculum astrologorum...8°, Basileæ (1596).
Not H; M/E3 – 38; Tr/NQ.16.195 ['Isaac Newton' on fly-leaf
under the signature 'Sir Henry Somers'].

1363 PURCHAS, Samuel. Purchas his pilgrimage. Or Relations of the world and the religions observed in all ages. . . In 4 partes. . . F°, London, 1613.
H; M/J9 – 13; ?

QUERCETANUS, Josephus. *See* DU CHESNE, Joseph.

1364 QUINTILIANUS, Marcus Fabius. M. Fab. Quintiliani Declamationum liber. Cum ejusdem, ut nonnullis visum, Dialogo de causis corruptæ eloquentiæ. Quæ omnia notis illustrantur. 8°, Oxonii, 1675.
Not H; M/F2 – 6; Tr/NQ.7.66.

1365 RAMUS, Jonas. Ulysses et Outinus unus et idem. . . Ed. nova. . . aucta. 8°, Hafniæ, 1713.
H; M/E3 – 29; ?

1366 RAMUS, Petrus. Arithmeticæ libri II et geometriæ XXVII. . . dudum quidem a L. Schonero recogniti et aucti. . . (3 pts.) 4°, Francofurti, 1627.
Not H; M/E6 – 24; ?

1367 RANDOLPH, Thomas. Poems, with the Muses looking-glasse, Amyntas, Jealous lovers, Arystippus. . . 4th ed. inlarged. 5 pts. 8°, London, 1652.
Not H; M/G9 – 39; ?

1368 RAPHSON, Joseph. Analysis æquationum universalis seu Ad æquationes algebraicas resolvendas methodus generalis, et expedita. . . 4°, Londini, 1690.
H; M/E6 – 34; Butler Library, Columbia University, New York [inscribed 'To M^r. Isaack Newton wth. my most humble service. J. R[aphson].' on fly-leaf; bound with 299, 350, 350a, 818, 902, 1254].

1369 — Analysis æquationum universalis. . . Ed. 2^a, cui accessit appendix de infinito infinitarum serierum progressu. . . Cui etiam annexum est, De spatio reali seu ente infinito. . . (1 pts.) 4°, Londini, 1702.
H; not M; ?

1370 — Historia fluxionum, sive Tractatus originem & progressum peregregiæ istius methodi brevissimo compendio (et quasi synopticè) exhibens. 4°, Londini, 1715 [1714].
H; M/E6 – 16; Tr/NQ.16.173 [a first issue, without Newton's appendix].
Latin notes and corrections by Newton on pp. 1, 93–5, some of which were introduced into the English version; Sotheran **828** (1931) 3602; see pp. 23–4 above.

1371 — The history of fluxions, shewing in a compendious manner the first rise of, and various improvements made in that incomparable method. [Pp. 1–96.] [Published by Newton with an anonymous appendix by him: Epistolæ sequentes à D. Leibnitio cum amicis suis in Gallia & alibi communicatæ, ad controversiam præcedentam spectant...Pp. 97–123.] 4°, London, 1715[–1717].
H; M (Tracts) D6 – 4; Tr/NQ.16.197⁶.
Corrections and alterations by Newton (still unpublished though presumably made for 1718 reissue) in Appendix pp. 111–12, 114–15, 117, 123; Math. papers, III, 11 n. 29; see pp. 23–4 above.

1372 RARES expériences sur l'esprit minéral, pour la préparation et transmutation des corps metaliques...Par Monsieur D*** [de Respour]. Vol. 1. 8°, Paris, 1668.
H; M/G8 – 9; Tr/NQ.16.57; Ekins 5 – 0.

1373 RATRAMNUS, *Corbiensis*. Bertram or Ratram concerning the Body and Blood of the Lord, in Latin, with a new English translation ...2nd ed. corrected, and enlarged. 8°, London, 1688.
Not H; M/H1 – 26; ?

RAWLET, John. A dialogue betwixt two Protestants, in answer to a Popish Catechism...[By J. Rawlet.] 2 pts. 1685. *See* 512.

1374 RAY, John. Synopsis methodica avium & piscium; opus posthumum ...(Ed. by W. Derham.) (2 pts.) 8°, Londini, 1713.
H; M/E5 – 18; Tr/NQ.16.187.

1375 — Three physico-theological discourses, concerning I. The primitive chaos, and creation of the world. II. The general deluge, its causes and effects. III. The dissolution of the world, and future conflagration...3rd ed., illustrated...8°, London, 1713.
H; M/H4 – 19; ?

1376 — The wisdom of God manifested in the works of the Creation, in 2 parts...2nd ed., very much enlarged. 8°, London, 1692.
H; M/H4 – 18; Tr/NQ.7.42.

1377 REASONS offer'd against the continuance of the Bank. In a letter to a Member of Parliament. 8°, London, 1707.
Not H; M (Tracts) J2 – 38; Tr/NQ.16.194⁴.

1378 RECONDITORIUM ac reclusorium opulentiæ sapientiæque numinis mundi magni, cui deditur in titulum Chymica vannus...(*With* Commentatio de pharmaco catholico...[By J. de Monte-Snyder.]) (2 pts.) 4°, Amstelodami, 1666.
Not H; M/B1 – 25; Tr/NQ.16.80.
Many page and line references by Newton in margin of pt 2, pp. 12, 13, 15, 19, 20, 29, 48, 66, 68, 72, 75; several signs of dog-earing in pt 2.

1379 RECUEIL de diverses pièces, sur la philosophie, la religion naturelle, l'histoire, les mathématiques, &c. Par Mrs. Leibniz, Clarke, Newton, & autres autheurs célèbres. [Ed. by P. Des Maizeaux.] 2 vols. 12°, Amsterdam, 1720.
H; M/G7 – 18 & 19; Tr/NQ.9.84 & 85 [vol. 1 has p. 182 turned down, vol. 2 has several signs of dog-earing including marking of Newton's name in text of p. 32; see pp. 24, 27 above].

1380 — — Vol. 2, pp. 1–88 only. 12°, [Amsterdam, 1720].
Not H; not M; Portsmouth (1888); ULC/Adv.d.39.2.
Preliminary proof-sheets, with signatures A–D12, E1–4, differing in pagination from the published ed.; two small inserts by Newton to text on pp. 16 & 17 (not included in the published version) and many corrections by Des Maizeaux.

— Recueil d'observations faites en plusieurs voyages par ordre de sa Majesté pour perfectionner l'astronomie et la géographie... Par Messieurs de l'Académie Royale des Sciences. 1693. *See* 1253.

1381 REEVE, Thomas. A cedars sad and solemn fall. Delivered in a sermon ...At the funeral of James late Earl of Carlisle. 4°, London, 1661.
Not H; M/F5 – 1; Tr/NQ.8.74³ [bound with 1383].

1382 — England's backwardnesse, or A lingring party in bringing back a lawful King. Delivered in a sermon...4°, London, 1661.
Not H; M/F5 – 1; Tr/NQ.8.74² [bound with 1383].

1383 — England's restitution, or The Man, the Man of Men, the Statesman. Delivered in several sermons...4°, London, 1661.
Not H; M/F5 – 1; Tr/NQ.8.74¹ [bound with 1381 & 1382].

1384 REFLECTIONS upon learning, wherein is shewn the insufficiency thereof, in its several particulars...By a gentleman [T. Baker]. 4th ed. 8°, London, 1708.
H; M/E4 – 24; Tr/NQ.9.154.

1385 REFUTATIO libelli, quem Iac. Wiekus Iesuita anno 1590 Polonicè edidit, De Divinitate Filii Dei, & Spiritus Sancti...8°, 1594.
H; M/A3 – 12; Tr/NQ.9.54¹ [bound with 496].

REGLEMENT ordonné par le Roy pour l'Académie Royale des Sciences. Du 26. de Janvier 1699. 1699. *See* 1254 & 1255.

1386 REIGN (The) of King Charles: an history faithfully and impartially delivered and disposed into annals...[By H. L'Estrange.] F°, London, 1655.
Not H; M/J9 – 28; Tr/NQ.10.29 [p. 87 turned down].

1387 REINECCIUS, Reinerus. Chronicon Hierosolymitanum...Pars 2 continens Duorum priorum familiæ Luceburg, imperatorum historiam...4°, Helmæstadii, 1585.
H; not M; ?

1388 RELAND, Adrian. Antiquitates sacrae veterum Hebraeorum breviter delineatae. 8°, Trajecti Batavorum, 1708.
H; M/A3 – 17; ?

1389 — De nummis veterum Hebraeorum, qui ab inscriptarum literarum forma Samaritani appellantur, dissertationes quinque...(3 pts.) 8°, Trajecti ad Rhenum, 1709.
H; M/A3 – 16; Tr/NQ.10.89.

1390 RELIGIOUS courtship: being historical discourses, on the necessity of marrying religious husbands and wives only...[By D. Defoe]. 8°, London, 1722.
Not H; M/J6 – 32; ?

1391 REMARKS on Dr. Henry More's Expositions of the Apocalypse and Daniel, and upon his Apology: defended against his answer to them. 4°, London, 1690.
Not H; M/F5 – 9; Tr/NQ.7.47.

RÉMOND DE MONMORT, Pierre. See MONMORT, Pierre Rémond de.

1392 REPLY (A) to Dr. Waterland's Defense of his queries...By a clergyman in the country [J. Jackson]. 8°, London, 1722.
H; M/H3 – 24; ?

1393 REPORT (A) containing an essay for the amendment of the silver coins. [By W. Lowndes.] 8°, London, 1695.
H; M/J3 – 19; Tr/NQ.9.59 [p. 87 turned up].

1394 RERUM Sicularum scriptores ex recentioribus præcipui, in unum corpus nunc primum congesti, diligentiq; recognitione plurimis in locis emendati [by T. Fazello]...F°, Francofurti ad Moenum, 1579.
H; M/D8 – 9; Tr/NQ.11.14.

RESPOUR, P. M. de. Rares expériences sur l'esprit minéral, pour la préparation et transmutation des corps metaliques...Par Monsieur D*** [de Respour]. Vol. 1. 1668. See 1372.

1395 RETZ, Jean François Paul de Gondi, *Cardinal de*. Memoirs, containing all the great events during the minority of Lewis XIV, and administration of Cardinal Mazarin. Done out of French. 8°, London, 1723.
H; M/J4 – 34; Tr/NQ.9.148.

1396 REYHER, Samuel. Dissertatio de nummis quibusdam ex chymico metallo factis. 4°, Kiliæ Holsatorum, 1692.
H; M/A6 – 6; ?

RHENANUS, Johannes. Harmoniæ inperscrutabilis chymico-philosophicæ...decas i(–ii). [By J. Grasshoff and J. Rhenanus.] (2 pts.) 1625. See 740.

1397 — Opera chymiatrica, quæ hactenus in lucem prodierunt omnia à plurimis...8°, Francofurti, 1668.
Not H; M/D1 - 25; Tr/NQ.8.90; Ekins 4 - 0.

1398 RHODOMANUS, Laurentius. Poesis Christiana. Palæstinæ, seu Historiæ sacræ, libri IX...(*Greek.*) 4°, Francofurdi, 1589.
H; M/F6 - 12; Tr/NQ.10.68² [bound with 64].

1399 RICARD, Samuel. Traité général du commerce, plus ample et plus exact que ceux qui ont paru jusques à présent...4°, Amsterdam, 1700.
H; M/F6 - 32; Tr/NQ.10.67 [pp. 187, 204, 300, 343 turned down and several other signs of dog-earing; see p. 25 above and plate 4].

RICCI, Michele Angelo. Exercitatio geometrica de maximis & minimis...1668. *See* 1073.

1400 RICCIOLI, Giovanni Battista. Almagestum novum, astronomiam veterem novamque complectens...2 vols. in 1. F°, Bononiæ, 1651.
H; M/A8 - 8; ?

1401 RICHER, Edmond. Historia Conciliorum generalium, in quatuor libros distributa. 4°, Coloniæ, 1680.
Not H; M/A6 - 10; Tr/NQ.16.106 [on fly-leaf '1682. pret. 6ˢ 6ᵈ, valet 10ˢ' in Newton's hand].

1402 RIDER, Cardanus. Rider's British Merlin. [An almanac.] 12°, London, [1690?].
H (described as bound in 'Red Turkey, [with] Silver Clasps'); not M; ?

1403 RIDLEY, Mark. A short treatise of magneticall bodies and motions. 4°, London, 1613.
Not H; M/J3 - 36; Tr/NQ.9.67.
Manuscript sketches at beginning and end of book, possibly by Newton; a few signs of dog-earing.

1404 RIDLEY, *Sir* Thomas. A view of the civile and ecclesiasticall law... 3rd ed., by J. G[regory]. 8°, Oxford, 1662.
Not H; M/J8 - 13; ?

1405 RIPLEY, George. Opera omnia chemica, quotquot hactenus visa sunt...8°, Cassellis, 1649.
H; M/E3 - 25; Tr/NQ.10.149.
Notes by Newton, mainly references to alchemical books, on pp. 18, 101, 123, 171, 383, 401; several signs of dog-earing.

1406 — — [Another copy.]
H (one of the 6 'Books that has Notes of Sir Is. Newtons'); not M; almost certainly not the copy immediately above; ?

— Opuscula quædam chemica. G. Riplei Medulla philosophiæ chemicæ [etc.]...1614. *See* 1208.

1407 — Ripley reviv'd: or, An exposition upon Sir George Ripley's hermetico-poetical works...Written by Eirenæus Philalethes [i.e. G. Starkey]...8°, London, (1677–)1678.
H; M/E2 – 31; ?

1408 ROBERTI, Gaudenzio. Miscellanea Italica physico-mathematica. 4°, Bononiæ, 1692.
H; M/B1 – 10; Thame/986; ?

1409 ROBERTS, Lewes. The merchants mappe of commerce...4th ed. F°, London, 1700.
H; M/D8 – 8; ?

1410 ROBERTSON, William. Thesaurus Græcæ linguæ, in epitomen, sive compendium...4°, Cantabrigiæ, 1676.
Not H; M/C6 – 23; ?

1411 — Thesaurus linguæ sanctæ...sive, Concordantiale lexicon Hebræo-Latino-Biblicum...4°, Londini, 1680.
H; M/C6 – 22; Tr/NQ.8.7.

ROBERVAL, Gilles Personne de. Aristarchi Samii [or rather, G. P. de Roberval's] De mundi systemate, partibus, & motibus eiusdem, libellus...1644. See 78.

1412 ROE, Nathaniel. Tabulæ logarithmicæ, or Two tables of logarithmes: the first...by N. Roe. The other...by E. Wingate. (2 pts.) 8°, London, 1633.
Not H; M/E5 – 45; Tr/NQ.10.150.
Note by Newton on fly-leaf: 'Arcus radio æqualis, est 57ᵍ29577995. Arcus quadrantalis 1 57079, 63267, 94896, 61923.'

ROGERS, John. Sermons. See 1492.

1413 ROGERS, Thomas. The faith, doctrine, and religion, professed, and protected in the realme of England...Expressed in thirty nine Articles...8°, London, 1633.
Not H; M/H2 – 18; ?

1414 ROGERS, Woodes. A cruising voyage round the world: first to the South-Seas, thence to the East-Indies, and homewards by the Cape of Good Hope. Begun in 1708, and finished in 1711...8°, London, 1712.
Not H; M/J4 – 36; Tr/NQ.9.14 ['Benjamin Tudman his book... 1716' on fly-leaf and below this 'C Huggins', so almost certainly not from Newton's library].

1415 ROHAULT, Jacques. Physica. Latinè vertit, recensuit, & adnotationibus ex Illustrissimi I. Newtoni philosophià maximam partem haustis, amplificavit & ornavit S. Clarke. Ed. 3ᵃ...multùm aucta. 8°, Londini, 1710.
H; M/A4 – 5; Tr/NQ.16.190.

1416 ROLEWINCK, Werner. Fasciculus temporum. F°, [Cologne, *c.* 1485].
H; M/C2 – 5; Tr/NQ.11.32¹ [bound with 296; see p. 73 above].

1417 ROLFINCK, Werner. Chimia in artis formam redacta, VI libris comprehensa. 4°, Genevæ, 1671.
Not H; M/B1 – 21; Tr/NQ.16.82.

1418 ROMAN forgeries, or A true account of false records discovering the impostures and counterfeit antiquities of the Church of Rome. By a Faithful son of the Church of England [T. Traherne]. 8°, London, 1673.
Not H; M/H1 – 15; Tr/NQ.7.64.

1419 ROMISH (The) horseleech: or, An impartial account of the intolerable charge of Popery to this nation...[By T. Staveley.] 8°, London, 1674.
Not H; M/H1 – 5; Tr/NQ.16.156 [a few signs of dog-earing].

1420 RONAYNE, Philip. A treatise of algebra in two books...8°, London, 1717.
H; M/E5 – 8; ?

1421 ROOKE, *Sir* George. The life and glorious actions of the Right Honourable Sir George Rooke. 12°, London, 1707.
Not H; M/J8 – 56; ?

1422 ROSENCREUTZ, Christian, *pseud.* The hermetick romance: or, The chymical wedding. Written in High Dutch by C. Rosencreutz [i.e. J. V. Andreae]. Transl. by E. Foxcroft. 8°, [London], 1690.
Not H; M/E3 – 41; ?

1423 ROUBAIS, Jacques, *de Tourcoin.* A physical dissertation, concerning the cause of the variation of the barometer...wherein some mistakes in Sir Isaack Newton's system are rectify'd...Transl. into English. 8°, London, 1721.
Not H; M (Tracts) J3 – 25; Tr/NQ.16.114³.

1424 ROUS, Francis. Archæologiæ Atticæ libri VII. Seven books of the Attick antiquities...5th ed. corrected and enlarged...4°, Oxford, 1658.
Not H; M/F5 – 26; Babson 405 [bound with 673].

1425 ROWE, Jacob. Navigation improved: in two books. Book I. containing an exact description of the fluid quadrant for the latitude ...Book II. An essay on the discovery of the longitude, by a new invention of an everlasting horometer...8°, London, 1725.
Not H; M (Tracts) J8 – 11; Tr/NQ.10.119³.
— Tracts on the longitude, by Hobbs, Keith, Rowe, &c. 1709. *See* 1636.

ROYAL SOCIETY. Commercium epistolicum...jussu Societatis Regiæ in lucem editum. *See* 422–4.

— Philosophical transactions. *See* 1304–8.

RUDD, Thomas. Euclides, Elements of geometry: the first vi. books ...By T. Rudd. 1651. *See* 584.

RUGGLE, George. Ignoramus. Comœdia coram Rege Jacobo... [By G. Ruggle.] Ed. 4ª...1668. *See* 830.

1426 RULAND, Martin. Lexicon alchemiæ sive Dictionarium alchemisticum...4°, [Frankfurt], 1612.
 Not H; M/C1 – 14; Tr/NQ.16.102 [a few signs of dog-earing].

1427 — Progymnasmata alchemiæ, sive Problemata chymica, nonaginta & una quæstionibus dilucidata: cum Lapidis philosophici vera conficiendi ratione. 3 pts. 8°, Francofurti, 1607.
 H; M/D1 – 27; Stanford University Library, Stanford, Calif.; Thame/987.

1428 RULE (The) for finding Easter in the Book of Common-Prayer, explain'd and vindicated...[By R. Watts.] 8°, London, 1712.
 Not H; M/H1 – 25; ?

1429 RUSHWORTH, John. Historical collections of private passages of state, weighty matters in law, remarkable proceedings in five Parliaments....8 vols. F°, London, 1721.
 Not H; M/B5 – 12 to 19; ?

1430 RUSSIAN (The) Catechism, composed and published by order of the Czar...[Transl. from the Russian by J. T. Philipps.] 8°, London, 1723.
 Not H; M (Tracts) H1 – 17; ?

1431 RUTHERFURD, Samuel. The tryal & triumph of faith: or, An exposition of the history of Christs dispossessing of the daughter of the woman of Canaan. Delivered in sermons...4°, London, 1645.
 Not H; M/H2 – 32; Tr/NQ.8.67.

1432 RUTILIUS NAMATIANUS, Claudius. Itinerarium, integris Simleri, Castalionis, Pithoei [etc.]...animadversionibus illustratum. 12°, Amstelædami, 1687.
 Not H; M/G1 – 6; ?

1433 RUYSCH, Fredrik. Adversariorum anatomico-medico-chirurgicorum decas secunda. 4°, Amstelodami, 1720.
 H; M/E6 – 39; ?

1434 RYMER, Thomas. Fœdera, conventiones, literæ, et cujuscunque generis acta publica, inter Reges Angliæ...Vol. 1. F°, Londini, 1704.
 H; not M; ?

SABINUS, Georgius. P. Ovidii Metamorphosis, seu Fabulæ poeticæ: earumque interpretatio ethica, physica et historica Georgii Sabini. . .Ultima ed. 1593. *See* 1224.

SACHEVERELL, Henry. Sermons. *See* 1494.

1435 SALLENGRE, Albert Henri. Novus thesaurus antiquitatum Romanarum. Vols. 1, 2. F°, Hagæ-Comitum, 1716–18.
H; not M; ?

1436 SALLUSTIUS, *philosophus*. De diis et mundo. L. Allatius nunc primus è tenebris eruit, & Latinè vertit. (*Greek & Latin*.) 12°, Lugduni Batavorum, 1639.
Not H; M/E1 – 36; ?

1437 SALLUSTIUS CRISPUS, Gaius. C. Sallustius Crispus, cum veterum historicorum fragmentis. Ed. novissima. 8°, Amstelodami, 1675.
Not H; M/G2 – 11; ?

1438 SALMON, Thomas. A proposal to perform musick, in perfect and mathematical proportions. . .4°, London, 1688.
Not H; M (Tracts) J6 – 2; Tr/NQ.16.77[7].

SALMON, William. Bibliothèque des philosophes (chymiques). . . Par Le Sieur S. D. E. M. [i.e. W. Salmon.] 2 vols. 1672–8. *See* 221.

1439 — Medicina practica: or The practical physician. . .8°, London, 1707.
H; M/E5 – 40; ?

1440 SAMMES, Aylett. Britannia antiqua illustrata: or, The antiquities of Ancient Britain. . .Vol. 1. F°, London, 1676.
H; M/H9 – 4; ?

1441 SANCHONIATHON. Sanchoniatho's Phœnician history, transl. from the first book of Eusebius De præparatione Evangelica. With a continuation of Sanchoniatho's History by Eratosthenes Cyrenæus's Canon. . .By R. Cumberland. 8°, London, 1720.
H (2 copies); M (1 only) J4 – 7; Tr/NQ.9.16 [p. 483 turned down and a few other signs of dog-earing].

1442 SANDERSON, Robert, *Bp of Lincoln*. Logicæ artis compendium. 3ª hac ed. recognitum. . .(2 pts.) 8°, Oxoniæ, 1631.
Not H; not M; provenance not known; Tr/Adv.e.1.15 [on recto of title-page 'Isaac Newton', on verso 'Isaac Newton Trin Coll Cant 1661'; a few signs of dog-earing].

1443 — xxxvi. sermons. . .8th ed.; corrected and amended. Whereunto is now added the life of the. . .author, written by I. Walton. F°, London, 1689.
Not H; M/J9 – 4; ?

1444 SANDIUS, Christophorus. Nucleus historiæ ecclesiasticæ: cui præfixus est Tractatus de veteribus scriptoribus ecclesiasticis. (3 pts.) 8°, Cosmopoli [Amsterdam], 1669.
H; M/A1 – 20; Tr/NQ.9.17 [a few signs of dog-earing].

SANDIVOGIUS, Michael. *See* SENDIVOGIUS, Michael.

1445 SANGUIS naturæ, or, A manifest declaration of the sanguine and solar congealed liquor of nature. By Anonimus [i.e. C. Grummet]. 8°, London, 1696.
Not H; M/E2 – 38; Tr/NQ.16.172 [many signs of dog-earing].

1446 — — [Another copy.]
Not H; not M; provenance not known; Memorial Library, University of Wisconsin – Madison.
*Note in Newton's hand inserted: 'Sanguis Naturæ, at Sowles [the bookseller] a Quaker Widdow in White Hart Court at y*e* upper end of Lombard Street*'; many signs of dog-earing.

SAPPHO. Anacreontis et Sapphonis carmina. Notas & animadversiones addidit T. Faber...(*Greek & Latin.*) 1680. *See* 40.

— Les poësies d'Anacreon et de Sapho, traduites de Grec en François, avec des remarques par Mademoiselle Le Fevre. (*Greek & French.*) 1681. *See* 41.

1447 SARBIEWSKI, Maciej Kazimierz. Lyricorum libri IV...8°, Antverpiæ, 1624.
Not H; M/G1 – 42; ?

1448 SAUNDERS, Richard. Palmistry, the secrets thereof disclosed... As also that most useful piece of astrology...concerning elections ...2 pts. 12°, London, 1663.
Not H; M/E1 – 10; ?

1449 SAURIN, Jacques. Dissertations...on the most memorable events of the Old and New Testaments...Made English by J. Chamberlayne. 8°, London, 1720.
H; M/H2 – 10; ?

1450 — — [Another ed.] Vol. 1. [*No more published.*] F°, London, 1723.
H; M/F7 – 14; ?

1451 SAVARON, Jean. Chronologie des Estats Généraux, où Le Tiers Estat est compris, depuis l'an MDCXV. iusques à CCCCXXII. 8°, Paris, 1615.
H; M/G10 – 37; Tr/NQ.9.35.

1452 SAVERY, Thomas. The miners friend; or, An engine to raise water by fire, described...8°, London, 1702.
Not H; M (Tracts) J3 – 33; Tr/NQ.16.162¹.

1453 SAXO, *Grammaticus*. Danorum historiæ libri XVI. [Ed. by J. Oporinus.] F°, Basileæ, 1534.
H; M/B2 – 25; Tr/NQ.11.28.

1454 SCALIGER, Joseph Justus. Opus novum de emendatione temporum in octo libros tributum. F°, Lutetiæ, 1583.
H; M/C3 - 13; Tr/NQ.11.22 [a few signs of dog-earing].
Alternative versions of the Jewish (and other) names of months in the text written by Newton in the margin of p. 378: against the printed ' Tisri', ' Marcheschwan', ' Nisan', ' Ijar', he supplied ' Elhanim', ' Bul', ' Abib', ' Zif' respectively.

1455 SCAPULA, Johannes. Lexicon Græcolatinum novum...Ed. 2ª. F°, Basileæ, 1589.
H; M/A8 - 7; ?

1456 —— Ed. nova accurata. F°, Lugduni Batavorum, 1652.
H; M/B3 - 1; ?

1457 SCHAAF, Karl. Lexicon Syriacum concordantiale, omnes Novi Testamenti Syriaci voces...4°, Lugduni Batavorum, 1709.
H; M/B6 - 9; Tr/NQ.16.174.

1458 SCHEDIUS, Elias. De diis Germanis...syngrammata IV. 8°, Amsterodami, 1648.
H; M/A2 - 7; ?

1459 SCHEINER, Christoph. Oculus; hoc est, Fundamentum opticum: in quo ex accurata oculi anatome...radius visualis eruitur... 4°, Londini, 1652.
Not H; M/C1 - 6; Tr/NQ.8.49 [p. 225 turned down and a few other signs of dog-earing].

1460 SCHELHAMMER, Gunther Christoph. De nitro, cum veterum, tum nostro commentatio...8°, Amstelodami, 1709.
H; M/E3 - 27; ?

1461 SCHELSTRATE, Emmanuel. Antiquitas illustrata circa Concilia generalia et provincialia, decreta et gesta Pontificum, et præcipua totius historiæ ecclesiasticæ capita. 4°, Antverpiæ, 1678.
H; M/A6 - 12; Tr/NQ.8.55 [pp. 78, 79, 90, 474 turned up, p. 98 down, and many other signs of dog-earing].

1462 — Ecclesia Africana sub primate Carthaginiensi. 4°, Parisiis, 1679.
H; M/A6 - 11; Tr/NQ.8.129 [a few signs of dog-earing].

1463 — Sacrum Antiochenum Concilium pro Arianorum conciliabulo... 4°, Antverpiæ, 1681.
H; M/C1 - 24; Tr/NQ.8.41.

1464 SCHEUCHZER, Johann Jacob. Herbarium diluvianum. F°, Tiguri, 1709.
H; M/B4 - 12; ?

1465 — Ούρεσιφοίτης Helveticus, sive Itinera Alpina tria...4°, Londini, 1708.
H; not M; ?

1466 — Οὐρεσιφοίτης Helveticus, sive Itinera per Helvetiæ Alpinas regiones facta annis MDCCII–MDCCXI. 4 vols. in 2. 4°, Lugduni Batavorum, 1723.

H; M/F6 – 6 & 7; Tr/NQ.10.4 & 5 [a few signs of dog-earing; presentation inscription on fly-leaf 'Perillustri Viro Isaaco Newtono Equiti aurato & Societatis Regiæ Præsidi Itinera hæcce Alpina nomine Authoris Ea quâ par est, observantiâ humillimè offert Joh. Casparus Scheuchzerus Fil. Londini Calendis Julij MDCCXXIII'; plates opposite pp. 1, 65, 147 of vol. 1 'Sumptibus D. Isaaci Newton'].

1467 — Piscium querelae et vindiciae. 4°, Tiguri, 1708.

Not H; M (Tracts) J6 – 15; Tr/NQ.8.52¹ [presentation inscription 'Illustr. D. Newtono Societ. Regiæ Præsid.' at foot of title-page].

1468 — Specimen lithographiæ Helveticæ curiosæ, quo lapides ex figuratis Helveticis selectissimi æri incisi sistuntur & describuntur. 8°, Tiguri, 1702.

Not H; M (Tracts) H8 – 13; Tr/NQ.16.99³.

1469 SCHICKARD, Wilhelm. Horologium Hebræum, sive consilium, quomodo sancta lingua...apprehendi queat...(With Rota Hebræa...recusa denuò.) (2 pts.) 8°, Londini, 1639.

Not H; M/E2 – 35; Tr/NQ.7.6.

1470 SCHINDLER, Valentin. Lexicon pentaglotton, Hebraicum, Chaldaicum, Syriacum, Talmudico-Rabbinicum, & Arabicum... F°, Hanoviæ, 1612.

H; M/H9 – 7; ?

SCHLICHTING, Jonas. Commentarius in Epistolam ad Hebræos ...[By J. Schlichting.] 1634. See 421.

1471 SCHOOTEN, Frans van. Exercitationum mathematicarum libri v...Quibus accedit C. Hugenii Tractatus, de ratiociniis in aleæ ludo. 4°, Lugd. Batav., 1657.

H; M/C1 – 16; Tr/NQ.16.184.

Early Newton mathematical notes (late 1664?) on slips of paper now attached between pp. 178–9 and pp. 360–61, see Math. papers, I, 46, where the first note is reproduced.

1472 SCHREVELIUS, Cornelius. Lexicon manuale Græco-Latinum et Latino-Græcum. Ed. 2ª auctior...8°, Lugduni Batavorum, 1657.

H; M/A4 – 14; ?; bought by Newton *c.* 1665 for 5*s.* 4*d.*, see Trinity Notebook (MS.R.4.48ᶜ).

1473 SCHROEDER, Johann. Quercetanus redivivus, hoc est, Ars medica dogmatico-hermetica, ex scriptis J. Quercetani...Ed. 2ª. 3 vols. in 1. 4°, Francofurti, 1679.

Not H; M/B1 – 8; Thame/978; Sotheran **800** (1926) 11953, **804** (1927) 3811; ?

1474 SCIENCE (La) des medailles antiques et modernes...Nouv. éd., revuë, corrigée & augmentée considérablement par l'auteur [L. Jobert]. 8°, Amsterdam, 1717.

H; M/G8 – 28; Tr/NQ.10.118.

1475 SCRIPTORES rerum Brunsvicensium illustrationi inservientes, antiqui omnes et religionis reformatione priores...cura G. G. Leibnitii. 3 vols. F°, Hanoveræ, 1707–11.

H; M/H9 – 8 to 10; Tr/NQ.17.13 to 15 [a few signs of dog-earing in vol. 1].

1476 SCRIPTURE (The) doctrine of the most holy and undivided Trinity, vindicated from the misinterpretations of Dr. Clarke. [By J. Knight.] To which is prefixed a letter...by R. Nelson. 8°, London, 1714.

Not H; M/H3 – 19; ?

1477 SECRET (The) history of Europe. [By J. Oldmixon.] 2 vols. 8°, London, 1712.

H; M/J4 – 25 & 26; Tr/NQ.8.57 & NQ.16.182.

1478 SECRETS reveal'd: or, An open entrance to the shut-palace of the King...by...Anonymous, or Eyræneus Philaletha Cosmopolita [i.e. G. Starkey]...8°, London, 1669.

H (one of the 6 'Books that has Notes of Sir Is. Newtons'); not M; Portsmouth (1888); Sotheby July 1936/121; Memorial Library, University of Wisconsin – Madison.

Heavily annotated and corrected throughout by Newton with some margins completely filled with notes; many signs of dog-earing; Duveen, p. 470 & pl. XII; *see also* 838.

1479 SEDER OLAM: or, The order of ages. Wherein the doctrin is historically handled. Transl. out of Latin, by J. Clark, upon the leave and recommendation of F. M. Baron of Helmont. 8°, London, 1694.

Not H; M/E2 – 47; ?

1480 SEIDEL, Kaspar. Ἐγχειρίδιον τῆς Ἑλλάδος φωνῆς. Sive, Manuale Græcæ linguæ gnomologicum novum...Ed. 2ª, priori emendatior. 8°, Londini, 1653.

Not H; M/E3 – 47; Tr/NQ.10.141.

1481 SELDEN, John. De Diis Syris syntagmata II...Ed. juxta alteram ipsius autoris operâ emendatiorem auctioremque omnium novissima...operâ A. Beyeri. 8°, Amstelodami, 1680.

H; M/A3 – 15; Tr/NQ.9.77 [pt 1, p. 193 turned up, and several other signs of dog-earing].

1482 — De jure naturali et gentium, juxta disciplinam Ebræorum, libri VII...4°, Argentorati, 1665.

H; M/D5 – 14; ?

1483 — De synedriis & præfecturis juridicis veterum Ebræorum libri III. Ed. ultima priori correctior. 4°, Amstelædami, 1679.

H; M/D5 – 12; Tr/NQ.8.35 [pt 1, p. 148 and pt 2, p. 85 each with both corners turned, and a few other signs of dog-earing].

1484 — Uxor Ebraica, seu De nuptiis et divortiis ex jure civili, id est, divino & Talmudico, veterum Ebræorum, libri III...Ed. nova. 4°, Francofurti ad Oderam, 1673.

H; M/D5 – 13; Tr/NQ.8.20.

SENDIVOGIUS, Michael. Cosmopolite, ou Nouvelle lumière chimique...[By M. Sendivogius.] Dernière éd., revue et augmentée...2 vols. 1691. See 445.

1485 — A new light of alchymie...Also nine books of the nature of things, written by Paracelsus...Also a chymicall dictionary...Transl. out of the Latin...by J. F[rench]. 3 pts. 4°, London, 1650.

H; M/F5 – 36; Tr/NQ.16.130 [several signs of dog-earing].

— Novum lumen chymicum...[By M. Sendivogius.] 1639. See 1192.

1486 SENECA, Lucius Annaeus. Opera omnia, ab A. Schotto ad veterum exemplarium fidem castigata, Græcis etiam hiatibus expletis. Vol. 1. 8°, Genevæ, 1626.

Not H; M/A3 – 8; Tr/NQ.9.128 [a few signs of dog-earing].

1487 — Operum tomus 2, cum A. Schotti ad veterum exemplarium fidem castigatione...8°, [1626?].

Not H; M/A3 – 9; Tr/NQ.16.141 [a different issue from 1486].

1488 — L. Annæi Senecæ & P. Syri Mimi, forsan etiam aliorum, singulares sententiæ...studio & opera J. Gruteri...8°, Lugduni Batavorum, 1708.

H; M/F2 – 3; ?

1489 — L. & M. Senecæ Tragœdiæ, cum notis T. Farnabii. 12°, Amsterdami, 1645.

Not H; M/A1 – 25; Tr/Adv.e.1.12 ['Isaac Newton hunc librum
s d
tenet pret: 2. 6.' on fly-leaf]; Thame/972; not part of the Pilgrim Trust gift.

1490 — L. & M. Senecæ Tragœdiæ. Cum notis T. Farnabii. 12°, Amsterodami, 1656.

Not H; M/G2 – 18; ?

1491 SENEX, John. The English atlas, by J. Senex and J. Maxwell. F°, London, 1714.
> H; not M; ?

SERENUS, *Antinsensis*. De sectione cylindri & coni libri ii. [Ed. by E. Halley.] 1710. *See* 62.

1492 SERMONS by Berriman, Rogers, Hind, &c. 8°, 1722. *Not identified individually.*
> Not H; M/H4 – 24; ?

1493 — Sermons by Butler, Bird, Stillingfleet, &c. 4°, 1688. *Not identified individually.*
> Not H; M/J6 – 3; ?

1494 — Sermons by Dr Lupton, Sacheverell, &c. 8°, 1709. *Not identified individually.*
> Not H; M/H3 – 32; ?

1495 — Sermons by several hands. 3 vols. 4°, 1720. *Not identified individually.*
> Not H; M/B6 – 13 to 15; ?

1496 — Sermons by Sprat, Fothergill, &c. 8°, 1710. *Not identified individually.*
> Not H; M/H7 – 16; ?

1497 — Sermons in the time of Oliver Cromwell. 4°, 1644. *Not identified individually.*
> Not H; M/J6 – 30; ?

1498 — Sermons on charity. 2 vols. 4°, 1720. *Not identified individually.*
> Not H; M/H6 – 1 & 2; ?

1499 — Sermons printed for the benefit of the poor. 2 vols. 8°, 1707. *Not identified individually.*
> Not H; M/H7 – 26 & 27; ?

1500 SEVERAL proposals conducing to a farther Union of Britain: and pointing at some advantages arising from it. [By George MacKenzie, 1st Earl of Cromarty.] 4°, London, 1711.
> Not H; M (Tracts) J3 – 41; Tr/NQ.8.111⁸.

1501 SEWELL, George. The life and character of Mr. John Philips. 3rd ed. 12°, London, 1720.
> Not H; M/J8 – 25; Tr/NQ.8.109¹ [bound with 473, 1299 & 1527].

1502 — The tragedy of Sir Walter Raleigh. As it is acted at the Theatre-Royal in Lincoln's-Inn-Fields. 5th ed....12°, London, 1722.
> Not H; M (Plays) H8 – 9; Tr/NQ.10.83⁴.

1503 SEXTUS, *Empiricus*. Opera quæ extant...H. Stephano interprete ...Græcè nunc primùm editi...F°, Parisiis, 1621.
> H; M/C3 – 5; ?

1504 SHAKESPEARE, William. Hamlet, Prince of Denmark. A tragedy. As it is now acted by His Majesty's servants. 12°, London, 1718.
> Not H; M (Plays) H8 – 9; Tr/NQ.10.83¹.

1505 — The Tempest: or, The enchanted island. A comedy. First written by Mr. William Shakespear, & since altered by Sr. William Davenant, and Mr. John Dryden. 8°, London, 1710.

Not H; M (Plays) H8 – 19; Tr/NQ.10.92⁴.

SHARROCK, Robert. De finibus virtutis Christianæ. The ends of Christian religion. . .By R. S[harrock]. LL.D. 1673. *See* 494.

1506 SHELTON, Thomas. Tachy-graphy. The most exact and compendious method of short and swift writing. . .8°, London, 1660.

Not H; M/E2 – 48; ?

1507 SHERLOCK, Thomas, *Bp of London*. The use and intent of prophecy in the several ages of the world: in six discourses. . .8°, London, 1725.

Not H; M/H6 – 24; ?

SHERLOCK, William. The notes of the Church, as laid down by Cardinal Bellarmin; examined and confuted [by W. Sherlock, S. Freeman, S. Patrick, etc.]. 1688. *See* 154.

1508 — A practical discourse concerning death. 8°, London, 1710.

Not H; M/H3 – 30; ?

1509 SHORT (A) account of Dʳ Bentley's humanity and justice, to those authors who have written before him: with an honest vindication of T. Stanley. . .In a letter to the Honourable Charles Boyle. . . 8°, London, 1699.

H; M/E4 – 30; Tr/NQ.9.57 [p. 1 has signs of dog-earing].

1510 — A short introduction of grammar, generally to be used. Compiled and set forth for the bringing up of all those that intend to attain to the knowledge of the Latine tongue. . .[By W. Lily and J. Colet.] (2 pts.) 12°, Oxford, 1692.

Not H; M/J7 – 47; Tr/NQ.10.140.

1511 SIDONIUS APOLLINARIS, *St, Bp of Clermont*. Opera. I. Savaro multò quàm antea castigatius recognovit, & librum commentarium adiecit. 4°, Parisiis, 1599.

H; M/A7 – 5; Tr/NQ.8.44 [pt 2, p. 123 has signs of dog-earing].

1512 SIGONIUS, Carolus. Historiarum de Occidentali Imperio libri xx. . .F°, Francofurti, 1593.

H; M/A9 – 4; Tr/NQ.11.13¹ [bound with 1513; a few signs of dog-earing].

1513 — Historiarum de Regno Italiae libri xx. . .F°, Francofurti, 1591.

Not H; M/A9 – 4; Tr/NQ.11.13² [bound with 1512; a few signs of dog-earing].

SIKE, Henricus. Evangelium Infantiæ. Vel Liber apocryphus de infantia Servatoris. Ex manuscripto edidit, ac Latina versione & notis illustravit H. Sike. (2 pts.) (*Arabic & Latin.*) 1697. *See* 59.

SIMON, Richard. Comparaison des cérémonies des Iuifs, et de la discipline de l'Église. Par le Sieur de Simonville [i.e. R. Simon]. [c. 1681.] See 945.

1514 — Critical enquiries into the various editions of the Bible...Translated into English, by N. S. 4°, London, 1684.
H; M/C6 – 26; ?

1515 — A critical history of the Old Testament...Translated into English, by a person of quality. (4 pts.) 4°, London, 1682.
H; M/H2 – 1; Tr/NQ.16.143 ['Is. Newton Oct 3. (80) pret 8ˢ.' on fly-leaf (the meaning of '(80)' is unclear, the date of the book's publication rules out 1680); a few signs of dog-earing].

1516 — A critical history of the text of the New Testament...4°, London, 1689.
H; M/H2 – 2; Tr/NQ.8.32 [pt 2, p. 154 turned down and a few other signs of dog-earing].

1517 — Lettres choisies, où l'on trouve un grand nombre de faits anecdotes de litérature. 12°, Amsterdam, 1700.
H; M/G6 – 28; Tr/NQ.9.1.

1518 SKINNER, Thomas. The life of General Monk: late Duke of Albemarle...To which is added a preface...by W. Webster. 8°, London, 1723.
Not H; M/J4 – 33; Tr/NQ.7.57.

1519 SLARE, Frederick. An account of the nature and excellent properties and vertues of the Pyrmont waters...8°, London, 1717.
H; M/E4 – 38; ?

1520 — Experiments and observations upon Oriental and other Bezoar-stones, which prove them to be of no use in physick...8°, London, 1715.
H; M/J7 – 35; Tr/NQ.9.60 ['Ab Authore' (not in Newton's hand) on fly-leaf].

1521 SLEIDAN, Johann. De quatuor monarchiis libri III. Cum notis H. Meibomi & G. Horni. Ed. prioribus correctior & emendatior. 12°, Cantabrigiæ, 1686.
Not H; M/G2 – 20; Tr/NQ.9.165.

1522 — De statu religionis et reipublicæ, Carolo Quinto, Cæsare commentarij. 8°, 1556.
Not H; M/D1 – 17; ?

1523 SLOANE, Sir Hans. A voyage to the Islands Madera, Barbados, Nieves, S. Christophers and Jamaica, with the natural history... of the last of those Islands...2 vols. F°, London, 1707–25.
H (vol. 1 only); M (2 vols.) C3 – 11 & 12; ?

239

1524 SLUSE, René François de. Mesolabum, seu Duæ mediæ proportionales inter extremas datas per circulum et per infinitas hyperbolas, vel ellipses et per quamlibet exhibitæ...4°, Leodii Eburonum, 1668.
H; M/E5 – 16; Tr/NQ.9.40.

1525 SMALRIDGE, George, *Bp of Bristol.* Twelve sermons preach'd on several occasions. 8°, Oxford, 1717.
Not H; M/H4 – 32; ?

SMEDLEY, Jonathan. An hue and cry after Doctor S—T [Swift]; occasion'd by a true and exact copy of part of his own Diary... [By J. Smedley.] 1714. *See* 815.

1526 SMIGLECIUS, Martinus. Logica, selectis disputationibus & quæstionibus illustrata...4°, Oxonii, 1658.
Not H; M/C1 – 22; ?

1527 SMITH, Edmund. Works...to which is prefix'd, A character of Mr. Smith, by Mr. Oldisworth. 3rd ed., corrected. 12°, London, 1719.
Not H; M/J8 – 25; Tr/NQ.8.109⁴ [bound with 1501].

1528 SMITH, George. The vanity of conquests, and universal monarchy: being a succinct account of all the great conquerors and heroes both of ancient and modern ages...8°, London, 1705.
H; M/J3 – 21; Tr/NQ.9.9.

1529 SMITH, John, †*1631.* The sea-mans grammar and dictionary...In 2 pts. Now much amplified and enlarged. 4°, London, 1692.
H; M/J3 – 39; ?

SMITH, John, *engraver.* Novem Fabulae, Melanograph ex Titiano de Amoribus Deorum. [The Loves of the Gods, plates, after Titian, engraved by J. Smith.] [*c.* 1720.] *See* 1617.

1530 SMITH, John, *fl. 1656.* The mystery of rhetorick unveil'd. Wherein above 130 of the tropes and figures are severally derived from the Greek into English...8°, London, 1683.
Not H; M/J8 – 22; Tr/NQ.9.4 [a few signs of dog-earing].

1531 SMITH, Samuel. Aditus ad logicam, in usum eorum qui primò academiam salutant. Ed. 6ª. 12°, Oxoniæ, 1649.
Not H; M/E1 – 46; Tr/NQ.9.166¹ [wants title-page; bound with 293].

1532 SMITH, Thomas. De Græcæ Ecclesiæ hodierno statu epistola. Ed. nova, auctior & emendatior. 8°, Trajecti ad Rhenum, 1698.
H; M/F4 – 16; Tr/NQ.9.155.

SMYTH, George. *See* SMITH, George.

1533 SOAREZ, Cyprianus. De arte rhetorica libri III, ex Aristotele,

Cicerone, et Quintiliano præcipue deprompti...8°, Coloniæ, 1604.

Not H; M/E1 – 2; ?

SOCINUS, Faustus. De Iesu Christi Filii Dei natura sive essentia ...disputatio, adversus A. Volanum...Secundò edita. [*Preface signed* F.S., i.e. F. Socinus.] 1627. *See* 495.

1534 — Defensio animadversionum F. Socini...adversus G. Eutropium. 8°, Racoviæ, 1618.

H; M/A3 – 19; ?

1535 SOLUTIO problematis Paschalis. [By F. Bianchini.] 4°, (Romæ, 1703).

H; M/E6 – 7; Tr/NQ.10.31.

1536 SOME reasons for an European state, proposed to the powers of Europe, by an universal guarantee, and an annual congress, senate, dyet, or parliament...4°, London, 1710.

Not H; M (Tracts) J3 – 41; Tr/NQ.8.111³.

1537 — Some remarks upon a pamphlet [by A. A. Sykes], entitled, The case of Dr. Bentley farther stated and vindicated, &c....By the author of the Full and impartial account, &c. [i.e. C. Middleton]. 8°, London, 1719.

Not H; M (Tracts) J3 – 7; Tr/NQ.7.1⁵.

1538 SOPHOCLES. Tragœdiæ VII. Unà cum omnibus Græcis scholiis ad calcem adnexis. Ed. postrema. (*Greek & Latin.*) 8°, Cantabrigiæ, 1673.

H; M/F3 – 15; Tr/NQ.9.42 [a few signs of dog-earing].

1539 SOUPIRS (Les) de l'Europe &c. or, The groans of Europe at the prospect of the present posture of affairs. In a letter from a gentleman at the Hague to a Member of Parliament. Made English from the original French [of J. Dumont]. 8°, [London], 1713.

Not H; M (Tracts) J3 – 14; Tr/NQ.9.20⁴.

1540 SPANHEIMIUS, Fridericus. Dubia Evangelica...3 vols. in 2. 4°, Genevæ, 1655–8.

H; M/A7 – 6 & 7; ?

1541 — Historia imaginum restituta, præcipuè adversos Gallos scriptores nuperos L. Maimburg et N. Alexandrum. 8°, Lugduni Batavorum, 1686.

H; M/A3 – 3; Tr/NQ.8.139 [p. 263 turned down and several other signs of dog-earing].

SPARROW, Anthony, *Bp of Norwich*. A collection of articles, injunctions, canons, orders, ordinances...of the Church of England ...[By A. Sparrow.] 4th impr....1684. *See* 411.

SPECIMEN sapientiæ Indorum veterum...Nunc primum Græce ...cum versione nova Latina, opera S.G. Starkii. 1697. *See* 874.

1542 SPECTATOR (The). [By J. Addison, Sir R. Steele and others.] 8 vols. 8°, London, 1712–15.

H (7 vols.); M (8 vols.) H8 – 22 to 29; ?

1543 SPEECHES in Parliament in 1640 & 1641. 4°, 1641. *Not identified.*

Not H; M/F5 – 33; ?

1544 SPEED, John. A prospect of the most famous parts of the world. 8°, London, 1646.

Not H; M/G9 – 58; ?

1545 SPENCER, John. De legibus Hebræorum ritualibus et earum rationibus, libri III. F°, Cantabrigiæ, (1683–)1685.

H; M/D8 – 5; Tr/NQ.17.18 [p. 782 turned down and a few other signs of dog-earing].

1546 SPENER, Jakob Karl. Historia Germaniæ universalis et pragmatica breviter ac perspicue exposita...2 vols. 8°, Lipsiæ, 1716–17.

H; M/F3 – 17 & 18; ?

1547 — Notitia Germaniae antiquae, ab ortu Reipublicae ad regnorum Germanicorum in Romanis provinciis stabilimenta...2 vols. 4°, Halæ Magdeburgicæ, 1716.

H; M/C1 – 25 & 26; ?

1548 SPERLING, Otto. Dissertatio de nummis non cusis tam veterum quam recentiorum. 4°, Amstelædami, 1700.

H; M/A6 – 4; ?

1549 SPRAT, Thomas, *Bp of Rochester*. The history of the Royal-Society of London, for the improving of natural knowledge. 4°, London, 1667.

H; M/F5 – 17; Tr/NQ.9.28 [a few signs of dog-earing]; bought by Newton in 1667 for 7s., see Fitzwilliam Notebook.

— Sermons. *See* 1496.

1550 SQUIRE, John. A plaine exposition upon the first part of the second chapter of Saint Paul his second Epistle to the Thessalonians. Wherein it is plainly proved, that the Pope is the Antichrist... 4°, London, 1630.

Not H; M/H3 – 36; ?

1551 STANLEY, Thomas. Historia philosophiæ Orientalis. Recensuit, ex Anglica lingua in Latinam transtulit...J. Clericus. 8°, Amstelodami, 1690.

H; M/A3 – 13; ?

1552 — The history of philosophy: containing the lives, opinions, actions and discourses of the philosophers of every sect...3rd ed....F°, London, 1701.

H; M/B3 – 6; ?

STANYAN, Temple. An account of Switzerland. Written in the year 1714 [by T. Stanyan]. 1714. *See* 3.

STARK, Sebastian Gottfried. Specimen sapientiæ Indorum veterum. Id est Liber ethico-politicus pervetustus...Nunc primum Græce ...cum versione nova Latina, opera S. G. Starkii. 1697. *See* 874.

STARKEY, George. Enarratio methodica Trium Gebri medicinarum, in quibus continetur Lapidis philosophici vera confectio. Autore Anonymo sub nomine Æyrenæi Philalethes [i.e. G. Starkey?]. 1678. *See* 554.

— Introitus apertus ad occlusum regis palatium: autore Anonymo Philaletha [i.e. G. Starkey]...nunc primum publicatus, curante J. Langio. 1667. *See* 838.

— Liquor Alcahest, or A discourse of that immortal dissolvent of Paracelsus & Helmont...[By G. Starkey.] [2nd ed.?] 1684. *See* 961.

— The marrow of alchemy...By Eirenæus Philoponos Philalethes [i.e. G. Starkey]. 1654. *See* 1034.

— A true light of alchymy. Containing, I. A correct edition of the Marrow of alchymy (by Eirenæus Philoponos Philalethes [i.e. G. Starkey])...1709. *See* 1644.

— Philadelphia, or Brotherly love to the studious in the hermetick art... Written by Eyreneus Philoctetes [i.e. G. Starkey?]. 1694. *See* 1296.

1553 — Pyrotechny asserted and illustrated, to be the surest and safest means for art's triumph over nature's infirmities...8°, London, 1658. Not H; M/G9 – 14; ?

— Ripley reviv'd: or, An exposition upon Sir George Ripley's hermetico-poetical works...Written by Eirenæus Philalethes [i.e. G. Starkey]...(1677–)1678. *See* 1407.

— Secrets reveal'd: or, An open entrance to the shut-palace of the King...by...Anonymous, or Eyræneus Philaletha Cosmopolita [i.e. G. Starkey]...1669. *See* 1478.

STATUTES (The) at large...from Magna Charta until this time. By J. Keble. 2 vols. 1695. *See* 1262.

STAVELEY, Thomas. The Romish horseleech...[By T. Staveley.] 1674. *See* 1419.

1554 STEARNE, John, *Bp of Clogher*. Tractatus de visitatione infirmorum, seu de eis parochorum officiis...Ed. 2ᵃ. 12°, Londini, 1704. Not H; M/D1 – 29; ?

1555 STEELE, *Sir* Richard. An account of the fish-pool: consisting of a description of the vessel so call'd...by Sir Richard Steele, and J. Gillmore. 8°, London, 1718. Not H; M (Tracts) J3 – 33; Tr/NQ.16.162⁴.

— The Spectator. [By J. Addison, Sir R. Steele and others.] 8 vols. 1712–15. *See* 1542.

1556 STEPHANUS, *Byzantinus*. De urbibus, quem primus T. de Pinedo
...Latii jure donabat, et observationibus...illustrabat. His
additæ...collationes J. Gronovii...(*Greek & Latin*.) F°,
Amstelodami, 1678.
H; M/F7 – 10; ?

1557 STEPHANUS, Henricus. Concordantiæ Græcolatinæ Testamenti
Novi, nunc primùm plenæ editæ...F°, [Geneva], 1600.
H; M/F7 – 24; ?

1558 — Thesaurus Græcæ linguæ...4 vols. F°, [Geneva], 1572.
H; M/B3 – 16 to 19; ?

1559 STEPHANUS, Robertus. Dictionarium nominum propriorum
virorum, mulierum, populorum...8°, Coloniae Agrippinae,
1576.
Not H; M/A2 – 8; Tr/NQ.16.10 ['Isaac Newton Trin: Coll:
Cant: 1661' on verso of title-page].

1560 STERNHOLD, Thomas. The whole Book of Psalms collected into
English metre by T. Sternhold, J. Hopkins, and others...8°,
Cambridge, 1661.
Not H; not M; Tr/Adv.d.1.10³ [bound with 240, where see note
on the provenance of the volume].
6 lines of Biblical notes and references in Newton's hand on end-paper.
STEVENS, John. The history of Persia...written in Arabick, by
Mirkond...translated into Spanish, by A. Teixeira...and now
render'd into English. By J. Stevens. 1715. *See* 1084.
STEWART, Robert. Tracts, Mathematical by Bernoulli, Stewart &c.
2 vols. 1713. *See* 1630.

1561 STILLINGFLEET, Edward, *Bp of Worcester*. An answer to several
late treatises, occasioned by a book entituled A discourse concerning the idolatry practised in the Church of Rome...Pt 1.
8°, London, 1673.
Not H; M/H3 – 39; Tr/NQ.16.168 [on 1st fly-leaf 'Is. Newton
pret. 2ˢ. 6ᵈ.', on 2nd fly-leaf 'I. B. [Isaac Barrow?] ex dono
eximij authoris'].

1562 — A discourse concerning the idolatry practised in the Church of
Rome, and the danger of salvation in the communion of it...8°,
London, 1671.
Not H; M/H1 – 42; Babson 410 ['Is. Newton pret 2ˢ 6ᵈ' written
by Newton on fly-leaf above the scored-through inscription
'Donum eruditissimi Authoris']; Thame/975.

1563 — A discourse in vindication of the doctrine of the Trinity...8°,
London, 1697.
Not H; M/H4 – 28; ?

1564 — Irenicum. A weapon-salve for the Churches wounds...2nd ed....
4°, London, 1662.
Not H; M/F5 – 23; ?

1565 — Origines sacræ: or A rational account of the grounds of natural and
reveal'd religion. 7th ed....F°, Cambridge, 1702.
Not H; M/A9 – 1; Peter Laslett, Trinity College, Cambridge.
— Sermons. *See* 1493.

1566 STIRRUP, Thomas. The description and use of the universall quadrat
[*sic*]...4°, London, 1655.
H; not M; ?

1567 STONE, Edmund. A new mathematical dictionary...8°, London,
1726.
H; M/E5 – 23; ?

1568 STOUGHTON, John. Choice sermons preached upon selected
occasions...4°, London, 1640.
Not H; M/F5 – 28; Tr/NQ.8.66¹ [bound with 1569 & 1571].

1569 — xi. choice sermons, preached upon selected occasions, in Cambridge
...4°, London, 1640.
Not H; M/F5 – 28; Tr/NQ.8.66² [bound with 1568].

1570 — Felicitas ultimi sæculi: epistola in qua, inter alia, calamitosus ævi
præsentis status serio deploratur...8°, Londini, 1640.
Not H; M/A1 – 22; Tr/NQ.16.63.

1571 — Seaven sermons, preached upon severall occasions...4°, London,
1640.
Not H; M/F5 – 28; Tr/NQ.8.66³ [bound with 1568].

1572 STRABO. Rerum geographicarum libri xvii. I. Casaubonus recensuit...(*Greek & Latin.*) F°, Lutetiæ Parisiorum, 1620.
H; M/C3 – 14; ?

1573 — Rerum geographicarum libri xvii. Accedunt huic editioni, ad
Casaubonianam iii expressæ, notæ integræ G. Xylandri [etc.].
(*Greek & Latin.*) Subjiciuntur Chrestomathiæ Græc. & Lat. 2 vols.
F°, Amstelædami, 1707.
H; M/B2 – 17 & 18; Tr/NQ.11.2 & 3 [vol. 2, p. 720 turned
down and a few other signs of dog-earing].

1574 STRADA, Famiano. Eloquentia bipartita. Pars prior Prolusiones
academicas exhibet...altera Paradigmata eloquentiæ...pro-
ponit...12°, Oxoniæ, 1662.
H; M/G1 – 18; Zeitlin & Ver Brugge, Los Angeles ['Isaac
Newton' in his hand on fly-leaf].

1575 STREETE, Thomas. Astronomia Carolina: a new theory of the
cœlestial motions...2nd ed. corrected. To which are added some
lunar and planetary observations...4°, London, 1710.
H; M/E6 – 41; Tr/NQ.9.95.

Newton added '48''' in margin of p. 44 as correction for '41''' in the printed text of the last line but one.

STRONG, Martin. An essay on the usefulness of mathematical learning... [By M. Strong.] 1701. *See* 577.

1576 STUART, Alexander. Dissertatio medica inauguralis de structura et motu musculari... 4°, Lugduni Batavorum, 1711.

Not H; M (Orationes) C6 – 19; Tr/NQ.10.35³.

STUKELEY, William. An account of a Roman temple, and other antiquities, near Graham's Dike in Scotland. [By W. Stukeley.] [1720.] *See* 2.

1577 SUETONIUS TRANQUILLUS, Gaius. Duodecim Caesares, ex Erasmi recognitione. 2 pts. 8°, Parisiis, 1543.

Not H; M/F3 – 26; ?

1578 — Opera, & in illa commentarius S. Pitisci... Ed. 2ª priori ornatior & limatior. 2 vols. 4°, Leovardiæ, 1714–15.

H; M/F6 – 3 & 4; Tr/NQ.10.3 [vol. 1 only].

1579 SUICERUS, Johannes Casparus. Thesaurus ecclesiasticus, e Patribus Græcis ordine alphabetico... 2 vols. F°, Amstelædami, 1682.

H; M/F7 – 22 & 23; Tr/NQ.11.11 & 12 ['Isaac Newton. pret £ s 1. 12 Oct 3. 1682' on fly-leaf; vol. 1 has col. 90 turned down, 95 up, and a few other signs of dog-earing].

1580 SUIDAS. Lexicon, Græce et Latine. Textum... purgavit, notisque perpetuis illustravit: versionem Latinam A. Porti... correxit; indicesque... adjecit L. Kusterus. 3 vols. F°, Cantabrigiæ, 1705.

H; M/C4 – 17 to 19; ?

1581 — Suidas, nunc primum integer Latinitate donatus... opera & studio A. Porti... 2 vols. F°, Coloniæ Allobrogum, 1619.

H; M/C3 – 2 & 3; ?

1582 SULLY, Henry. Description abregée d'une horloge d'une nouvelle invention, pour la juste mesure du temps sur mer... 4°, Paris, 1726.

Not H; M (Tracts) D6 – 4; Tr/NQ.16.197¹.

1583 — Règle artificielle du tems. Traité de la division naturelle & artificielle du tems, des horloges & des montres... 8°, Paris, 1717.

Not H; M (Tracts) H8 – 17; Tr/NQ.16.161¹.

1584 SULPICIUS SEVERUS. Opera omnia. Cum lectissimis commentariis accurante G. Hornio. Ed. 3ª... 8°, Amstelodami, 1665.

H; M/A4 – 18; ?

1585 SWIFT, Jonathan. A tale of a tub. Written for the universal improvement of mankind. To which is added, An account of a battel between the antient and modern books in St. James's Library. 4th ed. corrected. 8°, London, 1705.

Not H; M/E4 – 26; Tr/NQ.8.113.

SYKES, Arthur Ashley. A modest plea for the baptismal and scripture-notion of the Trinity...[By A. A. Sykes.] 1719. *See* 1089.

1586 SYLBURGIUS, Fridericus. Etymologicon magnum; seu Magnum grammaticæ penu...repurgatum, perpetuis notis illustratum... (2 pts.) F°, [Heidelberg], 1594.
H; M/H9 – 3; ?

1587 SYMSON, Patrick. The historie of the Church since the dayes of Our Saviour Iesus Christ, untill this present age. 4°, London, 1624.
Not H; M/F5 – 37; ?

SYNESIUS, *alchemist.* Philosophie naturelle de trois anciens philosophes renommez: Artephius, Flamel, & Synesius...Dernière éd. ...1682. *See* 1309 & 1310.

SYNESIUS, *Bp of Ptolemais.* Opera quæ extant omnia. Nunc denuò Græcè et Latinè coniunctim edita. Interprete D. Petavio...1631. *See* 476.

1588 TABLE (A) of logarithms, for numbers increasing in their natural order, from an unit to 10000...2nd ed., corrected. 8°, London, 1705.
Not H; M/G9 – 3; Tr/NQ.16.144² [bound with 858].

TACHENIUS, Otto. La lumière sortant par soy même des tenebres ...[Variously attributed to M.-A. Crassellame and O. Tachenius.] 1687. *See* 1003.

1589 — Otto Tachenius his Hippocrates chymicus...Transl. into English by J. W. 4°, London, 1677.
H; M/F5 – 19; Tr/NQ.16.103 [several signs of dog-earing].

1590 TACITUS, Gaius Cornelius. Annalium...sive Historiæ Augustæ libri xvi qui supersunt...recogniti...per B. Rhenanum. F°, Basileæ, 1533.
Not H; M/C2 – 4; ?

1591 — C. Cornelii Taciti quæ exstant. M. Z. Boxhornius recensuit & animadversionibus nonnullis illustravit...Ed. nova et auctior. 12°, Amstelodami, 1661.
Not H; M/A2 – 21; ?

1592 — Opera quæ exstant, integris J. Lipsii...commentariis illustrata. J. F. Gronovius recensuit...8°, Amstelodami, 1685.
H; M/F4 – 9; ?

1593 TALLENTS, Francis. A view of universal history...[Chronological tables.] F°, London, 1700.
Not H; M/B5 – 3 [bound with 1281]; ?

TARAPHA, Franciscus. Rerum Hispaniae memorabilium annales ...a J. Vasaeo et F. Tarapha...1577. *See* 1673.

1594 TATLER (The). 4 vols. 8°, London, 1710–11.
H (vols. 1, 2); M (vols. 1–4) H8 – 30 to 33; ?

1595 TAVERNIER, Jean Baptiste. Collections of travels through Turky
into Persia, and the East-Indies...Vol. 1. F°, London, 1684.
H; M/C2 – 3; Thame/964; ?

1596 TAYLOR, Brook. Linear perspective: or, A new method of repre-
senting justly all manner of objects as they appear to the eye in
all situations...8°, London, 1715.
Not H; M (Tracts) J3 – 24; Tr/NQ.9.125⁴.

1597 — Methodus incrementorum directa & inversa. 4°, Londini, 1715.
Not H; M (Tracts) J6 – 1; Tr/NQ.16.76³.

1598 TAYLOR, Jeremy, *Bp of Down and Connor*. Antiquitates Christianæ:
or, The history of the life and death of the Holy Jesus: as also the
lives...of His Apostles. 2 pts. (Pt 1 by J. Taylor, pt 2 by W. Cave.)
8th ed. F°, London, 1684.
Not H; M/H9 – 11; ?

1599 — The rule and exercises of holy living...(& The rule and exercises
of holy dying.) 7th ed. 2 pts. 8°, London, 1663.
Not H; M/H1 – 8; ?

1600 TEINTURIER (Le) parfait, ou Instruction nouvelle & générale pour
la teinture des laines, et manufactures de laine, comme aussi
pour les chapeaux...[By T. Haak]. 8°, Leyde, 1708.
H; M/G6 – 26; Tr/NQ.7.67.

1601 TEMPLE, *Sir* William. An introduction to the history of England.
2nd ed. corrected and amended. 8°, London, 1699.
Not H; M/J4 – 2; Tr/NQ.9.157.

1602 TERENTIUS, Publius. Comœdiæ VI. Ex recensione Heinsiana:
cum annotationibus T. Farnabii...12°, Amstelædami, 1651.
H; M/G1 – 4; ?

1603 — Comœdiæ VI. Interpretatione & notis illustravit N. Camus...in
usum Serenissimi Delphini...8°, Londini, 1709.
Not H; M/F4 – 26; ?

1604 — Publii Terentii Afri Comoediae, Phaedri Fabulae Aesopiae, Publii
Syri et aliorum veterum Sententiae, ex recensione et cum notis
R. Bentleii (et G. Faerni). (2 pts.) 4°, Cantabrigiae, 1726.
H; M/F6 – 1; Tr/NQ.8.2.

1605 TERTULLIANUS, Quintus Septimius Florens. Opera ad vetustis-
simorum exemplarium fidem locis quamplurimis emendata, N.
Rigaltii observationibus & notis illustrata...F°, Lutetiæ, 1634.
H; M/D8 – 3; Tr/NQ.11.17 [several signs of dog-earing].

1606 TESAURO, Emanuele, *conte*. Patriarchæ, sive Christi servatoris
genealogia, per mundi ætates traducta. 8°, Londini, 1651.
Not H; M/E2 – 22; Tr/NQ.16.41¹ [bound with 1622].

1607 TEXTE (Le) d'alchymie, et le Songe-verd. 8°, Paris, 1695.

H; M/G7 – 42; Tr/NQ.16.147 [many signs of dog-earing]; Ekins 2 – 6.

1608 THEATRUM chemicum, præcipuos selectorum auctorum tractatus de chemiæ et lapidis philosophici antiquitate, veritate, jure, præstantia, & operationibus, continens...[Ed. by L. Zetzner etc.] 6 vols. 8°, Argentorati, 1659–61.

H; M/D1 – 4 to 9; Thame/978; bought by Newton in 1669 for £1 8s. 0d., see Fitzwilliam Notebook; ?; Sotheran **800** (1926) 12098.

*Numerous annotations, textual corrections and references by Newton in vols. 1, 3–6, with vols. 3 & 5 described as very copiously annotated; very many pages turned down (see Sotheran **800**).*

1609 THEODORETUS, *Bp of Cyrus.* Opera omnia...cura & studio I. Sirmondi. 5 vols. (Vol. 5 cura & studio J. Garnerii.) (*Greek & Latin.*) F°, Lutetiæ Parisiorum, 1642–84.

H; M/F7 – 17 to 21; Tr/NQ.17.4 to 8 [vol. 1 has pp. 128, 395 turned down and a few other signs, vol. 3, p. 540 down, vol. 5, pp. 296, 306 down].

1610 THEODOSIUS II, *Emperor of the East.* Codex Theodosianus cum perpetuis commentariis I. Gothofredi...Opus posthumum... opera et studio A. Marvillii. 6 vols. in 4. F°, Lugduni, 1665.

H; M/B3 – 12 to 15; Tr/NQ.17.9 to 12 [vol. 1 has pp. lxxvi, lxxxvi, ci turned up, clv down, vol. 6, p. 123 down].

THEODOSIUS, *Tripolita.* Theodosii Sphærica: methodo nova illustrata, & succinctè demonstrata. Per I. Barrow. 1675. *See* 76.

1611 THEOPHYLACTUS, *Abp of Ochrida.* Institutio Regia. Ad Porphyrogenitum Constantinum interprete P. Possino...(*Greek & Latin.*) 4°, Parisiis, 1651.

Not H; M/A7 – 13; ?

1612 THIRTY FOUR conferences between the Danish missionaries and the Malabarian Bramans, or Heathen priests, in the East Indies, concerning the truth of the Christian religion...Transl. out of High Dutch by Mr. Philipps. 8°, London, 1719.

H (2 copies); M (1 only) H4 – 25; Tr/NQ.7.44.

1613 THOU, Jacques Auguste de. Catalogus Bibliothecæ Thuanæ a P. & I. Puteanis [i.e. P. & J. Dupuy]...2 vols. 8°, Parisiis, 1679.

H; M/D1 – 12 & 13; ?

1614 — Doctorum virorum Elogia Thuanea. Operâ C. B[arksdale]. 8°, Londini, 1671.

Not H; M/G2 – 2; ?

1615 THUCYDIDES. De Bello Peloponnesiaco libri VIII. Ex interpretatione
L. Vallæ, ab H. Stephano iterum recognita...8°, Francofurdi,
1589.
Not H; M/F2 – 17; Tr/NQ.10.116 [several signs of dog-earing].

1616 TILLOTSON, John, *Abp of Canterbury*. A sermon preached at the
funeral of the Reverend Mr Thomas Gouge, the 4th of Novemb.
1681...8°, London, 1682.
Not H; M/H1 – 48; ?

1617 TITIAN. Novem Fabulae, Melanograph ex Titiano de Amoribus
Deorum. [The Loves of the Gods, plates, after Titian, engraved
by John Smith.] F°, [*c.* 1720].
H; not M; ?

1618 TOLLIUS, Jacobus. Fortuita. In quibus, præter critica nonnulla;
tota fabularis historia Græca, Phoenicia, Ægyptiaca, ad chemiam
pertinere asseritur. 8°, Amstelædami, 1687.
H; not M; ?

1619 TOMBEAU (Le) de la pauvreté. Dans lequel il est traité clairement
de la transmutation des metaux...Par un philosophe inconnu
[i.e. d'Atremont]...12°, Paris, 1673.
Not H; M/G2 – 25; Tr/NQ.16.109 [a few signs of dog-earing].
— Le tombeau de Semiramis nouvellement ouvert aux sages...1689.
See 511.

1620 TORNIELLUS, Augustinus. Annales sacri et profani, ab orbe
condito, ad eundem Christi Passione redemptum...2 vols. in
1. F°, Francofurti, 1611.
H; M/A8 – 5; ?

1621 TOWERSON, Gabriel. An explication of the Catechism of the
Church of England. 4 vols. (Vol. 1, 2nd ed.) F°, London, 1685–8.
Not H; M/H9 – 19 to 22; ?

1622 TOZER, Henry. Christus: sive Dicta & facta Christi: prout à quatuor
Evangelistis sparsim recitantur...8°, Oxoniæ, 1634.
Not H; M/E2 – 22; Tr/NQ.16.41² [bound with 1606].

1623 TRACTATUS aliquot chemici singulares summum philosophorum
arcanum continentes. 1. Liber de principiis naturæ, & artis
chemicæ, incerti authoris. 2. Johannis Belye Angli tractatulus
novus, & alius Bernhardi Comitis Trevirensis, ex Gallico versus.
Cum fragmentis E. Kellæi, H. Aquilæ Thuringi, & J. I. Hollandi.
3. Fratris Ferrarii tractatus integer...4. Johannis Daustenii
Angli Rosarium...[Ed. by L. Combachius.] 8°, Geismariæ,1647.
H; M/E3 – 8; Tr/NQ.10.147 [a few signs of dog-earing].
At the end of the preface (p. 15) signed 'L. C.' Newton added 'id est Lud.
Combachius. Vide Bibl. Chem. p. 64' (a reference to Borel's 'Bibliotheca
chimica', no. 246 in the present catalogue).

1624 — Tractatus de alchemia, author. varii. 8°, 1572. *Not identified individually.*
Not H; M/F3 – 34; ?

1625 TRACTS in divinity by Dr Clarke and others. 4 vols. 8°, 1706. *Not identified individually.*
Not H; M/H7 – 21 to 24; ?

1626 — Tracts in divinity viz. the Russian Catechism &c. 8°, 1695. *Only the Russian Catechism identified, see no. 1430.*
Not H; M/H1 – 17; ?

1627 — Tracts in history, viz. Claims of the people of England, &c. 8°. *Not identified individually.*
Not H; M/H8 – 15; ?

1628 — Tracts historical & mathematical, viz. Kalendar menses, Var. genl., &c. F°, 1721. *Not identified individually.*
Not H; M/D6 – 1; ?

1629 — Tracts, Law, viz. Laws of pledges & pawns, agt. Papists, &c. 8°, 1723. *Not identified individually.*
Not H; M/J3 – 23; ?

1630 — Tracts, Mathematical by Bernoulli, Stewart &c. 2 vols. 4°, 1713. *Not identified individually.*
Not H; M/D5 – 22 & 23; ?

1631 — Tracts, Medical, chiefly on the small pox. 8°, 1724. *Not identified individually.*
Not H; M/J3 – 26; ?

1632 — Tracts on antiquities found in England. 8°, 1713. *Not identified individually.*
Not H; M/J3 – 5; ?

1633 — Tracts on coin and the publick accounts. 8°, 1695. *Not identified individually.*
Not H; M/J3 – 18; ?

1634 — Tracts on medals in Latin. 8°, 1671. *Not identified individually.*
Not H; M/H8 – 21; ?

1635 — Tracts on the Bangorian controversy. [A theological dispute originating in 1716 in which Benjamin Hoadly, Bishop of Bangor, played a leading part.] 4 vols. 8°, 1718. *Not identified individually.*
Not H; M/H7 – 17 to 20; ?

1636 — Tracts on the longitude, by Hobbs, Keith, Rowe, &c. 4°, 1709. *Only the tract by William Hobbs identified.*
Not H; M/D5 – 21; ?

1637 — Tracts on the longitude & latitude. 8°, 1720. *Not identified individually.*
Not H; M/D5 – 20; ?

TRAHERNE, Thomas. Roman forgeries...By a Faithful son of the Church of England [T. Traherne]. 1673. *See* 1418.

TRAITÉ de la lumière...Par C. H[uygens].... 1690. *See* 822 & 823.

1638 TREATISE (A) of the description and use of the globes. 12°, London, [1705?].
Not H; M/E2 – 44; Tr/NQ.16.202.

1639 TREATISES on Church government. 4°, 1647. *Not identified individually.*
Not H; M/J6 – 31; ?

1640 TRELCATIUS, Lucas. Locorum communium S. Theologiæ Institutio per epitomem...12°, Londini, 1608.
Not H; M/G1 – 1; Sotheran **789** (1924) 5731 where it is described as 'With auto. "Isaac Newton Trin Coll: Cant 1661" on flyleaf'; ?

TRENCHARD, John. The Independent Whig. [By J. Trenchard and T. Gordon.] January 20, 1720 – January 4, 1721. *See* 832.

1641 TRES tractatus de metallorum transmutatione. Quid singulis contineatur, sequens pagina indicat. Incognito auctore. Adjuncta est Appendix Medicamentorum Antipodagricorum & Calculifragi...Nunc primum in lucem edi curavit M. Birrius. 8°, Amstelodami, 1668.
Not H; M/E2 – 33; Tr/NQ.10.144 [many signs of dog-earing].

TRINITY COLLEGE, *Cambridge*. See CAMBRIDGE. *University. Trinity College.*

1642 TRIOMPHE (Le) hermetique, ou La pierre philosophale victorieuse ...[By A. T. de Limojon de Saint-Didier.] 8°, Amsterdam, 1689.
H; M/G6 – 27; Tr/NQ.16.123.
Below the author's name printed at end of book (p. 153) as the anagram 'DIVES SICUT ARDENS, S***' *Newton appended his solution* 'S. E. Sanctus Didierus'; *many signs of dog-earing.*

1643 TRUE (A) and impartial account of the present differences between the Master (Richard Bentley) and Fellows of Trinity College in Cambridge, consider'd. In a letter to a gentleman sometime member of that Society. 8°, London, 1711.
Not H; M (Tracts) J3 – 7; Tr/NQ.7.1².

1644 — A true light of alchymy. Containing, I. A correct edition of the Marrow of alchymy (by Eirenæus Philoponos Philalethes [i.e. G. Starkey])...II. The errors of a late tract called, A short discourse of the quintessence of philosophers...III. The method and materials pointed at, composing the Sophick Mercury...12°, London, 1709.
Not H; M (Tracts) H8 – 14; Tr/NQ.16.95².

1645 — The true state of England. Containing, lists of the Privy-Council; of the King's houshold; of the houshold of the Prince and Princess of Wales...4°, London, 1726.
Not H; M (Tracts) J3 – 14; Tr/NQ.9.20¹.

TRUMAN, Joseph. The great propitiation: or, Christ's satisfaction ...[By J. Truman.] 1669. See 696.

1646 TRUTH (The) acknowledged: or, Sufficient proof to disabuse the publick, of the misrepresentations and false reports which have been maliciously spread abroad against the work of Mr. P. R. Fremont...[Transl.] 4°, London, 1722.
Not H; M (Orationes) C6 – 19; Tr/NQ.10.35¹¹; see also 1680.

1647 TRYALS, for high treason. 4°, 1700. Not identified.
Not H; M/C6 – 27; ?

1648 ṬUGHRĀ'Ī, Ḥusain ibn 'Alī, al-. Lamiato'l Ajam, carmen Tograi, poetæ Arabis doctissimi; unà cum versione Latina, & notis... operâ E. Pocockii...(2 pts.) 8°, Oxonii, 1661.
H; M/A2 – 18; Tr/NQ.16.26 [a few signs of dog-earing].

1649 TURRETINUS, Johannes Alphonsus. Nubes testium pro moderato et pacifico de rebus theologicis judicio, et instituenda inter Protestantes concordia...4°, Genevæ, 1719.
H; M/C6 – 6; ?

1650 TWO letters to the Reverend Dr. Bentley, Master of Trinity-College in Cambridge, concerning his intended edition of the Greek Testament. Together with the Doctor's Answer, and some account of what may be expected from that edition...[By J. Craven?] 8°, London, 1717.
Not H; M (Tracts) J3 – 7; Tr/NQ.7.1⁴.

1651 ULSTADT, Philip. Cœlum philosophorum, seu Liber de secretis naturæ. Adcessit I. A. Campesij Directorium summæ summarum medicinæ...12°, Lugduni, 1572.
Not H; M/E1 – 41; ?

1652 — Cœlum philosophorum, seu Liber: de secretis naturæ, id est: quomodo non solum è vino, sed etiam ex omnibus metallis... Ed. emendatior...12°, Augustæ Trebocorum, 1630.
Not H; M (Tracts) H8 – 18; Memorial Library, University of Wisconsin – Madison [bound with 1685]; Thame/974.

1653 URBIGERUS, Baro, pseud. Aphorismi Urbigerani, or Certain rules, clearly demonstrating the three infallible ways of preparing the grand elixir or circulatum majus of the philosophers...8°, London, 1690.
Not H; M (Tracts) H8 – 14; Tr/NQ.16.95⁵.

1654 USHER, James, *Abp of Armagh*. Annales. In quibus, Præter Mac-
cabaicam et Novi Testamenti historiam...continetur chronicon
...F°, Londini, 1654.
H; M/C2 – 11; Tr/NQ.10.2.

1655 — Annales Veteris Testamenti, a prima mundi origine deducti: una
cum rerum Asiaticarum et Ægyptiacarum chronico...F°,
Londini, 1650.
H; M/C2 – 10; Tr/NQ.10.1 [pp. 58, 104, 147, 506, 527 turned
down and several other signs of dog-earing].

1656 — An answer to a challenge made by a Iesuite in Ireland [i.e.
W. Malone]...3rd ed., corrected...(3 pts.) 4°, London, 1631.
Not H; M/F5 – 32; Tr/NQ.16.155.

1657 — A body of divinitie, or The summe and substance of Christian
religion...4th ed., corrected...F°, London, 1653.
Not H; M/J9 – 18; ?

VAILLANT, Jean. *See* FOY-VAILLANT, Jean.

VALENTINE, Basil. *See* BASILIUS VALENTINUS.

1658 VALERIANO, Giovanni Pietro. Hieroglyphica, sive De sacris
Ægyptiorum aliarumcӡ gentium litteris, commentariorum libri
LVIII....Ed. novissima...4°, Coloniæ Agrippinæ, 1631.
Not H; M/B6 – 10; Tr/NQ.10.55.

1659 VALERIUS FLACCUS, Gaius. Argonautica. I. B. Pij carmen ex
quarto Argonauticon Apollonij. Orphei Argonautica innominato
interprete. 8°, (Venetiis, 1523).
H; M/F2 – 31; Tr/NQ.10.80 [p. 30 turned down and a few other
signs of dog-earing].

1660 — Argonautica. N. Heinsius...ex vetustissimis exemplaribus recensuit & animadversiones adjecit. 12°, Trajecti Batavorum, 1702.
H; M/G2 – 13; Tr/NQ.9.53 [p. 11 turned down].

1661 VALERIUS MAXIMUS. Dictorum factorumque memorabilium,
libri IX...Operâ & industriâ J. Min-Ellii. 12°, Roterodami,
1681.
H; M/G2 – 12; ?

1662 — Valerius Maximus cum selectis variorum observat: et nova
recensione A. Thysii. 8°, Lugd. Batavorum, 1660.
H; M/F2 – 10; Tr/NQ.7.40 [pp. 456, 764 turned down and a
few other signs of dog-earing].

1663 VALLA, Laurentius. De collatione Novi Testamenti libri II. Ab
interitu vindicavit, recensuit, ac notis addidit J. Revius. 8°,
Amstelodami, 1630.
Not H; M/A2 – 19; Tr/NQ.10.134.

VALLEMONT, Pierre le Lorrain de. Ductor historicus: or, A short
system of universal history...Partly transl. from the French of

M. de Vallemont, but chiefly composed anew by W. J. M.A. [i.e. T. Hearne.] 1698. *See* 541.

1664 VALOR beneficiorum: or, A valuation of all ecclesiastical preferments in England and Wales...12°, London, 1695.
Not H; M/J8 – 43; ?

1665 VANSLEB, Johann Michael. The present state of Egypt; or, A new relation of a late voyage into that Kingdom...1672 and 1673... Englished by M. D. 8°, London, 1678.
Not H; M/J8 – 17; ?

1666 VARENIUS, Bernhard. Descriptio Regni Japoniæ et Siam...8°, Cantabrigiæ, 1673.
H; M/D1 – 3; Tr/NQ.7.45 [p. 244 turned up and a few other signs of dog-earing].

1667 — Geographia generalis...emendata, &...aucta & illustrata. Ab I. Newton. Ed. 2ª auctior & emendatior. 8°, Cantabrigiæ, 1681.
H; M/D1 – 2; ?

1668 — Geographia generalis...Adjecta est appendix...a J. Jurin. 8°, Cantabrigiæ, 1712.
H; M/D1 – 1; ?
VARET, Alexandre Louis. The nunns complaint against the Fryars ...[By A. L. Varet.] 1676. *See* 1195.

1669 VARIGNON, Pierre. Eclaircissemens sur l'Analyse des infiniment petits (de M. le Marquis de L'Hôpital). 4°, Paris, 1725.
H; not M; ?

1670 — Nouvelle mécanique ou statique dont le projet fut donné en M.DC.LXXXVII. Ouvrage posthume. 2 vols. 4°, Paris, 1725.
H (3 copies); M (1 only) E6 – 2 & 3; Tr/NQ.10.53 & 54.

1671 VARII historiæ Romanæ scriptores, partim Græci, partim Latini, in unum velut corpus redacti...4 vols. 8°, [Geneva], 1568.
H; M/F2 – 21, F1 – 4, F2 – 22 & 23; Tr/NQ.16.1, 1a, 2 & 3 [vol. 3 imperfect, vol. 3 has pp. 1108, 1134 turned down, vol. 4 p. 1555 down and these 2 vols. have a few other signs of dog earing].

1672 VARRO, Marcus Terentius. Opera quæ supersunt. In lib. de ling. Lat. cõiectanea J. Scaligeri...Ed. 3ª, recognita & aucta. (4 pts.) 8°, [Geneva], 1581.
H; M/F3 – 27; ?

1673 VASEO, Juan. Rerum Hispaniae memorabilium annales...a J. Vasaeo et F. Tarapha...8°, Coloniæ, 1577.
H; not M; ?

1674 VAUGHAN, Rice. A discourse of coin and coinage...12°, London, 1675.
H; M/G9 – 11; ?

VAUGHAN, Thomas. Anima magica abscondita: or A discourse of the universall spirit of nature...By Eugenius Philalethes [i.e. T. Vaughan]. 1650. *See* 47.

— Anthroposophia theomagica: or A discourse of the nature of man and his state after death...By Eugenius Philalethes [i.e. T. Vaughan]. 1650. *See* 50.

— The fame and confession of the Fraternity of R : C: commonly, of the Rosie Cross...By Eugenius Philalethes [i.e. T. Vaughan]. 1652. *See* 605.

— Lumen de lumine: or A new magicall light discovered, and communicated to the world, by Eugenius Philalethes [i.e. T. Vaughan]...1651. *See* 1002.

1675 VENETTE, Nicolas. Traité des pierres qui s'engendrent dans les terres & dans les animaux, où l'on parle exactement des causes qui les forment dans les hommes...12°, Amsterdam, 1701.
H; M/G2 – 32; Tr/NQ.8.58 [a few signs of dog-earing]; Ekins 3 – 6.

VERATIUS, Job. Conciones et orationes ex historicis Latinis excerptæ ...[Ed. by J. Veratius.] 1686. *See* 426.

1676 VERGILIUS MARO, Publius. Opera. Mauri Seruii Honorati grammatici in eadem commentarii...Castigationes & varietates Virgilianae lectionis, per I. Pierium Valerianum. F°, Parisiis, 1532.
H; M/C2 – 12; Tr/NQ.11.31 [several signs of dog-earing].

1677 — Opera; T. Pulmanni studio correcta. P. Manutii annotationes... 8°, Lugduni Batavorum, 1595.
Not H; M/F1 – 22; ?

1678 — Opera. Cum notis T. Farnabii. 12°, Amsterodami, [1640?].
H; M/E1 – 6; ?

1679 — Opera. Interpretatione et notis illustravit C. Ruæus...ad usum Serenissimi Delphini. Juxta editionem novissimam Parisiensem. 8°, Londini, 1696.
Not H; M/F4 – 31; ?

1680 VÉRITÉ (La) reconnüe: ou Preuve suffisante pour désabuser le public, du mauvais caracter, & faux rapport qui a eté malicieuse-ment divulgué partout contre l'œuvre de Mr. P. R. Fremont... 4°, Londre, 1722.
Not H; M (Orationes) C6 – 19; Tr/NQ.10.35^{10}; *see also* 1646.

1681 VERTOT, René Aubert de. Histoire des révolutions arrivées dans le gouvernement de la République Romaine. 3e éd. augmentée...
3 vols. 12°, La Haye, 1724.
Not H; M/G8 – 35 to 37; Tr/NQ.10.128 to 130.

1682 — Histoire des révolutions de Portugal. 12°, Amsterdam, 1712.
H; M/G8 – 38; Tr/NQ.10.77.

1683 VETERUM mathematicorum Athenæi, Bitonis, Apollodori...et aliorum opera, Græce et Latine pleraque nunc primum edita. F°, Parisiis, 1693.
H; M/C5 – 10; Babson 411; Sotheran **804** (1927) 3812.

1684 VIALART, Charles, *Bp of Avranches*. Geographia sacra, sive Notitia antiqua dioecesium omnium patriarchalium, metropoliticarum, et episcopalium veteris Ecclesiæ...Accesserunt...notæ et animadversiones L. Holstenii. 2 pts. F°, Amstelædami, 1703–4.
H; M/B5 – 11; ?

1685 VIGANI, Giovanni Francesco. Medulla chymiæ, variis experimentis aucta, multisq; figuris illustrata. 8°, Londini, 1683.
Not H; M (Tracts) H8 – 18; Memorial Library, University of Wisconsin – Madison [bound with 18 & 1652; a few signs of dog-earing]; Thame/974.
Annotations by Newton in margins of pp. 8 & 9.

1686 VIGER, François. De præcipuis Græcæ dictionis idiotismis. 11ª ed.
...12°, Londini, 1647.
Not H; M/E1 – 7; Tr/NQ.16.25 ['Ch Huggins' on fly-leaf, probably not from Newton's library].

1687 VIGNIER, Nicolas. Rerum Burgundionum Chronicon...Ex Bibliotheca historica N. Vignierij. 4°, Basileæ, 1575.
H; M/J6 – 8; Tr/NQ.16.192 ['I. Newton. pret 1ˢ.' in Newton's hand on fly-leaf; a few signs of dog-earing].

1688 VINCENT, *St, of Lérins*. Adversus profanas omnium novitates hæreticorum commonitorium. Cum notis S. Baluzii. 12°, Cantabrigiæ, 1687.
H; M/G1 – 17; Stanford University Library, Stanford, Calif. [inscribed by Newton on fly-leaf 'Is. Newton. Donum amicissimi Authoris Præfationis'; p. 93 turned down and a few other signs of dog-earing]; Thame/975.

1689 VINCENT, Levin. Catalogus et descriptio animalium volatilium, reptilium, & aquatilium...4°, La Haye, 1726.
Not H; M/E6 – 17; Tr/NQ.10.42³ [bound with 1690].

1690 — Descriptio pipae, seu Bufonis aquatici Surinamensis...4°, Haarlem, 1726.
H; M/E6 – 17; Tr/NQ.10.42¹ [bound with 1689 & 1691].

1691 — Description abregée des planches, qui representent les cabinets &
quelques-unes des curiosités, contenuës dans le Théâtre des
merveilles de la nature de L. Vincent. 4°, Harlem, 1719.
Not H; M/E6 – 17; Tr/NQ.10.42² [bound with 1690].

1692 VINDICATION (A) of the new theory of the earth from the ex-
ceptions of Mr. Keill and others...[By W. Whiston.] 8°,
London, 1698.
Not H; M (Tracts) J3 – 24; Tr/NQ.9.125⁵.

1693 VINDICIAE veterum scriptorum, contra J. Harduinum...[By
M. V. de la Croze.] 12°, Roterodami, 1708.
H; M/A3 – 24; Tr/NQ.16.13 [a few signs of dog-earing].

1694 VLACQ, Adriaan. Trigonometria artificialis: sive Magnus canon
triangulorum logarithmicus... Cui accedunt H. Briggii Chiliades
logarithmorum viginti...F°, Goudæ, 1633.
H; M/B2 – 2; ?; Newton thanked Halley for the book and sent
him 8s. for it, 17 Oct. 1695 (Correspondence, IV, 173).

1695 VOSSIUS, Gerardus Joannes. De historicis Græcis libri IV; ed.
altera, priori emendatior, & multis partibus auctior. 4°, Lugduni
Batavorum, 1651.
H; M/B1 – 14; Tr/NQ.7.54 [pp. 41, 429 turned up and several
other signs of dog-earing].

1696 — De historicis Latinis libri III. Ed. altera, priori emendatior, & duplo
auctior. 4°, Lugduni Batavorum, 1651.
H; M/B1 – 15; Tr/NQ.16.179 [a few signs of dog-earing].

1697 — De theologia Gentili, et physiologia Christiana; sive De origine
ac progressu idololatriæ...liber I, et II. 4°, Amsterdami, 1641.
H; M/C1 – 12; Tr/NQ.8.46² [bound with 1019; very extensively dog-eared with 112 pages still turned and several similar
signs].

1698 — Oratoriarum institutionum libri VI. Ed. 2ª ab autore recognita
& altera parte aucta. 8°, Francofurti, 1617.
Not H; M/F1 – 24; Tr/NQ.16.170.

1699 VOYAGE de l'Arabie Heureuse par l'Océan Oriental, & le Détroit
de la Mer Rouge. Fait par les François pour la première fois,
dans les années 1708, 1709 & 1710...[By J. de la Roque.] 12°,
Amsterdam, 1716.
H; M/G7 – 6; Tr/NQ.9.147.

1700 VOYAGES and discoveries in South-America. The first up the River
of Amazons...by C. d'Acugna. The second up the River of
Plata...by Mons. Acarete. The third from Cayenne into Guiana
...By M. Grillet and Bechamel. Done into English from the
originals...8°, London, 1698.
H (2 copies); M (1 only) J4 – 38; Tr/NQ.9.156.

1701 VREAM, William. A description of the air-pump, according to the late Mr. Hawksbee's best and last improvements...8°, London, 1717.

Not H; M/E5 – 25; Tr/NQ.16.91² [bound with 505].

1702 WAFER, Lionel. A new voyage and description of the Isthmus of America. 8°, London, 1699.

Not H; M/J4 – 35; ?

WAKE, William, *Abp of Canterbury*. The genuine Epistles of the Apostolical Fathers...Translated and publish'd...by William [Wake], Lord Archbishop of Canterbury. 3rd ed. 1719. *See* 67.

1703 — The principles of the Christian religion explained: in a brief commentary upon the Church-Catechism. 3rd ed. corrected... 8°, London, 1708.

Not H; M/H3 – 21; Tr/NQ.8.138 [p. 50 turned down and a few other signs of dog-earing].

1704 WALKER, William. English examples of the Latin syntaxis...8°, London, 1683.

Not H; M/D1 – 24; ?

1705 — A modest plea for infants baptism...12°, Cambridge, 1677.

Not H; M/G9 – 6; Babson 412 [inscribed by Newton on fly-leaf 'Is. Newton. Ex dono Rⁿᵈⁱ Authoris']; Thame/975.

1706 — A treatise of English particles...With a praxis upon the same. 8th ed. 8°, London, 1683.

H; M/F2 – 12; Tr/NQ.9.7.

1707 — Βαπτισμῶν διδαχή, the doctrine of baptisms: or, A discourse of dipping and sprinkling...8°, London, 1678.

Not H; M/H2 – 16; ?

1708 WALLER, William. An essay on the value of the mines, late of Sir Carbery Price. 8°, London, 1698.

H; M/E5 – 44; Thame/987; ?

1709 WALLIS, John. Mechanica: sive, De motu, tractatus geometricus. 3 pts. 4°, Londini, 1669–71.

H; M/A6 – 17; Tr/NQ.16.149.

Long mathematical notes by Newton in margins of pp. 15, 36; Newton thanked Collins for sending him the 2nd pt, Jan. 1669/70 and 11 July 1670 (*Correspondence*, I, 16, 31); Sotheran **828** (1931) 3602.

1710 — Opera mathematica. 3 vols. F°, Oxoniæ, 1693–9.

H; M/A9 – 7 to 9; Whipple Science Museum Library, Cambridge University [vols. 1 & 2 only].

Minute corrections in Newton's hand in vol. 2, pp. 392–3, see *Math. papers*, VII, xvii n. 37, 171n. 2; vol. 2, p. 391 turned down to mark reference to Newton's name in text; Thame/992 (3 vols).

WANSLEBEN, Johann Michael. *See* VANSLEB, Johann Michael.

1711   WARD, Seth, *Bp of Salisbury*. Idea trigonometriæ demonstratæ, in usum juventutis Oxoniensis. Item Prælectio de cometis. Et Inquisitio in Bullialdi Astronomiæ Philolaicæ fundamenta. 4°, Oxoniæ, 1654.

H; M/E5 – 36; Tr/NQ.8.71.

1712   WARNEFRIDUS, Paulus, *Diaconus*. De gestis Langobardorum libri vi. Ad MS. & veterum codicum fidem editi. 8°, Lugduni Batavorum, 1595.

H; M/F1 – 16; Tr/NQ.16.9¹ [bound with 13 & 675].

1713   WASER, Caspar. De antiquis mensuris Hebræorum; quarum S. Biblia meminerunt, libri iii...4°, Heidelbergæ, 1610.

Not H; M/A6 – 3; Tr/NQ.9.12² [bound with 1714; p. 89 was formerly turned down].

1714   — De antiquis numis Hebræorum, Chaldæorum et Syrorum: quorum S. Biblia & Rabbinorum scripta meminerunt, libri ii ...4°, Tiguri, 1605.

Not H; M/A6 – 3; Tr/NQ.9.12¹ [bound with 1713; 'Is. Newton pret 3' [?] on fly-leaf].

1715   WATERLAND, Daniel. The case of Arian-subscription considered: and the several pleas and excuses for it particularly examined and confuted. 8°, Cambridge, 1721.

Not H; M/H3 – 35; ?

1716   — A critical history of the Athanasian Creed...4°, Cambridge, 1724.

H; not M; ?

1717   — Eight sermons preach'd at the Cathedral Church of St. Paul... 8°, Cambridge, 1720.

H; M/H6 – 37; ?

WATTS, Robert. The rule for finding Easter in the Book of Common-Prayer, explain'd and vindicated... [By R. Watts.] 1712. *See* 1428.

WEBER, Friedrich Christian. The present state of Russia...from the year 1714, to 1720... [By F. C. Weber.] Translated from the High-Dutch. 2 vols. 1722–3. *See* 1347.

1718   WEBSTER, John. Metallographia: or, An history of metals...4°, London, 1671.

H; M/F5 – 22; Tr/NQ.16.150 [many signs of dog-earing].

WEGHORST, Henrik. Regis Christiani Quinti Jus Danicum, Latine redditum ab H. Weghorst. 1698. *See* 374.

1719   WEIDENFELD, Johann Seger. De secretis adeptorum, sive De usu spiritus vini Lulliani libri iv...4°, Londini, 1684.

Not H; M/E6 – 43; Tr/NQ.16.104 [several signs of dog-earing].

1720   WEIR, James. The ready accomptant; or Book-keeping reform'd... 4°, London, 1700.

H; M/J3 – 37; Tr/NQ.8.79.

1721 WELCHMAN, Edward. xxxix articuli Ecclesiæ Anglicanæ, textibus è Sacra Scriptura depromptis confirmati, brevibusque notis illustrati...Ed. 4ª auctior & emendatior...8°, Oxonii, 1724.
Not H; M/H2 – 7 [described as 'interleaved']; ?

1722 WELLS, John. The practical Sabbatarian; or, Sabbath holiness crowned with superlative happiness. 4°, London, 1668.
Not H; M/J6 – 28; ?

1723 WENDELIN, Marcus Friedrich. Christianæ theologiæ libri ii...Ed. novissima emendatior. 12°, Lugduni Batavorum, 1656.
Not H; M/E1 – 40; ?

1724 WERNER, George Christoph. Inventum novum, artis et naturæ connubium, in copulatione levitatis cum gravitate & gravitatis cum levitate...4°, Augustæ, 1670.
Not H; M (Tracts) D5 – 24; Tr/NQ.16.79².

WHARTON, Henry. The pamphlet entituled, Speculum ecclesiasticum [of T. Ward]...[By H. Wharton.] 1688. See 1234.

1725 WHEAR, Degory. Relectiones hyemales de ratione & methodo legendi utrasque historias, civiles & ecclesiasticas...8°, Cantabrigiæ, 1684.
Not H; M/E3 – 9; Tr/NQ.16.12.

1726 WHISTON, William. Astronomical principles of religion, natural and reveal'd. In 9 pts...8°, London, 1717. [Dedicated to Newton.]
H; M/E4 – 13; Tr/NQ.9.152.
— A course of mechanical, optical, hydrostatical, and pneumatical experiments. To be perform'd by F. Hauksbee; and the explanatory lectures read by W. Whiston. [1713.] See 745.

1727 — An essay on the Revelation of Saint John...To which are added two dissertations...4°, Cambridge, 1706.
H; M/F5 – 3; Tr/NQ.8.73 [several signs of dog-earing].

1728 — An essay towards restoring the true text of the Old Testament... 8°, London, 1722.
H; not M; ?

1729 — An historical preface to Primitive Christianity reviv'd...8°, London, 1711.
H; M/H7 – 10; ?

1730 — The longitude and latitude found by the inclinatory or dipping needle. 4°, [London, 1719?].
Not H; M (Tracts) J3 – 25; Tr/NQ.16.114¹.

1731 — The longitude and latitude found by the inclinatory or dipping needle...To which is subjoin'd Mr. Robert Norman's New attractive, or Account of the first invention of the dipping needle. (2 pts.) 4°, London, (1720–)1721.
Not H; M (Tracts) J3 – 25; Tr/NQ.16.114².

1732 — A new method for discovering the longitude both at sea and land
...By W. Whiston and H. Ditton. 2nd ed.: with great additions,
corrections, and improvements. 8°, London, 1715.
Not H; M (Tracts) J8 – 11; Tr/NQ.10.119[1].

1733 — Prælectiones astronomicæ Cantabrigiæ in Scholis Publicis habitæ.
Quibus accedunt Tabulæ plurimæ astronomicæ Flamstedianæ
correctæ, Halleianæ, Cassinianæ, et Streetianæ. 8°, Cantabrigiæ,
1707.
H; M/E4 – 12; Tr/NQ.7.43.

1734 — Primitive Christianity reviv'd...4 vols. 8°, London, 1711.
H; M/H7 – 5 to 8; ?

1735 — Sermons and essays upon several subjects...8°, London, 1709.
H; M/H7 – 12; ?

1736 — A short view of the chronology of the Old Testament, and of the
harmony of the Four Evangelists. 4°, Cambridge, 1702.
H; M/F5 – 2; ?

1737 — Three essays, I. The Council of Nice vindicated from the Athanasian heresy. II. A collection of ancient monuments relating to
the Trinity...III. The liturgy of the Church of England reduc'd
nearer to the primitive standard. 8°, London, 1713.
H; M/H7 – 11; ?

1738 — Tracts. 3 vols. 8°, London, 1710. *Not identified individually.*
Not H; M/H7 – 13 to 15; ?
— A vindication of the new theory of the earth from the exceptions
of Mr. Keill and others...[By W. Whiston.] 1698. *See* 1692.
WIEKUS, Jacobus. Refutatio libelli, quem Iac. Wiekus Iesuita anno
1590 Polonicè edidit, De Divinitate Filii Dei, & Spiritus Sancti
...1594. *See* 1385.

1739 WILCOX, Thomas. Works...containing an exposition upon the
whole booke of David's Psalmes...F°, London, 1624.
Not H; M/J9 – 12; ?

1740 WILKINS, John, *Bp of Chester.* Ecclesiastes: or, A discourse concerning the gift of preaching as it falls under the rules of art. 3rd
ed. 8°, London, 1651.
Not H; M/H1 – 49; ?

1741 WILLIS, Thomas. Pathologiæ cerebri, et nervosi generis specimen.
In quo agitur de morbis convulsivis, et de scorbuto. 12°, Amstelo-
dami, 1668.
Not H; M/G2 – 17; ?

1742 WILSON, George. A compleat course of chymistry. Containing near
three hundred operations...8°, London, 1700.
H; M/E5 – 38; Tr/NQ.16.98.

1743 WING, Vincent. Astronomia Britannica: in qua per novam, con-
cinnioremq; methodum, hi quinq; tractatus traduntur. (2 pts.)
F°, Londini, 1669.

H; M/A8 – 4; Tr/NQ.18.36.

*Index with page references to parts of the book made by Newton on 1st fly-leaf
a further note on 2nd fly-leaf, and on 2 closely written rear end-papers
Newton added (in about 1670) extensive notes concerning the book and
other related topics; see p. 19 above.*

1744 — Harmonicon cœleste: or, The cœlestial harmony of the visible
world...F°, London, 1651.

H; M/J9 – 20; Butler Library, Columbia University, New York;
Sotheran **786** (1923) 3976.

*Numerous notes, corrections and additions in Newton's hand on many pages
and a loose leaf closely written by him on both sides.*

WINGATE, Edmund. Ludus mathematicus: or, The mathematical
game...By E. W[ingate]. 1681. *See* 993.

1745 — Mr. Wingate's Arithmetick...9th ed., very much enlarged...by
J. Kersey. 8°, London, 1694.

H; M/E5 – 35; Tr/NQ.8.95.

— Tabulæ logarithmicæ, or Two tables of logarithmes: the first...
by N. Roe. The other...by E. Wingate. (2 pts.) 1633. *See* 1412.

1746 WINIFRED, *St.* The life and miracles of St. Wenefrede, together
with her litanies...[By W. Fleetwood.] 8°, London, 1713.

Not H; M/J3 – 8 [bound with 732]; ?

1747 WINSTANLEY, William. England's worthies: select lives of the
most eminent persons...8°, London, 1684.

Not H; M/J7 – 44; ?

1748 WINTER-EVENING conference between neighbours. [By J.
Goodman.] 3rd ed. corrected. 3 pts. 8°, London, 1686.

Not H; M/H7 – 28; ?

WINTERTON, Ralph. Poetæ minores Græci...Accedunt etiam
Observationes R. Wintertoni in Hesiodum. 1684. *See* 1333.

1749 WINWOOD, *Rt Hon. Sir* Ralph. Memorials of affairs of State in the
reigns of Q. Elizabeth and K. James I....3 vols. F°, London,
1725.

H; M/H9 – 13 to 15; ?

1750 WOLLEBIUS, Johannes. Compendium theologiæ Christianæ...Ed.
novissima...12°, Oxoniæ, 1655.

Not H; M/E1 – 43; ?

1751 WOODWARD, John. Naturalis historia telluris illustrata et aucta
...8°, Londini, 1714. [Dedicated to Newton.]

H; M/G10 – 20; ?

1752 — The natural history of the earth, illustrated, inlarged, and defended. Written originaly [*sic*] in Latin: and now first made English by B. Holloway...8°, London, 1726.
H; M/E5 – 21; Tr/NQ.9.15.

1753 — The state of physick and of diseases...8°, London, 1718.
H; M/J7 – 29; ?

1754 WOODWARD, Josiah. An account of the rise and progress of the religious societies in the City of London...3rd ed., enlarged. 12°, London, 1701.
Not H; M/G9 – 31; ?

1755 WORTHINGTON, John. The great duty of self-resignation to the Divine Will. 8°, London, 1675.
Not H; M/H1 – 3; Tr/NQ.8.123 [pp. 56, 111 turned down and a few other signs of dog-earing].

1756 WOTTON, *Sir* Henry. Reliquiæ Wottonianæ, or, A collection of lives, letters, poems...[Ed. by I. Walton.] 12°, London, 1651.
Not H; M/G9 – 20; ?

1757 WOTTON, William. Reflections upon ancient and modern learning. 2nd ed., with large additions. With A dissertation upon the Epistles of Phalaris [etc.]. By Dr. Bentley. (2 pts.) 8°, London, 1697.
H; M/E4 – 31; Tr/NQ.10.107 [pt 2, p. 29 turned down and a few other signs of dog-earing].

1758 XENOPHON. De Cyri institutione libri VIII...cum Latinâ inter-interpretatione J. Leunclavii...8°, Londini, 1698.
H; M/F2 – 4; ?

1759 — Hieron, ou Portrait de la condition des rois. En grec et en françois. De la traduction de P. Coste. 8°, Amsterdam, 1711.
H; M/G6 – 49; Tr/NQ.16.42.

1760 YARWORTH, William. Introitus apertus ad artem distillationis; or The whole art of distillation practically stated...8°, London, 1692.
H; M/E4 – 39; ?
— Mercury's Caducean rod...By Cleidophorus Mystagogus [i.e. W. Yarworth]. 1702. *See* 1138.
— A philosophical epistle, discovering the unrevealed mystery of the three fires of the Sophi. [*Signed* Cloidophorus Mystagogus, i.e. W. Yarworth.] [*c*. 1702.] *See* 1302 & 1303.

1761 YOUNG, Edward. Sermons on several occasions...3rd ed. 2 vols. 8°, London, 1720.
Not H; M/H5 – 29 & 30; ?

1762 ZAHN, Godofredus Andreas. Dissertatio de origine, progressu et dignitate medicinæ...12°, Vesaliæ, 1708.

H; M/G1 – 8; ?

ZETZNER, Lazarus. Theatrum chemicum...[Ed. by L. Zetzner etc.] 6 vols. 1659–61. *See* 1608.

1763 ZOSIMUS, *Panopolitanus.* Historiæ novæ libri VI, nunquam hactenus editi...F°, Basileæ, [1576].

H; M/D8 – 12; Tr/NQ.11.8 ['Isaac Newton. pret 3ˢ. 6ᵈ.' on fly-leaf; many signs of dog-earing].

THE LYMINGTON SALE
OF BOOKS AND MSS, *c.* 1750

John Conduitt died in May 1737; his wife Catherine lived less than two years longer, and on her death in January 1739 the Newton material which they had acquired (manuscripts and a few annotated books) passed to her daughter, also named Catherine. In 1740 the latter married John Wallop, who afterwards became Viscount Lymington when his father was created the first Earl of Portsmouth in 1743. Lymington died in November 1749 while his father was still alive, and it was his eldest son John who in 1762 succeeded as the second Earl.[1] The death of Lymington was followed shortly afterwards by a sale of his books and manuscripts: the *Catalogue* (probably issued in 1750, though no year is given) states that it is

> *of the Valuable Library of Books and Manuscripts of the Right Hon. Lord Lymington, deceased; which will be sold by Auction...on Tuesday, February 27, in Poland-Street...consisting of about Four Thousand Volumes...with a compleat Collection of the Journals of the House of Commons...in near Three Hundred Volumes MSS, heretofore the Property of John Conduit, Esq.... John Heath, Auctioneer.*

In spite of the mention here of John Conduitt, I believe that no Newton items were included in the sale both from my examination of the contents of the catalogue and above all because it would have broken the terms of an agreement legally binding upon Conduitt and his family which he made with Newton's heirs in 1727 not to dispose of his papers without their prior consent.

Conduitt had in 1720 bought the house and estate at Cranbury Park in Hampshire, and Newton is said to have stayed there with the Conduitts. After Conduitt's death his widow Catherine sold Cranbury to Thomas Lee Dummer, from whose son the estate passed to William Chamberlayne and so down to his descendants, who are the present owners. It is, of course, possible that Conduitt took certain of his Newton items down to his country

---

[1] For further details of the family, and the descent of the manuscripts in particular, see D. T. Whiteside's general introduction to Newton, *Math. papers*, I (1967), xx–xxv.

home, but of this there is no record. There are, however, at the present time seven volumes in Cranbury Park which, hearsay has it, once belonged to Newton. Each of the books has a hole cut in its fly-leaf or end-paper deliberately removing the signature, but they bear none of the usual signs of Newton's ownership. While their association with Conduitt's family is not in question, I would require much stronger evidence before being willing to include them in a catalogue of Newton's library.[1]

[1] I gratefully acknowledge the courtesy of Major N. Chamberlayne-Macdonald in showing me his library at Cranbury Park in September 1976.

THE LEIGH & SOTHEBY SALE
OF NEWTON BOOKS, 1813

———

In asserting on page 41 above that Newton's books remained essentially intact at Barnsley Park from 1778 (or thereabouts) until 1920, I am aware of the existence of the puzzling *Catalogue of the Library of the late Mrs. Anne Newton, containing the Collection of the Great Sir Isaac Newton. To which is added, Part of the Library of Tycho Wing, Esq. Deceased. Which will be sold by Auction, by Leigh and Sotheby...On Monday, March 22, 1813, and Five following Days...* This lists 1363 individual lots, of which the last 102 are described as the property of Tycho Wing; but 441 of the preceding 1261 items were published after Newton's death and so cannot be his.[1] Twelve works are claimed to bear the 'Autograph of Sir I. Newton', namely: Lot 22, '*Historical Remarques upon the Revolutions in the United Provinces, 1675*'; Lot 29, 'Besongne [*sic*] (M. N.) *Etat de la France*, 2 vol., Paris, 1665'; Lot 56, '*Geographical Description of the World, 1671*'; Lot 62, '*Julian the Apostate, being a short Account of his Life* [By S. Johnson], 1682'; Lot 125, '*La Clef des Coeurs*, Paris, 1673'; Lot 162, '*Dictionarium, Teutsch, Franc. Lat.*, 1669'; Lot 336, 'Mackenzie's (Sir Geo.) *Moral Essay, preferring Solitude to publick Employment*, 1685'; Lot 372, Barrow's '*Euclidis Elementa*, 1659'; Lot 381, 'Henrion (D.) *Usage du Compas de Proportion*, Rouen, 1564'; Lot 407, 'Des-Cartes (Renati) *Opera Philosophica*, Amst. Elz., 1664'; Lot 768, 'Ryff (P.) *Quaestiones Geometricæ*, Oxoniæ, 1665'; Lot 772, 'Gale (T.) *Opuscula Mythologica, Gr. Lat.*, Cantab., 1671'. These volumes are titles which one might perhaps have found on Newton's shelves, but only Gale's *Opuscula* is in the Huggins List, and this is also recorded in the Musgrave Catalogue with the letter 'B' at the end of the entry, which indicates that it was removed in the late eighteenth century to Barnsley Park (though its present whereabouts is not known). None of the twelve books are extant, as far as I can discover, so that I have been unable directly to examine their 'Newton' autographs. Though, moreover, the circumstances of the sale cannot at present be established exactly, there is clear evidence both to infer that the signatures were those of Sir John Newton (1651–1734), rather than of

---

[1] The auctioneers' annotated copy which is now in the British Library shows that the purchasers were mostly booksellers and that the sale fetched a total of £495. 3s. 6d.

his distant kinsman Sir Isaac, and also to conclude that 'the late Mrs. Anne Newton' was Anne Newton (*née* Bagshawe) who died in June 1811, widow of Michael Newton, nephew and heir of Sir John's son Michael.[1]

The will of Anne Newton[2] 'of Upper Harley Street in the Parish of Mary-le-Bow, Middlesex, Relict of Michael Newton', dated 5 June 1811, was proved on 11 July 1812. There were no children of the marriage, and the long document goes into considerable detail about various bequests she wished to make, including (it is here significant to notice) some to cousins in the Bagshawe family: this establishes that she was indeed the same Anne Bagshawe who had married Michael Newton. Her executrix, a Miss Martha Harley, was directed that 'my house, furniture, plate, jewels and linen [are] to be sold', but there is no instruction in the will regarding the library – in fact, there is no reference there to it at all. I assume, nonetheless, that Miss Harley arranged for the books in Mrs Newton's house to be auctioned within a few months of probate being granted. As for the 'part of the Library of Tycho Wing' sold along with them, my guess is that it came to Leigh and Sotheby from a different source and that it was a mere convenience to sell off the two collections together.

In making a brief reference to this 1813 sale, I must in fairness add, Zeitlinger has conjectured[3] (mistakenly in my opinion) that 'Anne Newton probably was a descendant of John Newton, Sir Isaac's nephew and heir-at-law'. When he died in 1737 this John Newton willed his estate to his two sisters,[4] and from what is known of his character, while he was very anxious to obtain any money coming to him from Isaac Newton, he was unlikely to have been concerned about his uncle's books to do other than swiftly sell them off. Zeitlinger further remarked that 'the books belonging to Mrs Newton were chiefly theological and classical, those on science [being] few and only of minor importance'. The judgement is demonstrably not in accord with the wide variety of the titles listed in the sale catalogue; whoever it was to whom these books may have once belonged, he does their range of subject less than justice. We are told by Zeitlinger, finally, that 'several other books from Newton's library were sold during this period'. If this was so, I have not been able to discover whence they came, what they were, or who bought them.

Augustus De Morgan, writing in about 1852 of this Leigh and Sotheby sale, observed that 'it seems to have excited no curiosity' and offered the

---

[1] See *Miscellanea genealogica et heraldica*, N.S., i (London, 1874), 169–71. It would, of course strengthen the evidence if I were able to locate in the 1813 sale catalogue the three books bearing Sir John Newton's authenticated signature already mentioned (p. 4 above), but they are not to be found there.

[2] Public Record Office, London, PROB./11, vol. 1535, fols. 395–8.

[3] *Newton's library* (Cambridge, 1944), p. 13.

[4] C. W. Foster, 'Sir Isaac Newton's family', *Reports and Papers of the Architectural Societies of the County of Lincoln...*xxxix (1928), 59.

suggestion that the books were the two or three hundred which had once belonged to Newton's stepfather and were later given away by Newton: 'this explains how these books, Newton's by inheritance and not by use, remained in the name of Newton'.[1] But the mathematical books which the 1813 sale catalogue asserted to carry Newton's signature hardly seem the sort a parish priest like the Reverend Barnabas Smith would have been likely to acquire, and, much more damaging to De Morgan's conjecture, all except one of the titles which I have listed were published after Smith died. De Morgan was by no means impressed with the claims of the sale catalogue, however. 'It also appears that the purchasers were all booksellers: a sale which, was believed to be of Newton's books would', he goes on, 'in 1813, have drawn the bibliomaniacs together like vultures to a carcase.' A few years later, in the course of some correspondence published in *Notes and Queries* on Newton's family, a letter by De Morgan of 19 October 1861, referring to the 1813 sale and its claims to have contained 'the Collection of the Great Isaac Newton', adds, 'I do not know what the pretext for this description is...But there is every reason to suppose that Newton's books went to Mrs Conduitt.'[2] Since De Morgan was manifestly unaware of the agreement between John Huggins and John Conduitt regarding the purchase of Newton's books in 1727 by the former, his mistaken supposition that the collection passed by right of inheritance to the Conduitt family is understandable. Nor is it entirely wrong, for, as we have seen, John Conduitt in then gaining custodianship of Newton's loose manuscript writings also acquired the six books with notes by Newton which were withheld from the sale of his library to Huggins.

[1] A. De Morgan, *Newton: his friend and his niece* (London, 1885; facs. repr. 1968), pp. 152–3. (The work was published posthumously, the author having died in 1867.)
[2] *Notes and Queries*, 2nd ser., xii (19 Oct. 1861), 315.

BOOKS WITH NOTES BY NEWTON
(Excluding price-indications and inscriptions)

Catalogue no.	Author or title
84	Ars chemica
90	Artis auriferae
93	Ashmole
103	Aurifontina chymica
110	Bacon
122	Barrow
128	Basilius Valentinus
142	Bayer
174	Bernoulli
188	Bible, *English*
189	Bible, *English*
195	Bible, *Greek, Old Testament*
264	Boyle
265	Boyle
377	Chrysostom
488	Dary
507	Descartes
513	Dickinson
554	Enarratio methodica
581	Euclid
584	Euclid
585	Euclid
594	Everard
598	Fabre
605	Fame and confession
657	Geber
718	Groningius
733	Halifax

Catalogue no.	Author or title
740	Harmoniæ inperscrutabilis
810	Howard
889	Knatchbull
904	La Hire
990	Lucretius
1000	Lull
1031	Mariotte
1045	Mayer
1049	Mayer
1072	Mercator
1084	Mirkhond
1110	More
1115	More
1138	Mystagogus
1153	Newton: Analysis
1154	Newton: Arithmetica
1155	Newton: Universal arithmetick
1157	Newton: Opticks. Ed. 1
1158	Newton: Opticks. Ed. 1
1159	Newton: Opticks. Ed. 2
1162	Newton: Optice. Ed. 1
1163	Newton: Optice. Ed. 1
1167	Newton: Principia. Ed. 1
1168	Newton: Principia. Ed. 1
1169	Newton: Principia. Ed. 2
1170	Newton: Principia. Ed. 2
1189	Norwood

Catalogue no.	Author or title	Catalogue no.	Author or title
1192	Novum lumen	1454	Scaliger
1209	Origenes	1471	Schooten
1220	Oughtred	1478	Secrets reveal'd
1248	Médailles sur Louis Le Grand	1560	Sternhold
		1575	Streete
1301	Philosophiæ chymicæ	1608	Theatrum chemicum
1370	Raphson	1623	Tractatus aliquot chemici
1371	Raphson		
1378	Reconditorium	1642	Triomphe hermetique
1380	Recueil de diverses pièces	1685	Vigani
1403	Ridley	1709	Wallis
1405	Ripley	1710	Wallis
1406	Ripley	1743	Wing
1412	Roe	1744	Wing
1446	Sanguis naturæ		

BOOKS RECORDED IN THE HUGGINS LIST BUT NOT IN THE MUSGRAVE CATALOGUE

(Excluding the six books with notes by Newton withdrawn from the 1727 sale)

Catalogue no.	Author or title
190	Bible, *English*
218	Bible, *Concordance*
237	Boizard
288	Brandt
298	Brisson
313	Burnet
383	Cicero
387	Clarke
400	Clemens, *Romanus*
435	Connoissance des temps
474	Cyprian, *St*
486	Dampier
529	Discourse on judicial authority
559	Enquiry [King]
572	Essay d'analyse [Monmort]
658	Geber
690	's Gravesande
703	Gregory, *St*
826	Hypothesis physica [Leibniz]
944	Leo, *Africanus*
971	Locke
974	Locke

Catalogue no.	Author or title
975	Locke
1023	Malebranche
1055	Mead
1059	Medulla historiæ [Howell]
1089	Modest plea [Sykes]
1093	Moivre
1107	Moore
1108	Morden
1117	Morgagni
1133	Musschenbroek
1145	Nepos
1146	Neptune françois
1149	New state of England
1161	Newton: Opticks. Ed. 3 (2 copies)
1164	Newton: Optice. Ed. 2
1171	Newton: Principia. Ed. ultima
1198	Observations [Clarke]
1259	Parliament: Acts
1266	Paterculus
1332	Pococke
1369	Raphson
1387	Reineccius

Catalogue no.	Author or title	Catalogue no.	Author or title
1402	Rider	1617	Titian
1434	Rymer	1618	Tollius
1435	Sallengre	1669	Varignon
1465	Scheuchzer	1673	Vaseo & Tarapha
1491	Senex & Maxwell	1716	Waterland
1566	Stirrup	1728	Whiston

BOOKS WITH PROVENANCE
OTHER THAN THE
HUGGINS OR MUSGRAVE COLLECTIONS

Catalogue no.	Author or title	Catalogue no.	Author or title
174	Bernoulli	1163[a]	Newton: Optice. Ed. 1
188[b]	Bible, *English*	1167[a]	Newton: Principia. Ed. 1
189[a]	Bible, *English*	1169	Newton: Principia. Ed. 2
240[b]	Book of Common Prayer	1380	Recueil de diverses pièces
506[a]	Descartes	1406[a]	Ripley
507	Descartes	1442	Sanderson
629	Fox Morcillo	1446	Sanguis naturæ
793	Homerus	1478[a]	Secrets reveal'd
1158	Newton: Opticks. Ed. 1	1560[b]	Sternhold

[a] Recorded in Huggins List but withdrawn from the sale.
[b] Bound together in one volume.

LIST OF AUTHORITIES

Ashton, J. *The Fleet, its river, prison, and marriages.* London, 1888.
Babson Collection. *A descriptive catalogue of the Grace K. Babson Collection of the Works of Sir Isaac Newton... in the Babson Institute Library, Babson Park, Mass.* New York, 1950. *Supplement... compiled by H. P. Macomber.* Babson Park, 1955.
Bibliotheca alchemica et chemica: an annotated catalogue of printed books on alchemy, chemistry and cognate subjects in the library of Denis I. Duveen. London, 1949 (repr. 1965).
Birch, T. 'Newton', *Biographia Britannica...* vol. v (London, 1760), pp. 3210–44.
— 'Newton', *A general dictionary, historical and critical...* vol. vii (London, 1738), pp. 776–802.
Björnståhl, J. J. *Briefe aus seinen ausländischen Reisen...* Vol. 3. Rostock and Leipzig, 1781.
Bloxam, J. R. *A register of... Saint Mary Magdalen College... The Demies...* Vol. 3. Oxford, 1879.
Bodleian Library, Oxford, MS New College 361 (Ekins Papers), vol. ii. Fol. 47, Book-list in Newton's hand of 34 works, *c.* 1698.
— — Fol. 78, 'Books for Mr Newton', *c.* 1702–5, in an unknown hand.
— MS Rawl.D.878, fols. 39–59, 'A Catalogue of the bookes of Dr Isaac Barrow sent to S. S. by Mr Isaac Newton... 1677'.
— Wykeham–Musgrave deposited deeds, C.41.
Brewster, Sir D. *Memoirs of the life, writings, and discoveries of Sir Isaac Newton.* 2 vols. Edinburgh, 1855.
Cambridge University Library. MS. Add. 4000, Notebook of Newton, *c.* 1699.
Catalogue (A) of the Portsmouth Collection of Books and Papers written by or belonging to Sir Isaac Newton... Cambridge, 1888.
Catalogue (A) of the Valuable Library of Books and Manuscripts of the Right Hon. Lord Lymington, deceased; which will be sold by Auction... on Tuesday, February 27, in Poland-Street... John Heath, Auctioneer. [London, 1750?]
Chinnor Parish Registers. MSS.
Cohen, I. B. *Introduction to Newton's 'Principia'.* Cambridge, 1971.

Country Life, xxiii (2 May 1908), pp. 630–36, and cxvi (2 Sept. 1954), pp. 720–23, (9 Sept. 1954), pp. 806–9.

Craig, *Sir* J. *Newton at the Mint*. Cambridge, 1946.

De Morgan, A. *Newton: his friend and his niece*. London, 1885 (facs. repr. 1968).

de Villamil, R. *Newton: the man*. London [1931] (repr. New York, 1972).

— 'The tragedy of Sir Isaac Newton's library', *Bookman*, lxxi (1926–7), pp. 303–4.

Dobbs, B. J. T. *The foundations of Newton's alchemy, or 'The Hunting of the Greene Lyon'*. Cambridge, 1975.

Duveen, Denis I., see *Bibliotheca alchemica et chemica*...

Edleston, J. (ed.), *see* Newton, *Correspondence of Sir Isaac*...

English, J. S. '...*And all was light': the life and work of Sir Isaac Newton*. Lincolnshire Library Service, 1977.

Feisenberger, H. A. 'The libraries of Newton, Hooke and Boyle', *Notes and Records of the Royal Society of London*, xxi (1966), pp. 42–55.

Fitzwilliam Museum, Cambridge. Notebook of Newton, *c.* 1665–9.

Forbes, E. G. 'The library of the Rev. John Flamsteed, F.R.S., first Astronomer Royal', *Notes and Records of the Royal Society of London*, xxviii (1973–4), pp. 119–43.

Foster, C. W. 'Sir Isaac Newton's family', *Reports and Papers of the Architectural Societies of the County of Lincoln, County of York* [etc.] xxxix (1928), pp. 1-62.

Foster, J. *Alumni Oxonienses...1500–1714...*Vol. 2. Oxford, 1892.

— *Alumni Oxonienses...1715–1886...*Vol. 3. Oxford, 1888.

Gaskell, P. and R. Robson. *The Library of Trinity College, Cambridge: a short history*. Cambridge, 1971.

Gray, G. J. *A bibliography of the works of Sir Isaac Newton*. Cambridge, 1888 (2nd ed., revised and enlarged, 1907).

Harrison, J. 'Newton's library: identifying the books', *Harvard Library Bulletin*, xxiv (1976), pp. 395–406.

Harrison, J. and P. Laslett. *The Library of John Locke*. Oxford, 1965 (2nd ed. 1971).

Historical Manuscripts Commission. *Report on the manuscripts of the late Reginald Rawdon Hastings, Esq....*Vol. 1. London, 1928.

Huggins List, 1727. (British Library Reference Division, MS.Add.25, 424.)

Huish, M. 'Newton's library returns home', *The Librarian and Book World and Curator*, xxxiii, 12 (1944), p. 190.

— 'Where are Newton's books?' *The Librarian and Book World and Curator*, xxxii, 2 (1942), p. 20.

Jukes, H. A. L. (ed.). *Articles of enquiry addressed to the clergy of the Diocese of Oxford at the primary visitation of Dr. Thomas Secker, 1738*. (Oxfordshire Record Society, xxxviii.) Oxford, 1957.

King, Peter, *7th Lord*. *The life of John Locke with extracts from his journals and common-place books*. New ed. 2 vols. London, 1830.

King's College, Cambridge. Keynes MS 2, Newton's Common Place Book.
— Keynes MS 127A, Documents relating to the Settlement of Newton's Estate.
— Keynes Papers, Box 6. Correspondence etc. of J. M. Keynes relating to the Newton Collection.
Koyré, A. and I. B. Cohen. 'The case of the missing *tanquam*: Leibniz, Newton & Clarke', *Isis*, LII (Baltimore, 1961), pp. 555–66.
Leigh & Sotheby. *A Catalogue of the Library of the late Mrs. Anne Newton, containing the Collection of the Great Sir Isaac Newton...Which will be sold by Auction...on Monday, March 22, 1813, and Five following Days...*London, 1813.
Library of Sir Isaac Newton: presentation by the Pilgrim Trust to Trinity College Cambridge 30 October 1943. With Appendix: Newton's library and its discovery, by H. Zeitlinger. Cambridge, 1944.
Low, D. 'Sir Isaac Newton's library and Chinnor Rectory', *Buckinghamshire and Chiltern Life*, no. 26 (1974), pp. 30–31.
— *With all faults.* 'Tehran' [Chinnor, Oxford], 1973.
Macphail, I. *Alchemy and the occult: a catalogue of books and manuscripts from the Collection of Paul and Mary Mellon given to Yale University Library...*2 vols. New Haven, 1968.
Manuel, F. E. *Isaac Newton, historian.* Cambridge, 1963.
— *A portrait of Isaac Newton.* Cambridge, Mass., 1968.
Markham, A. A. *The story of Grantham and its church.* 18th ed. Gloucester, 1973.
Maude, T. *Wensley-Dale; or, rural contemplations: a poem.* 3rd ed. London, 1780.
Miscellanea genealogica et heraldica, N.S., I (1874), 169–71, 'Pedigree of Newton'.
More, L. T. *Isaac Newton: a biography, 1642–1727.* New York, 1934.
Morning Post (The). 8 February 1928. 'Newton's library discovered'.
— 7 January 1929. 'More Newton discoveries'.
Munby, A. N. L. *A Catalogue of the Manuscripts and Printed Books in the Sir Isaac Newton Collection forming Part of the Library bequeathed by John Maynard, Baron Keynes of Tilton to King's College, Cambridge.* 2 pts. (Pt 1 an annotated copy of the Sotheby Newton Catalogue; pt 2 a typescript catalogue.) London, 1936; Cambridge, 1949. (Copies in Cambridge University and King's College Libraries.)
Musgrave Catalogue, *c.* 1767. MS. (Trinity College, NQ.17.36.)
Neu, J. 'Isaac Newton's library: ten books at Wisconsin', *U.W. Library News*, xv, 4 (1970), pp. 1–10. Repr. without illustrations in *Friends of the Library Messenger, University of Wisconsin – Madison*, no. 15 (1974), pp. 1–5.
Newton, Isaac. *Correspondence.* Vols. 1–3 (*1661–75, 1676–87, 1688–94*) ed. H. W. Turnbull; vol. 4 (*1694–1709*), ed. J. F. Scott; vols. 5–7 (*1709–13, 1713–18, 1718–27*), ed. A. R. Hall and L. Tilling. Cambridge, 1959–77.
— *Correspondence of Sir Isaac Newton and Professor Cotes, including letters of other eminent men...with notes,* ed. J. Edleston. London, 1850.

Newton, Isaac. *Mathematical papers*, ed. D. T. Whiteside. Vols. 1– . Cambridge, 1967– .

— *Philosophiae naturalis principia mathematica, 3rd ed., 1726, with variant readings*, ed. A. Koyré and I. B. Cohen. 2 vols. Cambridge, 1972.

Nicolson, M. H. (ed.). *Conway letters: the correspondence of Anne, Viscountess Conway, Henry More, and their Friends, 1642–84*. London, 1930.

Notes and Queries, 2nd ser., XII (19 Oct. 1861), p. 315, letter of A. De Morgan.

Observer (The). 11 April 1943. 'Pilgrim Trust buys Newton Books lost for 200 years.'

Plot, R. *The natural history of Oxford-shire*. London, 1677.

Public Record Office, London. 'Inventory [Sir I. Newton]. PROB.3/26/66.'

— 'Will of Anne Newton, June, 1811', PROB./11, vol. 1535, fols. 395–8.

Royal (The) Society: Newton tercentenary celebrations, 15–19 July 1946. Cambridge, 1947.

Shapin, S. and S. Hill. 'The Turner Collection of the History of Mathematics at the University of Keele', *British Journal for the History of Science*, VI (1973), pp. 336–7.

Sotheby & Co. *Catalogue of the Newton Papers sold by Order of the Viscount Lymington...July 13th, 1936, and Following Day*. London, 1936.

Sotheran (Henry) & Co. *Bibliotheca chemico-mathematica: Catalogue of Works in many Tongues on Exact and Applied Science*, ed. H. Zeitlinger and H. C. Sotheran. 2 vols. and Suppl. 1–2. London, 1921–37.

— Catalogue nos. 804 (1927), 843 (1935), 865 (1940). London.

— *The Newton Library, for Sale by Henry Sotheran, Limited*. [Brochure.] London [1929].

Spargo, P. E. 'Newton's library', *Endeavour*, XXXI (1971), pp. 29–33.

Spence, J. *Anecdotes, observations, and characters, of books and men*...London, 1820.

Stanford University Library, Stanford, Calif., MS, Container 2, Folder 13. A list by Newton headed 'De scriptoribus chemicis'.

Stukeley, W. *Memoirs of Sir Isaac Newton's life*...ed. A. Hastings White. London, 1936.

Thame Park, Thame, Oxon. The Greater Portion of the Contents of the Mansion... Hampton & Sons...are...to sell the above...on Tuesday, January 13th 1920, and two following Days. London, 1920.

Times (The). 12 April 1943. 'Isaac Newton's library. Purchase by Pilgrim Trust'.

Trinity College, Cambridge. Add.MS.a.101. Class List and Catalogue, 1667.

— Add.MS.a.103a, b. Catalogue and Finding Lists, *c*. 1675–6.

— Add.MS.a.104. Class List and Catalogue, *c*. 1690.

— Add.MS.a.106, fol. 13v. List of Donors, *c*. 1680.

— MS.R.4.48ᶜ. Pocket-book of Newton with record of his accounts, *c*. 1661–8.

— NQ.17.37. 'Notes, Papers etc. relating to the Newton Library'.

Verey, D. *Gloucestershire*. 2nd ed. Vol. I. (*The buildings of England*.) London, 1970.

Victoria History of the Counties of England: Oxfordshire, vol. VIII. London, 1964.

Wallis, P. and R. Wallis. *Newton and Newtoniana, 1672–1975: a bibliography*. Folkestone, 1977.

Whiteside, D. T. 'Before the *Principia*: the maturing of Newton's thoughts on dynamical astronomy', *Journal for the History of Astronomy*, I (1970), pp. 5–18.

— 'Isaac Newton: birth of a mathematician', *Notes and Records of the Royal Society of London*, XIX (1964), pp. 53–62.

— 'Newton's early thoughts on planetary motion: a fresh look', *British Journal for the History of Science*, II (1964–5), pp. 117–37.

— 'Newton's marvellous year: 1666 and all that', *Notes and Records of the Royal Society of London*, XXI (1966), pp. 32–41.

Zeitlinger, H. 'A Newton bibliography'. In *Isaac Newton, 1642–1727: a memorial volume*, ed. W. J. Greenstreet for the Mathematical Association. London, 1927, pp. 148–70.

— *Newton's library*, see *Library of Sir Isaac Newton*. . .

INDEX

accountancy books, 73
acquisition dates, 2, 3, 80
Acta eruditorum, 49, 67, 68
Adams, H. M., 55, 56 n. 3
Addison, Joseph, 68
additions to the library after 1727, 35–6, 38, 45–6, 56, 57, 58, 71
advertisements of the library, 53–4
alchemical books and manuscripts, 8, 20, 41–2, 59, 60 n. 1, 64–5
anatomical books, 59, 63
Andrade, E. N. da C., 57 n. 2, 60 n. 1, 64–5
annotations in books
  by Newton, 8, 11, 14–24, 32, 33, 37, 43 n. 2, 50, 51, 80, 271–2
  by others, 24–5
antiquities, books on, 59
Arabic literature in the library, 72
Argonauts, books on, 70
arrangement of books in the library
  at Barnsley Park, 39–40, 49
  at Chinnor, 35, 37–9
astronomy, books on, 58, 59, 62, 65 n. 2
authors not represented in the library, 62, 72–3

Babson College, Mass., 2, 4 n. 2, 19, 80.
  Nos. 112, 142, 261, 442, 673, 674, 716, 717, 734, 1009, 1159, 1224, 1228, 1276, 1424, 1562, 1683, 1705
  MS 418 'Lib. Chem', 1 n. 1, 41–2
Baluze, Étienne, 3, 12
Barchas, S. I., Sonoita, Ariz., nos. 657, 820
Barnby, Bendall & Co., Ltd, Cheltenham, 54

Barnsley Park, Glos., 39–41, 48–9, 51–2
Barrow, Isaac
  annotations by, 10, 11
  library of, 6–7, 8, 11, 60, 61–3, 64, 70, 71–2, 73
  presentations to Newton, 3, 10
  presentations to Trinity, 63 n. 2
Barton, Catherine, *see* Conduitt, Catherine
Basilius Valentinus, 65
Bernard, Edward, 13
Bernoulli, Johann I, 14
biblical books, 19, 59, 66
bibliographical books, 69
biographies, 59
Birch, Thomas, 35 n. 2
Björnståhl, J. J., 37–8, 44 n. 2, 64 n. 6
blanket entries in Huggins List, 32–3, 45
book-lists made by Newton, 1, 8–9, 41–2
book-marking, *see* 'dog-earing'
bookplates
  Huggins, 36, 38
  Musgrave, 38, 39, 40, 45
  not used by Newton, 4
  Pilgrim Trust, 56
bookseller's account, a, 9
botanical books, 59
Boyle, Robert, 3, 11–12, 39, 51, 57 n. 2
British Museum, *aftw.* Library, 30, 55, 268 n. 1. Nos. 817, 1192
Brown University Library, Providence, R.I., no. 1091
Burnet, Thomas, 3, 12–13

California, University of, at Berkeley, Bancroft Library, no. 1115

California, University of, at Los Angeles, William Andrews Clark Memorial Library, no. 1255

California Institute of Technology, Pasadena, nos. 756, 878

Cambridge University Library, 6, 22, 23, 34, 39, 43 n. 1, 54–5, 57, 63, 64, 79. Nos. 506, 1158, 1162, 1167, 1169, 1189, 1380

Chamberlayne, John, 66

chemistry
  materials bought by Newton for experiments, 8
  see also alchemical books and manuscripts

Chicago, University of, Library, 49 n. 1. Nos. 276, 1086

Chinnor
  library, 35–8; visitors to, 35, 37–8
  living, 30, 31 n. 1, 35 n. 1, 36, 39
  village, 36 n. 5

chronology, books on, 59, 66, 70

Church Fathers, books of, 59, 66

Church history, books on, 59

Clark, Mr, apothecary, 5

Clarke, Samuel, 75

classical literature in the library, 37, 59, 60, 70

Cockburn, William, 64

coding in books, 41–3

Cohen, I. B., 52 n. 1

Collins, John, 3, 6, 11, 61 n. 2, 75

Columbia University Library, New York, 4 n. 2. Nos. 350, 350a, 770, 818, 902, 1254, 1368, 1744

commerce, see trade

condition of the books (1727), 31

Conduitt, Catherine, 28, 29, 31 n. 1, 33, 68, 69, 70, 73, 266, 270

Conduitt, Catherine, the younger, 266

Conduitt, John, 19 n. 3, 28, 29, 30, 31 n. 1, 33, 34, 266, 270

Congreve, William, 71

Conti, Antonio Schinella, 23

cooking, books on, 73

Cornell University Library, Ithaca, N.Y., no. 136

Cox, John, 19 n. 3

Cox, Joseph, 19 n. 3

Cranbury Park, Winchester, 266–7

cross-references in the catalogue, 79

currency, books on, 59, 72

Dale, Sir Henry, 55 n. 1

date of publication analysis, 77–8

De Morgan, Augustus, 3 n. 2, 269–70

de Villamil, Richard, viii, 51–3, 56, 57, 62 n. 4, 64, 71 n. 3

dedications to Newton, 10

density of content of books, 60

Des Maizeaux, Pierre, 24, 33 n. 3, 35

Descartes, René, 14–15, 57 n. 3

descent of Newton's books, 28–31, 33–4, 36–7, 39, 40–41, 48–57

diagrams drawn by Newton, 16

dictionaries in the library, 59, 68–9

discovery of the library (1928), 51–2, 53

dispersal of part of the library (1920), 48–51

Dobbs, B. J. T., 64 n. 1, 65 n. 3

'dog-earing' by Newton, 25–7, 45, 68, 72, 80; pls. 4, 5

Doyley, Oliver, 3, 13

duplicate copies of books, 33, 34, 48, 79

Duport, James, 5 n. 1, 6. No. 704

Duveen, Denis I., 34. Nos. 259, 419, 436, 752, 1034

economics books, 59, 72

Ekins List, Bodleian Library MS New College 361, vol. II, 9, 80. Nos. 237, 511, 531, 539, 540, 619, 635, 1242, 1263, 1316, 1372, 1397, 1607, 1675

English, books published in, 73

English literature in the library, 59, 68, 70–71

European literature in the library, 71–2

Fatio de Duillier, Nicolas, 75

Feisenberger, H. A., 62

Fitzwilliam Museum, Cambridge, Notebook of Newton, 7, 8

Fleet prison, 30, 31 n. 1

floristry books, 73

fluxion priority dispute, 23–4, 67 n. 4

Fontenelle, Bernard Le Bovier de, 14, 27

forgery of Newton's signature, 3–4

French, books published in, 32, 74, 75–6

French literature in the library, 71

Gale, Thomas, 4

Galileo, Galilei, 62

gardening books, 73

Gassendi, Pierre, 62

geography books, 59, 66

INDEX

Golding, Edward, 19 n. 3
grammars in the library, 59, 69
Grantham, Lincs.
King's School, 5, 74–5
St Wulfram's Church library, 5
Greek, books in, 74, 75
Green, Benjamin, 38
Greene, Robert, 13
growth of the library, 58 n. 1, 77–8

Hale Observatories, Pasadena, Calif., no. 526
Halifax, Charles Montague, *Earl of*, 15, 68, 71
Halley, Edmond, 25, 67 n. 4
Hampton & Sons, London, 48–50
Harley, Martha, 269
Harper, Lathrop C., New York, no. 501
Hebrew, books published in, 74, 76
Hill, A. V., 55 n. 1
history books, 59, 66–7
Honeyman, R. B., Jr, San Juan Capistrano, Calif., no. 594
Huggins, Charles, 24, 30, 34–6, 45, 58, 71. Nos. 117, 1686
Huggins, Jane, *see* Musgrave, Jane
Huggins, John, 30, 31 n. 1, 35 n. 2, 270
Huggins, John, jun., 30 n. 2, 31 n. 1, 36
Huggins, William, 30 n. 2, 36
Huggins List, 15 n. 4, 28, 30–34, 45, 46, 48, 52–3, 57, 71, 79, 268, 273–4
Humanists, books by, 70
Huntington Library, San Marino, Calif., 29
Hurstbourne Park, 19 n. 3
Huygens, Christiaan, 1–2, 3, 12, 75

identification of the books, 45, 57
imprint analysis, 76–8
incunabula in the library, 73
interleaved volumes, 22–3
inventory of Newton's effects, 29, 45, 57, 72 n. 2

Jewish National & University Library, Jerusalem, no. 1048
Jewish rites and customs, books on, 59, 66, 76
John Crerar Library, Chicago, no. 266
Johnson, Thomas, 67–8
Jones, William, 35 n. 2, 44 n. 2
Journal des sçavans, 49, 67, 68

journals in the library, *see* periodicals
Keele University Library, nos. 173, 265, 600, 699, 713, 1166
Keill, John, 67 n. 4, 75 n. 1
Kepler, Johann, 62
Keynes, *Sir* Geoffrey, 53 n. 2. Nos. 493, 793
Keynes, J. M., 53, 55 n. 1
King, Peter, *1st Lord*, 69 nn. 3–4
King's College, Cambridge, Keynes Collection, 2, 28 n. 2, 42 n. 1, 53 nn. 2–3, 55 n. 1, 65 n. 3. Nos. 322, 1317
Knox, G. D., 53 n. 1

language analysis, 74–5
Laslett, Peter, no. 1565
Latin language, 74–5
Latin literature, modern, 59, 70
Laughton, John, 4. No. 629
law books, 59
Leibniz, Gottfried Wilhelm von, 23, 24, 27
Leigh & Sotheby sale (1813), 41 n. 2, 268–70
literature, modern, 59, 70–72
locations of Newton's books, 80. *Libraries etc. owning Newton books are listed in this index and at the catalogue entries for the items*
Locke, John, 1, 4, 8 n. 2, 10 n. 2, 11, 69–70, 72, 75
logic, books on, 59
London publishing, 76
Lord's Prayer, 66
losses from the library, 47–8, 273–4
Lowy, B. A., New York, no. 444
Lucas, Henry, 63 n. 1
Lull, Raymund, 65
Lymington, John Wallop, *Lord*, 266
Lymington sale (*c.* 1750), 266–7

McKitterick, D. J., no. 1206
Macmillan, H. P. M., *Baron*, 54, 55 n. 1, 56
Manuel, F. E., 65
manuscripts in the library, 29, 33, 46, 270
mathematicians, classical Greek, 61
mathematics
books in the library, 46–7, 59, 60–63
books not in the library, 62–3

283

Thame Park (*cont.*)
 876, 954, 961, 967, 979, 992, 993,
 995, 1034, 1084, 1096, 1115, 1189,
 1244, 1296, 1304, 1408, 1427, 1473,
 1489, 1562, 1595, 1608, 1652, 1685,
 1688, 1705, 1708, 1710
Theatrum chemicum...6 vols. (1659–61), 8,
 16, 50, 52, 64 n. 4, 65
theological books, 59, 66
Thorp, T., Guildford, nos. 322, 1317
Tilson, Christopher, 30
tracts in the library, 32–3
 bound volumes of, 33, 38, 46–8, 61, 80
trade, books on, 72
transfer of books
 to Chinnor, 30, 34
 to Barnsley Park, 39–41
 to Thame Park, 48
 to London, 54
 to Cheltenham and Banbury, 54
 to Cambridge, 55
travel books, 59, 60, 66
Tregaskis, J., London, no. 259
Trevelyan, G. M., 54, 56
Trinity College, Cambridge, 1, 3, 6, 7–8,
 55–7
 Library, 4–5, 6, 63, 64
 Newton collection, 2, 8, 19, 25, 33, 40,
 45, 46, 47, 51, 55–7, 60, 65 n. 3,
 68, 80
 Pocket-book of Newton (*c.* 1661–8),
 7, 8
 Newton books in Trinity are shown at
 their catalogue entries
triplicate copies of books, 32–3

unidentified works in Musgrave catalogue, 46, 47

valuation of the library (1727), 29, 30;
 (1920), 52 n. 1; (1929), 53
Varignon, Pierre, 67
Vaughan, Thomas, 65
voyages, books on, 59, 60, 66

Walker, William, 3, 12
Wallis, John, 62–3
'wast books', 32, 33
Whipple Science Museum Library, Cam-
 bridge, nos. 273, 1710
White, F. P., 39
Whiteside, D. T., 61
Wing, Tycho, 268
Wisconsin, University of, at Madison,
 Memorial Library, 19, 34, 47. Nos.
 18, 229, 259, 419, 436, 659, 752,
 1034, 1446, 1478, 1652, 1685
Wren Library, *see* Trinity College, Cam-
 bridge, Library
wrestling, book on, 73–4
Wykeham, Georgina, 41 n. 1
Wykeham-Musgrave, Aubrey Wenman,
 41 n. 1
Wykeham-Musgrave, Henry Wenman,
 48, 52, 53

Yale Medical Library, New Haven,
 Conn., Fulton Collection, no. 271
Yale University, Beinecke Library, nos.
 605, 1116

Zeitlin & Ver Brugge, Los Angeles,
 no. 1574
Zeitlinger, Heinrich, 2, 8, 27, 41 n. 2, 50,
 51, 52, 56 n. 3, 62, 65 n. 2, 269
zoology books, 59